Ilyushin IL-18/-20/-22

A Versatile Turboprop Transport

Yefim Gordon and Dmitriy Komissarov

AeroFax
An imprint of
Ian Allan Publishing

Ilyushin IL-18/20/22:
A Versatile Turboprop Transport

ISBN 1 85780 157 1

Published by Midland Publishing
4 Watling Drive, Hinckley, LE10 3EY, England
Tel: 01455 254 490 Fax: 01455 254 495
E-mail: midlandbooks@compuserve.com

Midland Publishing and Aerofax are imprints of
Ian Allan Publishing Ltd

Worldwide distribution (except North America):
Midland Counties Publications
4 Watling Drive, Hinckley, LE10 3EY, England
Telephone: 01455 254 450 Fax: 01455 233 737
E-mail: midlandbooks@compuserve.com
www.midlandcountiessuperstore.com

North American trade distribution:
Specialty Press Publishers & Wholesalers Inc.
39966 Grand Avenue, North Branch, MN 55056
Tel: 651 277 1400 Fax: 651 277 1203
Toll free telephone: 800 895 4585
www.specialtypress.com

Design concept and layout
© 2004 Midland Publishing and
Stephen Thompson Associates

Printed in England by
Ian Allan Printing Ltd
Riverdene Business Park, Molesey Road,
Hersham, Surrey, KT12 4RG

Title page: **A beautiful air-to-air study of an IL-18B (possibly CCCP-75690, c/n 189001205?) as it cruises over cumulus clouds. The IL-18 is an elegant airliner.** Il'yushin OKB

Contents

Below: **IL-18V CCCP-75554 (c/n 184007501) operated by Aeroflot's Central Directorate of International Air Services was preserved as a 'gate guard' at Moscow/Shermet'yevo-1 – a fitting tribute to the sterling service rendered by the type on Aeroflot's international routes.**
Yefim Gordon

Introduction

The Fourth of July is a meaningful date for many Russians as well as for Americans – albeit in a totally different way. On 4th July 1957 the Il'yushin IL-18 turboprop airliner took off on its maiden flight. Originally called *Moskva*, since it was designed and mass-produced in Moscow (though this popular name was shortlived), and code-named *Coot* by NATO, this Soviet equivalent of the Lockheed L-188 Electra proved to be extremely successful, and a few are in passenger service to this day.

The IL-18 deserves to be called an outstanding design. For one thing, it came on the scene at exactly the right moment, which was no small thing; many an excellent aircraft has fallen victim to bad timing. For another, it was designed and built to a high engineering standard. Thirdly, it had quite a large production run – enough to fill the needs of both Aeroflot, the sole Soviet airline, and foreign customers. Finally, the IL-18 was easy to master for crews converting to the type and suited for operating in fairly spartan conditions. By combining these virtues the IL-18 brought world fame and recognition not only to the design bureau led by Sergey Vladimirovich Il'yushin but to the Soviet aircraft industry as a whole. Starting in the late 1950s, there was not a single major event in Soviet (and later CIS) history where the IL-18 was not involved, directly or indirectly. The roles filled by this remarkable aircraft

The IL-12 was the Il'yushin OKB's first airliner to achieve production status and one of Aeroflot's principal types in the years when the turboprop IL-18 was developed. Here, an IL-12 *sans suffixe* registered CCCP-Л3925 (ie, SSSR-L3925) is seen on the apron at Moscow-Vnukovo in 1958.
Yefim Gordon archive

Left: **The IL-12B was developed into the IL-14 which saw even larger-scale production and wider service with Aeroflot. This is an IL-14P in Aeroflot's 1973-standard blue/white colours. Judging by the non-standard registration CCCP-48098, this example belonged to one of the divisions of the Ministry of Aircraft Industry (MAP); IL-14s were normally registered in the 41***, 61*** and 91*** blocks.** via Avico Press

Below: **Apart from the IL-12 and IL-14, the Soviet airline scene in the immediate post-war years and the 1950s was populated largely by the Lisunov Li-2 (which the Il'yushin types were designed to replace). CCCP-06134 is a Li-2P passenger version with no cargo door to starboard – and, judging by the registration in the 06*** block, again an MAP aircraft.** Yefim Gordon archive

included Arctic and Antarctic research, development of sparsely populated regions, the whole range of civil aviation tasks and a wide range of military duties in Soviet/Russian Air Force (VVS – *Voyenno-vozdooshnyye seely*) and Soviet/Russian Naval Air Arm (AVMF – *Aviahtsiya voyenno-morskovo flota*) service. First and foremost, however, it was a civil aircraft, remaining one of the world's principal airliners for several decades.

In the Soviet Union the years after the Second World War were characterised by the rapid development of numerous industry branches; these were first and foremost the branches having military importance, including the aircraft industry. New types of combat aircraft were developed and fielded; modern airfields and airbases were constructed all over the country. At the same time the Soviet national economy was being rebuilt at a speedy pace; this applied not only to the western and central regions of the USSR which had been ravaged by the war but also to Siberia and the Soviet Far East – regions which had a great potential but were largely underdeveloped. This, in turn, required workforce to be delivered to remote areas of the country at short notice. Thus, the immediate post-war years had all the prerequisites for the development of civil aviation in the USSR as well as the Soviet air arm.

The first new Soviet airliners to enter mass production after the war emanated from the Il'yushin OKB (*opytno-konstrooktorskoye byuro* – experimental design bureau), alias OKB-240 (the number was a code allocated for security reasons). The twin-engined IL-12 (NATO codename *Coach*) and its successor, the IL-14 *Crate*, were powered by Shvetsov ASh-82 radial engines, offering a very modest seating capacity (27-40) and short range. True, even at that early stage there were attempts to create something bigger and better. In 1946 OKB-240 brought out an aircraft which, by some quirk of fate, was designated IL-18. It was a low-wing monoplane with conventional tail surfaces, a tricycle landing gear and four 2,300-hp Shvetsov ASh-73 18-cylinder two-row radials (the 2,400-hp supercharged ASh-73TK version planned originally was not yet available). For the first time in the Soviet Union, it featured a pressurised cabin – a feature which accounted for the unusual 'pistol bullet' nose profile strongly reminiscent of the Boeing Model 307 Stratoliner. With a crew of five or six, the aircraft could carry up to 66 passengers and 900kg (1,980 lb) of baggage over a maximum range of 2,800km (1,740 miles).

The 'first-generation' IL-18 entered flight test on 17th August 1946, completing the manufacturer's test programme on 30th July 1947 with good results, but did not progress beyond the prototype stage. This was because the aircraft was ahead of its time – not so much technically as 'marketologically', so to say. Passenger air traffic volumes were still small at the time, and Aeroflot's fleet of IL-12s and Lisunov Li-2 *Cab* (licence-built Douglas DC-3 derivative) airliners, of which there were plenty, could still cope with the existing demand.

The rival OKB-156 headed by Andrey Nikolayevich Tupolev undertook a similar attempt, deriving the Tu-70 (aka Tu-12) airliner from the Tu-4 *Bull* strategic bomber (a copy of the Boeing B-29 Stratofortress – but *that* story lies outside the scope of this book). However, Tupolev had no better luck – only a single prototype of the Tu-70 was built for much the same reason. Another reason was that large piston-engined airliners had limited range, which made them uneconomical to operate on transcontinental routes in the USSR with its vast expanse. Thus, by the time traffic volumes began to grow the time was ripe for a new generation of passenger aircraft.

In the 1950s the Western world had already begun to introduce turbojet-powered aircraft into airline service. In contrast, the new large-capacity airliners entering flight test in the Soviet Union in those days were predominantly powered by turboprops. For the two decades that followed, the USSR became a veritable 'wildlife refuge' for turboprop passenger aircraft. And fine aircraft they were! There was the Antonov An-10 *Cat* – the world's only four-turboprop airliner to utilise a high-wing layout; there was the majestic Tupolev Tu-114 *Cleat* – the largest, fastest and most powerful turboprop airliner ever (which combined turboprops with swept wings into the bargain). In short, the Soviet airline scene was well worth a look for the aviation enthusiast and expert alike.

Finally, there was the 'second-generation' IL-18 which was destined to become not just a highly successful and widespread aircraft but a classic aircraft of the 1950s and 1960s – a period often described as the golden age of civil aviation.

On what grounds do we draw such conclusions, you may ask? It was not so much a matter of numbers – though few civil aircraft designed in the 1940s and 1950s were lucky enough to be produced in more than 500 copies. Indeed, if we compare the IL-18 with the other four-turboprop airliners of the day (Vickers Viscount, Vickers Vanguard, Bristol Britannia/Canadair CL-44, Lockheed L-188 Electra, An-10 and Tu-114), each of them surpassed the IL-18 in some respect. For instance, the Viscount – another classic airliner – was much more popular, judging by the number of airlines that operated it (not to mention the fact that it was the 'founding father' of the four-turboprop airliner class). Yet, neither of the airliners listed above could compare with the IL-18 in terms of production volumes and *versatility*.

Yes! Versatility is the key word – as, incidentally, was often the case with Soviet aircraft. Quite apart from the many passenger versions differing in seating capacity and range, the IL-18 was converted into numerous testbeds and research aircraft which found use in civil and military programmes alike. Originally a purely civil design, it spawned a number of purely military derivatives operated by the Soviet Air Force and Soviet Navy under separate designations – the IL-20 electronic intelligence aircraft, the IL-22 airborne command post and the IL-38 anti-submarine warfare aircraft. Civil specialised versions were built, too – the IL-20RT space tracker aircraft (although this version obviously had military uses as well!), the IL-24N ice patrol aircraft and the IL-18DORR fishery reconnaissance aircraft. These and other versions are described in detail here.

The IL-18 was the first Soviet airliner to be certificated for ICAO Category I blind landing capability, decision altitude 60m (200ft), horizontal visibility 800m (2,600ft), which facilitated night and poor-weather operations considerably, one of the reasons for the type's success.

The IL-18 saw service with airlines and air forces in various parts of the world, earning a reputation as a well-built and reliable aeroplane. Indeed, it was probably the first Soviet airliner to achieve considerable export success. In the process of selling Il'yushin aircraft to foreign customers the Soviet foreign trade organisations accumulated a wealth of experience not only in concluding sales deals but also in filling out the supporting technical documentation conforming to international

The first aircraft to bear the IL-18 designation (and the first Il'yushin aircraft to feature a pressurised cabin) was this airliner powered by four Shvetsov ASh-73 radials which entered flight test on 17th August 1946. A fine aircraft in its day, the 'first-generation' IL-18 was killed off by poor timing – there just wasn't enough passenger traffic to warrant aircraft of this size. Il'yushin OKB

The Tu-70 (formerly known as the Tu-12) was the Tupolev OKB's competitor to the 'first-generation' IL-18 and suffered a similar fate for much the same reasons. Tupolev OKB

standards. The advent of the IL-18 also sparked a massive development of Aeroflot's international services.

Yet the *Soviet Electra*'s road to world recognition was long and arduous. Hopefully the facts which the authors have set forth in this book will give the Western reader an unbiased picture of the development, progressive refinement, production history and service career of the Il'yushin four-turboprop family.

Though no longer as common a sight as it used to be in the former days, the IL-18 is still occasionally encountered both in and outside the Commonwealth of Independent States (CIS) – sometimes quite a long way from home. The reasons for this longevity are the airliner's excellent field performance enabling it to use relatively short runways, the capable navigation suite and automatic landing system, the high power-to-weight ratio and durable structure. Also, the IL-18 has 1,000km (620 miles) longer range than the Tupolev Tu-154 *Careless* trijet medium-haul airliner; it is more manoeuvrable and has safer landing characteristics (this is *not* meant to say that the Tu-154 is unsafe! – *Auth.*). On the down side, the 40-year-old design has a much smaller seating capacity and less comfort. The propeller-driven IL-18's cruising speed, cruise altitude and rate of climb are 50% lower than the Tu-154's, while the number of crewmembers is higher. The IL-18's amazing revival in the 1980s and 1990s was due first and foremost to the downturn in passenger traffic which turned the Il'yushin turboprop into an indispensible machine.

Acknowledgements

The authors wish to express their gratitude to the following persons who contributed to the making of this book: Sergey S Tsvetkov whose authoritative monograph on the IL-18 published in the *Mir Aviatsii* (Aviation World) magazine assisted greatly in the making of this book; Anatoliy F Artem'yev whose publications on the IL-38 were also of great help; Rudolf A Teymurazov, Ivan G Faleyev, Lydia N Anghelova and Natal'ya I Titova of the CIS Interstate Aviation Committee who provided an insight into IL-18 operational details in the Soviet Union and CIS; Nigel A Eastaway of the Russian Aviation Research Trust, whose extensive photo archive yielded many valuable details; Viktor G Kravchenko, Dmitriy A Petrochenko and Yuriy Kirsanov for supplying additional photos and information; Aleksandr P Boonaryov (East Line Aviation Security) and Stanislav I Looschchik (Vnukovo Airport Aviation Security) who granted ramp access at Moscow-Domodedovo and Moscow/Vnukovo-1, allowing some valuable photos to be taken; Andrey A Yurgenson for making the line drawings illustrating the book.

Special thanks go to the editorial group of the *Soviet Transports* reference book (Peter Hillman, Stuart Jessop, Guus Ottenhof and Tony Morris) who kindly granted permission to use the materials they have amassed on the type's operation and supplied the authors with up-to-the-minute updates on IL-18 operations. Last but not least, thanks go to Dmitriy S Makovenko who printed many of the photos used here with excellent quality.

Russian Language and Transliteration

The Russian language is phonetic – pronounced as written, or 'as seen'. Translating into English gives rise to many problems and the vast majority of these arise because English is not a straightforward language, with many pitfalls of pronunciation!

Accordingly, Russian words must be translated through into a *phonetic* form of English and this can lead to different ways of helping the reader pronounce what he sees. Every effort has been made to standardise this, but inevitably variations will occur. While reading from source to source this might seem confusing and/or inaccurate but it is the name as *pronounced* that is the constancy, not the *spelling* of that pronunciation!

The 20th letter of the Russian (Cyrillic) alphabet looks very much like a 'Y' but is pronounced as a 'U' as in the word 'rule'.

Another example is the train of thought that Russian words ending in 'y' are perhaps better spelt out as 'yi' to underline the pronunciation, but it is felt that most Western speakers would have problems getting their tongues around this!

This is a good example of the sort of problem that some Western sources have suffered from in the past (and occasionally even today) when they make the mental leap about what they see approximating to an English letter.

IL-18 Junior or Moscow Was Not Built in a Day

CCCP-Л5811 (ie, SSSR-L5811), the prototype of the 'second-generation' turboprop-powered IL-18, as originally flown with large *Moskva* (Moscow) titles; this was the aircraft's popular name which unfortunately did not catch on.
Yefim Gordon archive

In the early 1950s Soviet civil aviation was still in short pants, so to say. There were many reasons for that; the turbulent pre-war period, not to mention the years of the Great Patriotic War when every available aircraft was in the hands of the military and the immediate post-war years when half the nation lay in ruins, left no time or resources for the furtherance of air transport. On the other hand, during the Stalin era there was no need for a large and modern airliner fleet and a well-developed airline network. The impoverished population was veritably chained to its place of residence by the passport registration system – in effect, a modern form of serfdom; and even if they chose to travel, people preferred the more traditional kind of public transport – the railroad. However, the liberalisation that followed after Stalin's death (the 'Khruschchov thaw'), together with the need to develop the remote regions of Siberia and the Soviet Far East, forced the government to turn its attention to civil aviation in general and fleet renewal in particular.

Until then, only two of the Soviet aircraft design bureaux had any experience of airliner design (that is to say, apart from those design teams which did not survive the war due to Stalin's purges or other reasons, such as the Kalinin and Putilov bureaux). The Tupolev OKB, which specialised in multi-engine heavy aircraft, looked set to win. The No 2 contender was the Il'yushin OKB which, though it had little experience of designing heavy aircraft, was the sole maker of passenger aircraft in the Soviet Union at the time. The IL-12 had just entered service, livening up a scene then populated solely by the Li-2/C-47, and the more advanced IL-14 derivative was already undergoing tests. However, there was also the up-and-coming OKB-473 headed by Oleg Konstantinovich Antonov.

Rather than design large piston-engined aircraft of the kind in service with Western airlines at the time (in the same class as the Lockheed L-049/L-349/L-1049 Constellation, Boeing Model 377 Stratocruiser and Douglas DC-7), the Soviet government and the aircraft industry decided to go for the more advanced gas turbine engines. The Tupolev OKB chose a different 'market niche' from the other contenders, developing aircraft which were not directly comparable with the subject of this book – the Tu-104 *Camel* medium-haul twinjet (the first Soviet jet airliner) and the long-haul Tu-114. To save time and effort Andrey N Tupolev chose an approach which seemed fairly logical at the time, taking the wings, empennage, powerplant and landing gear of a production bomber and mating them to a new fully pressurised fuselage of larger cross-section; the Tu-16 *Badger* twinjet medium bomber and the Tu-95 *Bear-A* four-turboprop strategic bomber acted as the 'donors'. The trick worked; the Tu-104 made its maiden flight on 17th June 1955, entering service just over a year later, on 15th September 1956 when the competitors were still unbuilt.

On 30th December 1955 the Soviet Council of Ministers issued a directive ordering Nikolay Dmitriyevich Kuznetsov (head of the OKB-276 engine design bureau based in Kuibyshev, Russia) and Aleksandr Gheorgiyevich Ivchenko (head of the OKB-478 engine design bureau in Zaporozhye, the Ukraine) to develop new turboprop engines designated NK-4 and TV-20 respectively. (The latter designation derived from ***toorbovintovoy*** ***[dvigatel']*** – turboprop engine – proved shortlived and was soon changed to AI-20 in keeping with the common practice of designating Soviet aircraft and engines after the OKB's founder.) The same document tasked General Designer (the official title of Soviet OKB heads) Sergey V Il'yushin with developing a new airliner designated IL-18 around these alternative powerplants, with the implication that the best one would be selected to power the production aircraft. This was, in fact, the first instance when a new civil aircraft was developed on a competitive basis in the Soviet Union.

Unlike the Tupolev jet, the 'IL-18 Jr' and Antonov's first fully-fledged airliner – the An-10 – were both designed from scratch (though they did draw heavily on experience accumulated with the respective bureaux' previous designs, the 'IL-18 Sr.' and An-8 *Camp* military transport). Interestingly, they were designed around the same engines. Unlike the Tu-104 which received a proven powerplant, the engines for the Il'yushin and Antonov airliners were developed in parallel with the aircraft designed to take them.

The reason why Il'yushin and Antonov opted for turboprops was that, while turbojet engines offered high speed, turboprops (which were the latest state of the art in aero engine design in the 1950s) had a lower fuel burn. Besides, turboprop aircraft were anticipated to have better field performance – no small thing in the Soviet Union where many airports had short runways with a less-than-perfect surface. On the minus side, turboprop engines turned out to be more difficult both to build and to maintain than pure jets.

Both aircraft had their strong points: the IL-18 offered higher fuel efficiency on long routes, while the An-10 could operate from a much wider network of airfields, including unpaved airstrips, and could be easily modified into a dedicated transport aircraft (which emerged as the An-12 *Cub*).

On 25th May 1956 the Council of Ministers issued a follow-up directive on the IL-18 powered by four NK-4 turboprops. According to this document the aircraft was to be submitted for State acceptance trials – to be held in parallel by the Main Directorate of the Civil Air Fleet (the branch of Aeroflot responsible for regular passenger and cargo services) and the Soviet Air Force – in October 1957.

To this day the reason for ultimate choice of engine type for the IL-18 and An-10 (as well as for the An-8 and An-12) remains obscure and is one of the dark pages of these two types' history. A state order for an engine to power a mass-produced four-engined aircraft meant a production run of several thousand copies – much too lucrative an order to let the competitors take it away. In Soviet times the appraisal of a new aircraft or weapons system by the Powers That Be was not always objective; quite often its fate was decided not by its design merits but by who had more political clout – the designer or his opponents pursuing their own ends. Thus, only an expert in aircraft propulsion could make an objective comparison of the NK-4 and AI-20; still, competitive juices ran riot. Thus, Sergey V Il'yushin was quoted as saying that under no circumstances would he let engines from Zaporozhye be installed on his aircraft (ie, the IL-18). Similarly, Oleg K Antonov whose OKB resided in Kiev, the Ukrainian capital, stated that, while the AI-20 lacked the finesse and sophistication of the NK-4, it was more rugged and reliable; a classic case of the 'not invented here' syndrome on both sides.

On 27th September 1955 Nikolay D Kuznetsov called a meeting of his design staff, informing them of the government orders concerning the NK-4. The people set to work with a will and the project was completed in record time – the first prototype was ready for bench testing just 96 days after the meeting. Kuznetsov succeeded in making the engine as lightweight as possible; rated at 4,000ehp for take-off, the NK-4 had a low specific fuel consumption (SFC) and was easy to manufacture, too.

The engine completed its manufacturer's test programme and subsequently the State acceptance trials without any major problems. The flight tests were held at the Flight Research Institute named after Mikhail M Gromov (LII – ***Lyot****no-is****sled****ovatel'skiy insti****toot***), involving one of the institute's Tu-4LL engine testbeds; serialled '22 Red', this aircraft (construction number 221203) had two NK-4 development engines replacing the outer pair of Shvetsov ASh-73TK radials. Interestingly, the No 1 engine was mounted in an overwing installation as envisaged for the low-wing IL-18, while the No 4 turboprop was in an underwing nacelle developed for the An-10!

Thereupon the NK-4 entered production at aero engine factory No 24 in Kuibyshev (now renamed back to Samara). However, production was unexpectedly terminated after more than 200 engines had been delivered. Shortly afterwards at a board meeting of the Ministry of Aircraft Industry (MAP – *Mini****sterstvo*** *aviatsi****on****noy pro****myshl****ennosti*)[1] someone suggested making a performance comparison of the NK-4 and AI-20. OKB-478 chief Aleksandr G Ivchenko brushed off the idea, saying this was pointless; the NK-4 was out of production anyway whereas the AI-20 was in production, so there was 'nothing to discuss', as he put it. Another engine designer, the renowned Academician Vladimir Yakovlevich Klimov who specialised in fighter engines, was next to take the floor and spoke up in defence of Kuznetsov's engine. 'There *is* room for discussion here' he said, 'The NK-4 is 200kg [440lb] lighter than the AI-20 and has a much lower SFC. In its concept the NK-4 is an engine of the future, while the AI-20 could easily have been designed ten years ago.' Yet Klimov did not succeed in tipping the scales in favour of Kuznetsov's engine; nor did Il'yushin himself, even though he said more than once that the NK-4 was better suited for the IL-18 than the AI-20.

The Kuznetsov NK-4 turboprop (and later the Ivchenko AI-20 turboprop) was put through its paces on this Tupolev Tu-4LL engine testbed ('22 Red', c/n 221203) operated by the Flight Research Institute (LII). Note that the two development engines are installed differently; the No 1 engine is the overwing installation developed for the IL-18, while the No 4 turboprop represents the underwing installation for the Antonov An-10. Yefim Gordon archive

A group of workers of the Moscow Machinery Plant (MMZ) No 30 poses beside the IL-18 prototype at Khodynka; note the broken line of the crease marking the joint between the pressurised fuselage and the unpressurised extreme nose. Yefim Gordon archive

The Il'yushin OKB had started work on a new large long-range airliner back in 1954 – ahead of the government directives to this effect, as was often the case. The work proceeded quickly, as the engineers put the experience gained with the piston-engined 'first-generation' IL-18 to good use. While the general trend in the Soviet Union those days was that new civil aircraft should be easily adaptable for military roles, the General Designer succeeded in convincing the government that the IL-18 should be an uncompromised airliner utilising the classic low-wing layout with a circular-section fuselage and a conventional tail unit which made it possible to create a highly efficient aircraft. At a very early design stage Sergey V Il'yushin set an efficiency target for his team – or rather what might be called an affordability target: the price of a ticket on a flight aboard the IL-18 should be on a par with a train ticket for a sleeping car. This target was met in the long run.

On 26th August 1956 the General Designer endorsed the advanced development project (ADP) of the IL-18. Construction of the prototype began in September at MAP's experimental plant No 247 located on the west side of Moscow's Central airfield named after Mikhail V Frunze, right next door to the Il'yushin OKB's premises. (In passing, it may be noted that the field – better known as Moscow-Khodynka – is located in a heavily populated area some 6.75km (4.2 miles) from the Kremlin.) In so doing the plant received assistance from MMZ No 30 'Znamya Trooda' (*Moskovskiy mashinostroitel'nyy zavod* – Moscow Machinery Plant No 30 'Banner of Labour'), a series production factory located at the east end of Khodynka which was then manufacturing IL-14s. At this point the NK-4 was already undergoing flight tests on the abovementioned Tu-4LL testbed, whereas the competing AI-20 did not even exist in hardware form yet.

As one would expect from an aircraft designed in those days, the IL-18 was of all-metal construction. The choice of the low-wing layout with conventional tail surfaces was dictated by the need to ensure maximum flight safety while keeping seat-mile costs and fuel burn down. The engineers paid special attention to the wing aerodynamics. In contrast to the swept-wing Tu-114, the IL-18 featured conventional straight wings of trapezoidal planform with an aspect ratio of 10 and a taper of 3. The wings' aerodynamic layout assured the aircraft a high cruising speed by the day's standards, as well as good stability at low speed and high angles of attack. The stabilisers were likewise of trapezoidal planform, while the fin featured moderate leading-edge sweep and an unswept trailing edge. The general arrangement and aerodynamic features of the IL-18 were optimised for high performance.

The engines were housed in four slender nacelles adhering to the wings' upper surface; the inboard nacelles also accommodated the main landing gear units. The required engine power was calculated in such a way as to make sure that enough power would be available to continue flight with one or two engines inoperative. In particular, the aircraft was to be able to continue take-off in the event of an engine failure or maintain cruise altitude with one engine dead with virtually no reduction in range. With two dead engines the IL-18 was required to continue level flight at lower altitude and speed, with a slight reduction in range. An electro-mechanical feathering system was developed, allowing the variable-pitch propellers to be feathered automatically or manually throughout the flight envelope, precluding the possibility of windmilling – a mode in which the propeller of a dead engine creates tremendous drag which may cause loss of control. A go-around (ie, aborting a landing approach) was possible at any moment with all four engines running or down to 50m (165ft) with one or two dead engines.

The 'second-generation' IL-18 had a tricycle landing gear, of course. However, in contrast to the piston-engined aircraft of 1946, which had twin wheels on each unit and an aft-retracting nose gear, all three units now retracted forward. In order to reduce runway loading the main units featured four-wheel bogies; these rotated through 90° (with the forward pair of wheels uppermost) before the struts retracted so that the bogies lay inverted beneath the inboard engines.

Like its predecessor and namesake, the IL-18 featured a pressurised fuselage, offering comfortable conditions for the passengers and crew throughout the altitude range; the pressurisation and air conditioning system used engine bleed air. Since the pressure differential at high altitude meant the fuselage structure would be subjected to recurring loads applied from within, the fuselage had to be designed in such a way as to ensure adequate fatigue strength. Metal fatigue and recurring loads were thrust into the spotlight in 1954 after two fatal accidents involving de Havilland DH.106 Comet 1 jet airliners owned by the British Overseas Airways Company (BOAC) – G-ALYP (c/n 06003) which crashed into the Mediterranean near Elba Island on 10th January and G-ALYY (c/n 06012) which crashed into the Tyrrhenian Sea off Stromboli, Italy, on 8th April. The investigation of these accidents revealed that both aircraft had disintegrated in mid-air due to explosive decompression caused by fatigue failure.

For safety's sake the engineers of OKB-240 introduced a number of completely new features into the IL-18's design and evolved methods of testing the complete fuselage and its components for fatigue resistance. The most important of these features were as follows. To minimise the consequences in the event of an engine fire the engines were mounted ahead of the wing leading edge, the jetpipes passing over the wings all the way to the trailing edge (as a bonus, the latter feature made it possible to use the engines' residual thrust to the full). All hot components of the powerplant, including the jetpipes, were separated from the airframe by firewalls made of titanium steel. The engine nacelles were designed to provide efficient ram air cooling of the engines and equipped with a powerful fire extinguishing system utilising a highly effective extinguishing agent.

All of the principal airframe subassemblies featured structural elements increasing their resistance to long-term recurrent loads; some of these elements were duplicated for extra reliability. Thus in designing the IL-18 Soviet engineers gained their first experience with fail-safe structures.

Another 'first' in Soviet airliner design was the provision of a weather radar in the fuselage nose. Apart from storm fronts, it could detect other air traffic ahead, high terrain, tall buildings and the like and perform some navigation tasks. Hence the 'IL-18 Jr' had a conventional nose profile with a stepped windscreen. This was obtained by simply grafting a 'thimble' onto the bullet-shaped front end of the pressure cabin, the unpressurised structure accommodating the radar and the nosewheel well. This 'nose job' resulted in a prominent crease curving down and aft from the windscreen – a fact which the Western press did not hesitate to mock at, making comments about 'unsophisticated aerodynamics'.

The comprehensive flight instrumentation and navigation suite enabled the IL-18 to operate at night and in adverse weather conditions. As already mentioned, this was the first Soviet airliner to have automatic landing approach capability, which further enhanced flight safety and extended the aircraft's operational envelope. The automatic approach system was built around the flight director system and autopilot; proving its worth in operation, it paved the way for a fully-fledged automatic landing system later fitted to other aircraft in Aeroflot service.

The electric system had a multi-channel layout and four-channel connection of the equipment to the distribution buses. This guaranteed that the equipment remained operational as long as at least one generator was serviceable.

A reliable and effective de-icing system permitted operation in icing conditions. Electric de-icing was provided for the wing and tail unit leading edges, propeller blades and spinners, engine air intakes, air conditioning system heat exchangers, pitot heads and flightdeck glazing.

The powerful wing high-lift devices gave the IL-18 good field performance, enabling it to use short runways – which may be inevitable if the aircraft has to make a refuelling stop on a long flight. The wheels were equipped with relatively low-pressure tyres, making operation from suitably prepared grass and dirt strips possible – in theory at least.

The high economic efficiency of the IL-18 anticipated by its creators was proved time and again by the type's long and successful operational career. The main criterion for assessing the aircraft's efficiency was the load ratio, the payload making up 46-49% of the take-off weight. This impressive figure was attained thanks to the rationally designed airframe structure, the use of new structural materials and advanced design and calculation methods. As a result, the IL-18's direct operating costs (DOC) were lower compared to other turboprop airliners in the same 'weight category', which was due not only to the high load ratio but also to the aircraft's good aerodynamics, powerful engines and high maintainability reducing the IL-18's fly-away price, repair and maintenance costs.

Despite the fact that the IL-18 was the Il'yushin OKB's first pressurised airliner to achieve production status (not counting the 'first-generation' IL-18 which was also pressurised!) and was developed in the late 1950s, it offered a level of comfort comparable in many respects to that of many airliners designed many years later. The air conditioning system maintained a cabin temperature of +20°C (+68°F) throughout the flight, regardless of the altitude and the season. Sea level pressure was maintained automatically up to 5,200m (about 17,000ft); at a flight level of 8,000m (16,250ft) the cabin pressure was equal to that at 1,500m (about 4,900ft) above sea level, and at 10,000m (32,800ft) the cabin pressure equalled 2,400m (7,875ft) ASL. The air in the cabins was exchanged completely in less than two minutes. The overhead baggage racks incorporated passenger service units (PSUs) with individual ventilation nozzles and reading lights. The comfortable seats featured reclining backs; a lunch tray made of polished wood was stowed in a pocket on the back of the seat in front to be mounted on the armrests when lunch was served.

Two entry doors were provided on the port side of the fuselage ahead and aft of the wings, allowing passengers to board or disembark via two mobile boarding ramps at once for quicker turnarounds. The doors were of the so-called plug type, opened by pushing inwards and sliding towards the nose (forward door) or the tail (rear door); thus the excess pressure in the cabin held the doors firmly shut in flight.

To provide catering in accordance with internationally accepted standards the IL-18 featured a galley, as well as coat closets and well-appointed toilets. The galley was located amidships, separating the two cabins which seated 10 and 65 passengers respectively. The baggage was stowed in two compartments under the cabin floor and a third, unpressurised compartment in the rear fuselage, all with doors on the starboard side; the prototype also featured a baggage stowage area at the 'main deck' level immediately aft of the forward entry door. The cabin walls were lined with heat insulation/soundproofing mats; the cabin trim (wall liners, bulkheads, seat upholstery etc) was made of locally produced materials with carefully chosen colours and decorative patterns.

The prototype of an all-new aircraft, especially an aircraft of this size and complexity, inevitably takes a lot of time to build, since it is virtually hand-made. Nevertheless, the work went right on schedule; registered CCCP-Л5811 (ie, SSSR-L5811 in Cyrillic characters), the first prototype was ready for rollout by June 1957. What looks right flies right, as the saying goes; well, the turboprop IL-18 certainly looked right. The Il'yushin engineers succeeded in creating an elegant aeroplane with clean lines and well-chosen proportions. The aircraft had 15 cabin windows and two overwing emergency exits on each side (2 windows+3+2 exits+1+7); the entry doors were located at the extremities of the cabin (ie, fore and aft of all the windows on the port side).

(Note: Under the Soviet civil aircraft registration system used in 1922-1958 the CCCP- country prefix was followed by a letter designating the agency to which a particular aircraft was assigned, plus up to four figures. In this case the operator designator is an L (for *lee**ney**nyy* [*samo**lyot***] – aircraft in airline service) denoting the Main Directorate of the Civil Air Fleet (GU GVF – ***Glahv**noye oopravl**en**iye grazh**dahn**skovo vozd**oosh**novo **flo**ta*) which operated scheduled passenger/cargo services. Cf. CCCP-Hxxx (the Cyrillic N) for the Main Directorate of the Northern Sea Route (GU SMP – ***Glahv**noye oopravl**en**iye **Se**vernovo mor**sko**vo pu**tee***) which included the Polar Aviation branch, CCCP-Cxxxx (the Cyrillic S) for the Osoaviakhim sports organisation running Soviet air clubs, CCCP-Axxxx for the agricultural division, CCCP-Kxxx (derived from ***krah**snyy krest* – red cross) for ambulance aircraft etc. *The rendering of the registrations as actually applied is used throughout.*)

After a period of ground checks the IL-18 commenced taxying tests and high-speed runs at Khodynka on 1st July 1957. Just three days later, on 4th July, a test crew captained by Vladimir Konstantinovich Kokkinaki (the Il'yushin OKB's chief test pilot since the early 1930s) took the airliner aloft for its maiden flight. The flight did not reveal anything untoward; the machine behaved as predicted and the performance figures measured during the first flight were close to the designers' estimates. (Interestingly, the flight ended at the same airfield; in later years the prototypes of Il'yushin aircraft taking off from Khodynka were ferried to another airfield straight away for safety reasons.) For the initial flight test period the first prototype was equipped with a long air data boom attached directly to the radome, which was still empty.

A few days later CCCP-Л5811 left its birthplace for good, making a short hop from Khodynka to Moscow-Vnukovo airport located about 30km (18.5 miles) south-west of the capital. There the prototype was formally unveiled to a Soviet government commission on 10th July together with the prototypes of two other new turboprop airliners – An-10 CCCP-У1957 (ie, SSSR-U1957; a highly unusual registration, it should be noted, since the code letter U was anything but official!) and Tu-114 CCCP-Л5611. This was a time when the Soviet aircraft design bureaux made a brief attempt to get away from the nondescript alphanumeric type designators, allocating popular names to newly-developed aircraft, as was customary in the West. After all, why should a Soviet airliner have a less stylish name than, say, the Bristol 175 Britannia or Boeing Model 377 Stratocruiser? Thus the IL-18 prototype proudly bore a large legend reading *Moskva* (Moscow), while the An-10 was inscribed *Ookraïna* (the Ukraine). Apparently someone decided that Aeroflot's new flagship deserved something more glamorous and the Tu-114 received the name *Rosseeya* (Russia) – though, unlike the other two aircraft, this name was not worn visibly by the prototype. (It should be noted that the names proved shortlived; at any rate, the production IL-18 was never referred to as the Moskva during its service career. A few Soviet aircraft developed in later years were luckier in this respect, receiving popular names which did catch on.)

Later the IL-18 was shown to foreign diplomats (as a first step towards attracting export orders) and Aeroflot representatives at Vnukovo. Other demonstrations took place at LII's airfield in Zhukovskiy (where the aircraft was inspected by representatives of the Soviet Army's General Headquarters) and at Kubinka airbase west of Moscow where the prototype was shown to Mohammad Zakir Shah, King of Afghanistan, and the Indian Minister of Defence.

There is evidence that the decision to launch series production of the IL-18 was made *before the airliner had even flown*! The aforementioned MMZ No 30 at Moscow-Khodynka was chosen to build the new airliner. There were several reasons for this. Firstly, the plant had already gained considerable experience with Il'yushin aircraft, having manufactured the IL-12 and IL-14 airliners, the IL-28 *Beagle* tactical bomber and the IL-28U *Mascot* trainer version of same. Secondly, the close proximity of the OKB allowed any problems arising during production to be quickly resolved.

On 20th July 1958 the IL-18 had its public debut when CCCP-Л5811 – again captained by Vladimir Kokkinaki – took part in the annual flypast at Moscow's Tushino airfield, the seat of the Central Air Club named after Valeriy Pavlovich Chkalov. Of course this did not escape the attention of foreign military attachés who always attended these public events. After this, the NATO's Air Standards Co-ordinating Committee (ASCC) allocated the reporting name *Coot* to the new Soviet airliner.

On 4th September 1958 General Designer Sergey V Il'yushin was awarded a gold medal at the World Expo fair in Brussels for the development of a new airliner. In 1960, barely three years after the airliner's maiden flight, a group of OKB-240 engineers headed by Il'yushin received the prestigious Lenin Prize, one of the highest awards in the Soviet Union, from the government in recognition of their work on creating the IL-18.

Many publications mention that a second prototype bearing the registration CCCP-Л5812 and the construction number 187000002 was completed at MMZ No 30 in 1957. However, no documentary proof or other evidence of its existence, such as photographs, has been found to date.

IL-18 CCCP-Л5811 caught by the camera as it becomes airborne at Moscow-Khodynka. Yefim Gordon archive

IL-18 *sans suffixe* CCCP-Л5811 cruises over the city it was born in and named after. Note the streaks extending aft from the Soviet flag on the tail which were a curious feature of the prototype's colour scheme. Sergey and Dmitriy Komissarov archive

Above: **СССР-У1957 (ie, SSSR-U1957), the prototype of the rival An-10 Ookraïna (The Ukraine). This type's service career was cut short by serious design flaws which resulted in several fatal crashes.** Yefim Gordon archive

Below and bottom: **Later in its flight test programme the IL-18 prototype received Aeroflot titles replacing the earlier Moskva titles and the aircraft type was indicated on the nose. Here the aircraft is seen taking off from the airfield of its birthplace, the Central Airfield named after Mikhail V Frunze (Moscow-Khodynka).** Yefim Gordon archive

The Civil Family

IL-14 production at MMZ No 30 began winding down gradually in 1957 as the plant began preparing to build the IL-18. MMZ No 30 (referred to in official documents as P/O Box 2402 and later as P/O Box A-1122) was the sole manufacturer of the IL-18 family, except that, as already mentioned, the prototype was built elsewhere.

True, plans were in hand to launch licence production in Czechoslovakia at the Avia factory near Prague. Moreover, this factory's engineering department was working on a radical redesign of the airliner – a 'baby IL-18' with an abbreviated fuselage, a shorter wingspan and only two AI-20 turboprops. Had this aircraft been built, it would have fitted nicely into the market niche populated today by such types as the Aérospatiale/Alenia ATR-72 or SAAB 2000. However, these plans fell victim to socialist planning and 'economic co-operation' between the member states of the Council for Mutual Economic Assistance (COMECON), the Eastern Bloc's equivalent of the EEC; the Avia factory's long involvement with aviation ended when the plant was reoriented towards utility vehicle production. (The use of quotation marks is intentional; in many cases this 'co-operation' looked more like tripping up the competitors! This phenomenon worked both ways; two Soviet aircraft – the Beriyev Be-30/Be-32 feederliner (NATO reporting name *Cuff*) and the Yakovlev Yak-30 (Yak-104) *Magnum* jet trainer – were similarly victimised in favour of the Czech Let L-410 Turbolet and Aero L-29 Delfin respectively, both of which were decidedly inferior to the Soviet designs.)

CCCP-Л5819 (ie, SSSR-L5819), the second production IL-18 (c/n 187000102), was one of the 20 aircraft representing the first production version designated IL-18A. Like the prototype, it carried Moskva titles initially (albeit in a different typeface) and featured streaks aft of the Soviet flag. Note the extreme fore and aft position of the entry doors. Yefim Gordon archive

IL-18A Medium-haul Airliner

As already mentioned, the decision to launch full-scale production was taken before the IL-18 had ever flown; this is evidenced by the fact that several production airframes were laid down at MMZ No 30 as early as the spring of 1957.

Designated IL-18A, the initial production version differed from the prototype mainly in interior layout. For one thing, the noise and vibrations generated by the powerful turboprops proved to be unexpectedly high, especially in the propellers' plane of rotation; for another, the prototype's 75 seats were not enough. Hence the rear wall of the forward cabin had to be moved forward to shift the rearmost row of seats out of the noisiest area, providing adequate comfort for the passengers. The baggage stowage area near the forward entry door gave way to additional seating and the seat pitch was reduced, increasing the number of seats in the forward cabin from 10 to 19; similarly, higher-density seating in the rear cabin provided 70 seats instead of the prototype's 65, bringing the total capacity to 89. This, in turn, brought about changes in the cabin window arrangement; for instance, IL-18A CCCP-Л5819 (c/n 187000102) had 15 windows to port (door+2+3+2 exits+1+7+door) and 17 windows to starboard (1+2+3+2 exits+1+7+1).

(A note must be made here on *construction numbers (manufacturer's serial numbers)*. The IL-18 used the same c/n system as Moscow-built IL-14s, which was fairly straightforward. For instance, IL-18V CCCP-75520 manufactured on 19th October 1963 is c/n 183006703 – ie, ***izdeliye*** (product) 18,[1] year of manufacture 1963, MMZ No 30 (the first digit is omitted for security reasons to confuse would-be spies), Batch 067, third aircraft in the batch (there were almost invariably five aircraft per batch, except Batch 000, which consisted of one (or possibly two) aircraft, and Batch 113, the last one for the civil production run, which had four). The c/n is usually stencilled on the fin and under the wing leading edge at the roots; IL-18s intended for the 'home market' carried the complete c/n, while export aircraft usually had only the batch number and number of the aircraft in the batch painted on (unless they were ex-Aeroflot aircraft). Thus the c/n 187000001 sometimes stated for the first prototype appears extremely doubtful, since the aircraft was *not* built by MMZ No 30!

Some sources in the Il'yushin OKB claim that the first production IL-18A (original registration unknown but possibly CCCP-Л5818, later

reregistered CCCP-75634; c/n 187000101) made its first flight on 26th October 1957. However, this seems doubtful; there is documentary evidence that the *second* production aircraft – the abovementioned IL-18A CCCP-Л5819 – was manufactured on *2nd* October 1957, which automatically means it flew at least once before that date!

(Note: The Soviet civil aircraft registration system was changed in 1958. The operator designator was deleted and the CCCP- prefix was now followed by a five-digit number. The first two digits were usually a sort of type designator introduced for flight safety reasons (this allowed air traffic control officers to identify the aircraft type by its registration and thus avoid placing excessive demands on the crew); this system is still in use today in Russia and most of the other CIS republics.[2] Under the new system the IL-18 was allocated the 75xxx registration block.)

Despite MMZ No 30's prior experience with Il'yushin aircraft, mastering the IL-18 proved to be a major challenge for the factory – just as it would be for Aeroflot a short while later. Quite apart from the obvious size in dimensions (the *Coot* was twice as big as the IL-14 it replaced on the production line!), the IL-18 represented a qualitatively new level in many respects. First of all, the new airliner had a huge pressure cabin occupying almost the entire fuselage; this placed high demands on manufacturing quality to make sure the fuselage structure was properly sealed. The provision of a pressure cabin, in turn, required the installation of pressurisation and air conditioning systems which were *terra incognita* for the plant (the IL-14 had only a simple cabin heater). The much more powerful engines caused much more powerful vibrations and ran on a new kind of fuel, kerosene; all of this created additional problems. Finally, the new-generation aircraft were chock-full of electric and electronic equipment and assorted automatic systems – by the day's standards, of course.

Airframe manufacturing started in the sheet metal department where the metal was cut and supplied to the shops specialising in various airframe subassemblies. Shop 31 manufactured the outer wings, Shop 34 was responsible for the tail unit, Shop 35 for the inner wings and centre section, and Shop 37 for the fuselage. All this and other components were mated in Shop 38, the final assembly shop which was the heart of the plant. All IL-18s rolled off a single final assembly line. True, there was a second final assembly line at MMZ No 30, but that was occupied by another famous aircraft – the Mikoyan/Gurevich MiG-21 *Fishbed* tactical fighter which the factory built in large numbers.

Of course every single aircraft underwent a pre-delivery test programme, and this was where the factory's location suddenly posed a major problem. By the early 1960s Khodynka found itself right in the middle of the steadily growing city, in a built-up area some 6.75km from the Kremlin (the official appellation 'Central airfield' was an apt one). The MiG-21s built by MMZ No 30 were dismantled upon completion and delivered by rail to the plant's flight test facility in Lookhovitsy south-east of Moscow.[3] (Flying them out of Khodynka was out of the question, both for security reasons and out of environmental concerns; imagine the ear-splitting roar a jet fighter produces during take-off!) However, the IL-18's large size made this approach impossible, and trucking the airliners in dismantled form all the way to Lookhovitsy was too troublesome. Hence all IL-18s and other aircraft based on the *Coot* made their first flights from downtown Moscow, passing literally between the rooftops as they climbed out from Khodynka, winging their way to Lookhovitsy. The pre-delivery flight tests and official acceptance of each aircraft took place there. With a few exceptions, the aircraft were fully painted before the first flight.

The first IL-18As were powered by Kuznetsov NK-4s, as were the very first An-10s. However, this basically good engine was plagued by teething troubles and needed a lengthy refinement process. Being in acute need of new turboprop airliners, Aeroflot could not afford to wait until the NK-4 had been perfected; therefore on 15th July 1958 the Council of Ministers ruled that all IL-18s manufactured from November 1958 onwards be powered by Ivchenko AI-20s due to the NK-4's unsatisfactory service test results. All IL-18s manufactured by then were to be re-engined with the new powerplant.

As already noted, it is hard to say why the AI-20 was selected over the NK-4 in the long run, but the whole affair was very probably a case of unfair play. In reality both engines were far from perfect in those days. The *official* story is that the Kuznetsov engine was less reliable and there were cases of compressor failure. However, the AI-20 (which allegedly did not suffer from this weakness) was prone to catching fire in the early years, which led to several fatal crashes. Also, the AI-20's time between overhauls (TBO) was initially barely a hundred hours.

When the first Ivchenko-powered IL-18s became available, Aeroflot started comparing the two engines in service. The airline's representatives had no objections against the AI-20; the engine was simple, and this fact alone was regarded as an asset.

The IL-18A put on weight as production proceeded, mainly due to the thicker fuselage skin introduced from c/n 188000203 (probably registered CCCP-75639) onwards.[4] The empty weight of different 'As varied from 29,450 to 30,579kg (64,925 to 67,414 lb) – a significant increase over the prototype's 28,000kg (61,730 lb). Likewise, the maximum take-off weight of initial production aircraft was 58,000-59,350kg (127,865-130,840 lb) versus the prototype's 59,000kg (130,070 lb).

Production of the A model was brief, the final example rolling off the line in mid-1958. Official sources say IL-18A production was limited to the first four batches (c/ns 187000101 through 188000405), which equals a production run of only 20 aircraft.

Several IL-18As were delivered to the Vnukovo United Air Detachment (UAD) of the Moscow Territorial Civil Aviation Directorate (CAD) which began route proving flights in January 1958. In keeping with the usual practice the *Coots* carried freight and mail at this stage. The airliners were a mix of Kuznetsov- and Ivchenko-engined examples, the former variety including c/ns 187000101 (CCCP-Л5818?), 188000201 (CCCP-Л5821) and 188000202 (possibly CCCP-75638); selecting the best engine for the IL-18 was one of the goals of the test programme. Based at Moscow-Vnukovo airport, the Vnukovo UAD was traditionally the first Aeroflot unit to evaluate new passenger types; the IL-18s were operated by the 65th Flight, one of its constituent sub-units.[5]

Intensive evaluation of the new airliner continued throughout 1958. Among other things, between 10th May and 20th August the IL-18 underwent joint trials at the Soviet Air Force State Research Institute named after Valeriy Pavlovich Chkalov (GK NII VVS – *Gosoo**dar**stvennyy krasnozna**myon**nyy na**ooch**no-is**sle**dovatel'skiy insti**toot** vo**yen**no-vozd**oosh**nykh seel*) at Chkalovskaya airbase about 30km (18.5 miles) east of Moscow.[6] This institution tested all new civil aircraft in those days, assessing possible military uses. The trials included 59 flights in which the *Coot* logged a total of 142 hours.

A new stage of the service trials began on 10th June 1958 when a mixed GKAT/Aeroflot crew including captain Vladimir K Kokkinaki (Il'yushin OKB) and first officer Boris A Anopov (Aeroflot) started flying the aircraft. Mikhail S Gol'dman was the OKB's engineer in charge at this stage of the tests.

On 13-27th December 1958 the airliner successfully underwent cold-weather trials in Yakutsk, one of the coldest places in Russia, operating at ambient temperatures down to –52°C (–62°F). Eduard I Kuznetsov, another Il'yushin OKB test pilot, headed the crew this time. 'Operation Cold Soak' continued on 15th January – 1st February 1959 when an IL-18A was used to test the propeller feathering system at low ambient temperatures in all flight modes. This part of the programme, which proceeded in Yakutsk, Irkutsk and Tiksi, was performed by captain Vladimir K Kokkinaki and first officer Eduard I Kuznetsov.

In 1958-59 the IL-18 demonstrated its capabilities for the first time – but far from the last time – when the third production IL-18A (CCCP-Л5820, c/n 187000103)[7] established a series of world speed and altitude records; these are described in detail in Chapter 4.

IL-18B Medium-haul Airliner

As the IL-18 entered production and service trials began, the OKB strove to improve flight safety and reliability and enhance passenger comfort. Increasing the airliner's seating capacity/payload and extending range was another area of prime concern, as this would improve profitability.

These reasons logically led to the development of the first major production version, the IL-18B. This was basically a high gross weight version of the definitive (Ivchenko-powered) IL-18A. The payload was increased from 12 tons (26,455 lb) to 14 tons (30,865 lb); the maximum take-off weight grew accordingly to 61.2 tons (134,920 lb), requiring the wing torsion box, fuselage structure and landing gear to be reinforced. The unpressurised fuselage nose was redesigned to incorporate a larger dielectric radome, increasing the overall length by 20cm (7⅞in) and resulting in a 'puffier' nose when seen head-on.

Bearing the out-of-sequence registration CCCP-75473, the IL-18B prototype (c/n 188000501) took to the air on 30th September 1958; the aircraft was powered by the initial production variant of the AI-20.[8] The new model allowed MMZ No 30 to fully master the *Coot*, and by the time the final 'Bs left the production line in December 1959 the plant was turning out four IL-18s per month.

It is reported that 65 IL-18Bs was manufactured, comprising batches 5 through 17, and the last one was CCCP-75709 (c/n 189001801). (Mechanically adding up c/ns 188000501 through 189001801 makes 66 units, but airframe No 189001302 is excluded because it was completed as an IL-18V; see page 17.) Some sources, however, say 61 or 63 copies were built and the last IL-18B was either c/n 189001701 or c/n 189001801.

On 12th January 1959 the Il'yushin OKB, GK NII VVS and the Civil Air Fleet Research Institute (NII GVF – *Na**ooch**no-is**sled**ovatel'skiy insti-**toot** Grazh**dahn**skovo voz**doosh**novo **flo**ta*)[9] located at Moscow-Sheremet'yevo airport began joint trials of IL-18B CCCP-75666 (c/n 188000705). More about this 'devil's aeroplane' (three sixes, that's a nice registration) can be found in Chapter 5. On 24th March a joint manufacturer's test/State acceptance trials programme got under way, in the course of which IL-18B CCCP-75673 (c/n 188000902) captained by Vladimir K Kokkinaki was used to verify the compass system, autopilot and astro-compass.

A major milestone was achieved on 20th April 1959 when Aeroflot finally launched revenue services with the type after more than a year of evaluation. On that day a Moscow Territorial CAD IL-18B (CCCP-75674, c/n 188000903) captained by Boris A Lakhtin, Hero of the Soviet Union, performed the Moscow/Vnukovo-1 – Adler service, while sister ship CCCP-75672 (c/n 188000901) captained by A Averkin flew from Moscow-Domodedovo to Alma-Ata.

On 11-19th June 1959 the IL-18B had its international debut when CCCP-75673 was displayed at the 23rd Paris Air Show. Once again the crew was headed by Il'yushin CTP Vladimir K Kokkinaki. Back at home in Moscow, IL-18B CCCP-75644 (c/n 188000303?) was put on show at the VDNKh fairground (***Vys**tavka dos-ti**zhe**niy na**rod**novo kho**ziay**stva* – National Economy Achievements Exhibition) in 1959 together with a Tu-104 registered CCCP-42394 (c/n 9350804?). Interestingly, both aircraft were later returned to service, giving place to a Tu-124V *Cookpot* short-haul airliner (CCCP-45052); in contrast, most of the aircraft visiting the fairground's central plaza were destined to fly no more.

As was the case with the A model, the IL-18B underwent various refinements in the course of production. Early-production 'Bs had 15 cabin windows to port (door+2+3+2 exits+1+7+door) and 17 to starboard (2+2+3+2 exits+1+7). However, soon after the type entered airline service it became obvious that the noise and vibrations created by the mighty turboprops were several times higher than the customary levels on piston-engined airliners. Quite apart from the discomfort, the OKB was evidently worried about fatigue life: the constant vibrations could affect the fuselage's strength in the area of the propellers. Anyway, the interior layout of the forward fuselage was changed again.

Outwardly the updated aircraft differed in having one window less on each side ahead of the wings – ie, door+1+3+2 exits+1+7+door on the port side and 2+1+3+2 exits+1+7 to starboard. It is not known for certain when the

An early-production IL-18B wearing the post-1958 registration CCCP-75669 (c/n 188000803?). This example is representative of the original interior layout and hence cabin window arrangement, with two windows aft of the forward entry door on the port side.
Yefim Gordon archive

A production IL-18B kicks up a minor snowstorm as it begins its take-off run. Note the almost unbroken window arrangement, the lack of the anti-collision light on top to the fin and the long strake aerial for the VHF communications radio along the forward fuselage top.
Yefim Gordon archive

This IL-18B with four windows to starboard in the forward cabin is CCCP-75666 (c/n 188000705), one of several examples completed in IL-18S VIP configuration. This aircraft was operated by the Soviet Air Force. Yefim Gordon archive

All IL-18s, including this 'B, were flown from Moscow-Khodynka to the plant's flight test facility at Tret'yakovo airfield in Lookhovitsy south of Moscow for pre-delivery tests. With no airstairs on site, the occupants were obliged to use rickety stepladders. With a few exceptions, all *Coots* were painted before they left the plant. Yefim Gordon archive

change was made, but Batch 12 appears most likely. At any rate, IL-18B CCCP-75684 (c/n 189001104) operated by the Moscow Territorial CAD/Vnukovo UAD/65th Flight still had the old window arrangement, whereas CCCP-75694 (c/n 189001304) had 14 windows to port and 16 to starboard.

Most of the already existing IL-18Bs were modified by sealing the appropriate (or rather inappropriate!) windows with metal plugs. Later, when these aircraft were due for an overhaul, the offending windows were removed altogether and the stringers and solid skin panels in the area were reinstated to bring the aircraft up to late IL-18B standard.[10] Some IL-18As underwent a similar conversion, too.

An IL-18B with the out-of-sequence registration CCCP-75699 (c/n 189001402)[11] manufactured on 31st August 1959 was the first *Coot* to be powered by improved AI-20 Srs 2 engines.[12] While having the same ratings as the original AI-20 Srs 1 – 4,000ehp for take-off, a nominal power rating of 3,400ehp at sea level and 2,800ehp at 8,000m (26,250ft) – the new version featured improved reliability. Previously manufactured IL-18s were refitted with AI-20 Srs 2s in due course.

The 'B was also the first variant to be exported. The first-ever export delivery of a *Coot* took place on 21st November 1959 when an IL-18B registered 202 (c/n 189001401) was handed over to the 'Chinese Aeroflot' – the Civil Aviation Administration of China (CAAC). This aircraft apparently left the production line with the old AI-20 Srs 1 turboprops but was refitted with Srs 2 engines before delivery. In addition to China (which bought three 'Bs), two IL-18Bs were delivered to the Czechoslovak flag carrier ČSA Československé Aerolinie.

IL-18S VIP Aircraft

The IL-18S ([*samo**lyot***-] *sa**lon*** – VIP aircraft) was developed as a VIP version of the IL-18B pursuant to a GKAT order dated 16th May 1958. At least two such aircraft were operated by the Soviet Air Force in an airline-style colour scheme with a 'lightning bolt' cheatline different from the pre-1973 Aeroflot IL-18 standard.[13] The first one, coded '001 Red' (ex-CCCP-75667?, c/n 189000801?), was an early-production IL-18B with 15 cabin windows to port and 17 to starboard. Marshal Andrey A Grechko, the then Defence Minister of the Soviet Union, used this aircraft during his travels around the country.

(Note: Unlike Western military aircraft which have *serials* allowing positive identification, since 1955 Soviet and CIS military aircraft have *tactical codes* which are usually simply the aircraft's number in the regiment operating it, rendering positive identification impossible. Three-digit tactical codes are rare and usually allocated to development aircraft only, in which case they still tie in with the construction number (manufacturer's serial number) or fuselage number (line number). With transports, however, the three-digit codes are now usually the last three of the aircraft's former civil registration.)

The ministerial aircraft status and the tactical code '001 Red' later passed to a Tu-104V VIP transport (c/n unknown). As for the IL-18S, it received a non-standard civil registration, CCCP-33569 (actually its former ATC callsign); this enabled it to fly abroad, including nations which would not allow a Soviet military aircraft to enter their airspace.[14] Still later this aircraft was reregistered CCCP-75479.

The other known SovAF IL-18S was quasi-civil and based on a late-production IL-18B with 14 cabin windows to port and 16 to starboard (c/n 189001603?). It was registered CCCP-75749 (out of sequence) but probably had overt military markings before that.

At least one IL-18S (CCCP-75705, c/n 189001702) belonged to Aeroflot's VIP unit – the 235th Independent Air Detachment (IAD) catering for the Soviet federal government. Unfortunately this aircraft was lost in a fatal crash near Kiev, the capital of the Ukrainian SSR, on 17th August 1960.

IL-18V Medium-haul Airliner

Operational experience accumulated with the first two production models led the Il'yushin OKB to develop a further refined version of the airliner. In Russian alphabetical sequence the new model was designated IL-18V.

Interestingly, it was quite some time before the suffix letters to the designation were introduced on the *Coot* – and it was exactly the IL-18V that 'kicked off' this practice by being sufficiently different from the preceding versions. Fact is, apart from the initial variations on the engine theme before the AI-20 was standardised, the early IL-18s featured only minor alterations between the versions or within the version. These mostly concerned the fuselage, being limited to some local reinforcement, moving the interior partitions to change the size of the cabins and galley, changing the number of toilets and adding or deleting a window or two. This may explain the fact that initially neither Aeroflot nor the manufacturing plant, nor even the Il'yushin OKB discerned between the versions in their official correspondence, referring to the aircraft simply as 'IL-18'. Even when the IL-18V appeared, the earlier versions were still referred to as 'IL-18 (interior layout 17A)'. And, in fact, the IL-18V itself was initially known as the 'IL-18 (interior layout 15A)' before the suffix letters became widespread. (It may be noted that the type was invariably painted on the actual aircraft as 'IL-18' (with no suffix letter); not until the 1990s did some airlines apply the full nose titles to a few *Coots*.

While being structurally identical to that of the IL-18A/B, the IL-18V's fuselage was redesigned to allow greater freedom of interior layout, permitting the separation of the cabin into different classes, and improve passenger comfort. Unlike previous versions, the IL-18V had three completely separate passenger cabins – one at the front between fuselage frames 6-14, one in the centre fuselage (frames 22-42) and one at the rear (frames 47-54). The entry vestibules were now located in between, not at the extremities of the cabin. The toilet immediately ahead of the forward cabin section on the IL-18A/B was eliminated – most of the passengers now had to use the two toilets on the starboard side of the forward vestibule, with a coat closet in between. A second 'cloakroom' was located on the port side just aft of the forward entry door, together with rescue equipment stowage.

The galley was relocated from the area near the wing leading edge (in the propellers' plane of rotation) to the rear vestibule, much to the joy of the flight attendants. Previously the poor stewardesses were forced to spend most of the time in the area where the noise level was at its highest, exceeding 110dB in some flight modes. The area of the galley was increased and the equipment fit became more comprehensive; another welcome change was the addition of a window in the galley compartment allowing the flight attendants to work in daylight conditions. Also, the deafening roar of the engines stopped the passengers from lingering in the forward lavatories too long – which other passengers waiting in line to go to the loo were grateful for! (In passing, it is really uncanny how passengers seem to get the urge to relieve themselves all together immediately after take-off!) The passengers in the rearmost cabin enjoyed the convenience of a separate compartment between frames 54-56 which accommodated a coat closet (to port) and a toilet (to starboard).

These changes resulted in a totally different window arrangement with 15 windows to port (3+door+2+2 exits+1+4+door+3) and 16 to starboard (3+2+2 exits+1+4+1+3). This arrangement was to remain almost unchanged until the end of production. The location of the door tracks also changed: the forward door now slid rearwards when opened and the rear door forwards.

Some minor changes were made to the pressurisation and air conditioning system; among other things, this resulted in the forward excess air spill valve being moved from the port side of the fuselage to the starboard side. Incidentally, the position of this valve enclosed by a small fairing with an outlet at the rear is the only way to tell an IL-18V from an IL-18B in a head-on view when the doors and windows are not visible.

The IL-18V prototype, again with an out-of-sequence registration (CCCP-75710, c/n 189001302) entered flight test on 10th December 1959.[15] Production of the new model began with CCCP-75711 (c/n 189001802). The IL-18V remained in production until December 1965 and a total of 334 was built in various configu-

Soon after the IL-18B entered service it became imperative to change the interior layout near the propellers' plane of rotation to enhance fatigue resistance and passenger comfort. The updated aircraft had one window less on each side ahead of the wings, as illustrated by CCCP-75665 (c/n 188000704?). This aircraft crashed near Alma-Ata on 22nd November 1966. Yefim Gordon archive

This view illustrates the old window arrangement with two pairs of windows in the forward cabin. Note the smoothly curving joint between the pressurised and unpressurised parts of the nose, a result of a 'puffier' extreme nose section being introduced on the B model. Il'yushin OKB

001

Opposite page:

Top and centre: **Several IL-18S VIP aircraft were delivered to the Soviet Air Force, wearing overt military markings and an unusual 'lightning bolt' cheatline. This example is seen during pre-delivery tests, hence the absence of a serial.** Il'yushin OKB

Bottom: **An in-service IL-18S serialled '001 Red' (ATC callsign CCCP-33569, c/n 188000801) in the same 'lightning bolt' colour scheme. This was the ministerial aircraft of Marshal Andrey A Grechko.** Boris Vdovenko

This page:

This shot of three *Coots* performing at the Domodedovo air fest in the summer of 1961 is interesting in that it shows the IL-18V prototype, CCCP-75710 (c/n 189001302) – farthest from camera – in a unique formation with an IL-18B (CCCP-75702, c/n 189001503, lead aircraft) and an IL-18A (CCCP-75645, c/n 188000304) Yefim Gordon archive

rations, making it the most widespread version of the *Coot*. (Note: Production of the IL-18E, the next model on the line, started in *September* 1965; however, the claim about the last 'Vs being manufactured in December may well be true because they may have been manufactured late due to complex outfitting jobs or the need to rectify defects. With few manufacture dates known, it is hard to say.)

Initially the IL-18V was a 78-seater, the three cabins accommodating twenty, fifty and eight passengers respectively. The forward and centre cabins had five-abreast tourist class seating (2+3 with an aisle offset to port) at 90cm (35½in) pitch. The rear cabin, however, was conceived as a first class cabin with two rows of four-abreast sleeperette seats at 112cm (44in) pitch; the seats featured reclining backs and extendable leg rests. However, a report summing up the checkout tests of an early-production IL-18V, CCCP-75713 (c/n 180001804),[16] killed off the idea. *'An aircraft with such an interior layout is unacceptable as a mass-produced aircraft serving domestic routes in the USSR because Aeroflot has common fares and the tickets are not subdivided into first-, second- and third-class tickets as far as prices are concerned* – the report went – *Therefore it is considered advisable to manufacture the IL-18V in an 84-seat all-tourist configuration, which will improve operating economics.'* [17]

Commenting on the first-class sleeperette seats, the report said, *'Despite the advantages they offer, using the seats creates a number of inconveniences; the extendable leg rest is too short,* [...] *when the seat back is tilted it may injure the legs of the passenger sitting in the seat behind. The leg rest locking mechanism is troublesome and unsafe; these mechanisms were out of order on almost all seats during the one-month period of operation.'*

Isn't that typical of the Soviet industry (not just aircraft industry) which could make a basically good product but show complete disregard for the details – and it's exactly the minor but irritating things that can drive you up the wall. Such cock-ups would be pardonable in the tourist or economy class, but not in the first class.

The report was duly considered, and from then on there were no more first class seats on Soviet airliners. (There you are. Read between the lines: So much for bourgeois excess in the form of first class travel, otherwise the Soviet people will get fat and lazy, period.) On the other hand, there could be no doubt that the rear cabin of the IL-18V offered the greatest comfort as far as noise and vibrations were concerned – a fact which was later taken into account when developing VIP versions for government use.

Nevertheless, the factory continued turning out IL-18Vs with the controversial clumsy sleeperette seats in the rear cabin for a while, and this mixed-class configuration was in use until Aeroflot set about increasing revenues from its *Coot* operations. The seat pitch in the centre cabin was reduced to 84cm (33in) to provide room for an eleventh row of seats; the rear cabin was reconfigured to feature three rows of ordinary tourist class seats – two rows five-abreast and a row of four seats (2+2) at the rear where the fuselage started narrowing. The first aircraft to be so configured was c/n 181003803 (presumably registered CCCP-75779), and most IL-18Vs were built and operated in this 89-seat layout.

Unlike the cabins, the three baggage compartments remained unchanged. Operational experience accumulated with the IL-18A/B made it possible to assess them objectively. The *Coot*'s baggage compartments were better equipped than those of the Tu-104A and An-10; however, the baggage loading doors were rather too small, allowing only very compact items to be carried. The abovementioned report concerning IL-18V CCCP-75713 No 1 stated that the baggage doors needed to be enlarged; yet this deficiency was never corrected – the size of the doors remained unaltered until the end of IL-18 production. To make matters worse the doors of the two underfloor baggage compartments opened inwards and upward, rendering part of the floor area unusable. Nothing could be done about it at the time – the Soviet aircraft engineers had not yet figured out how to design a pressure door that did not eat up part of the baggage space. Outward-opening pressure doors for the baggage compartments did not appear until 1965 when the Tupolev OKB brought out the Tu-154 trijet medium-haul airliner (which, incidentally, succeeded the IL-18 on many of Aeroflot's services).

Whereas longer range and higher capacity were the Il'yushin OKB's 'strategic' goals with respect to the IL-18 (which were attained by developing other versions described here), the more down-to-earth everyday work centered on enhancing passenger comfort and improving the avionics suite. Of course, the OKB also paid due attention to refining various structural components and systems – a natural process for an aircraft which has been in production for years.

Meanwhile, the aircraft industry took great pains to improve the reliability of the AI-20 engine which was alarmingly low at first. As noted earlier, the refined AI-20 Srs 2 was introduced from Batch 14 onwards in August 1959 when the IL-18B was still in production. Soon afterwards, in mid-1960, this engine gave way to the identically rated AI-20A (aka AI-20 Srs 3) incorporating further improvements; IL-18V DM-STB (c/n 180002001), the second *Coot* delivered to East Germany, was the first to feature the new engines.[18]

While the basic airframe design remained the same, subtle changes were introduced almost in every production batch in order to eliminate design shortcomings and manufacturing defects as they cropped up. Better flight instruments, new systems components and improved parts found their way to the production line. Since these modifications could affect performance and handling (usually adversely because aircraft tend to put on weight!), a single aircraft – known as an *etalon* (standard-setter) – was picked each year for checkout trials to see if the updated IL-18 still met the specifications. Known examples are the aforemen-

Above: **The IL-18V was the most numerous version of the *Coot*. This view of a production example manufactured on 26th August 1964 (CCCP-75558, c/n 184007505) illustrates the new door and window arrangement resulting from the different interior layout. This aircraft crashed near Frunze, the capital of the Kirghiz SSR, on 30th January 1976.** Yefim Gordon archive

Left: **IL-18Vs on the final assembly line in Shop 38 of MMZ No 30 at Moscow-Khodynka.** Yefim Gordon archive

tioned CCCP-75713 No1 (c/n 180001804) which was the *etalon* of 1960, an IL-18V with the out-of-sequence registration CCCP-75842 (c/n 182005201)[19] which was the *etalon* of 1962 and IL-18V CCCP-75523 (c/n 183006801)[20] which was the *etalon* of 1963.

IL-18V CCCP-75745 (c/n 181003001) manufactured on 18th April 1961 introduced an important feature – a TG-16 auxiliary power unit (APU) for self-contained engine starting and ground power supply.[21] Previously all IL-18s were provided with a bank of silver-zinc or lead-acid DC batteries for use as a ground power source. Developed by the Kazan' Machinery Design Bureau (*Ka**zahn**skoye pro**yekt**noye byu**ro** ma**shin**ostro**yen**iya*),[22] the APU was installed in the unpressurised rear baggage compartment; the exhaust was located just ahead of the starboard stabiliser and surrounded by a heat-resistant steel plate to protect the fuselage skin. On some aircraft the door of the rear baggage compartment featured a prominent 'elephant's ear' air intake allowing the APU to run in flight.

IL-18V CCCP-75881 (c/n 183006104) manufactured on 24th April 1963 featured a revised interior layout with high-density seating for 110 passengers (24+72+14). In 1966 the Ministry of Civil Aviation's aircraft overhaul plant No 402 at Moscow-Bykovo airport refitted IL-18V CCCP-75506 (c/n 183006404) with a 100-seat interior. The aircraft completed its trials programme on 16th December 1966, and most IL-18Vs undergoing refurbishment were progressively upgraded to this standard.

CCCP-75506 manufactured on 31st July 1963 was noteworthy in one more respect – it was the first *Coot* to be powered by the new AI-20K (AI-20 Srs 5) engines. IL-18V CCCP-75559 (c/n 184007703) manufactured on 15th October 1964 introduced an electro-pulse de-icing system. Two weeks earlier, on 1st October, Vladimir K Kokkinaki had made his last flight as a test pilot in the captain's seat of IL-18V CCCP-75549 (c/n 184007304).

On 25th May 1965 IL-18V CCCP-75544 (c/n 184007202)[23] began a series of demonstration flights in India (which did not lead to any *Coot* sales in that country, however). The aircraft was captained by Il'yushin OKB test pilot Aleksandr M Tyuryumin; D N Simanovich was the engineer in charge of the programme.

IL-18V/VIP Version (IL-18V 'Salon')

A VIP transport version of the IL-18V for government and military use was developed concurrently with the standard passenger version as a successor to the IL-18S. No separate designation is known (with a separate suffix letter, that is); however, some sources suggest the aircraft was designated IL-18V 'Salon', as simple as that.

At least 24 IL-18Vs were built as VIP aircraft. Known Soviet examples were registered CCCP-75530,* -75563, -75591*, -75716, -75717, -75742, -75764 (later sold to Algeria as 7T-VRA), -75774*, -75823, -75825*, -75873 and -75894*. IL-18V 'Salons' were also delivered to the Algerian Government (the abovementioned 7T-VRA), the Chinese People's Liberation Army Air Force (serials '50850 Red', '50851 Red' and '50854 Red'), the Czechoslovak Federal Government Flight (OK-BYP), the East German Air Force ('493 Black',[24] '497 Black', '499 Black', DM-STH* and DM-STI*), the Polish Air Force ('101 Red' No 1 [c/n 180002504], '102 Red' No 1 [c/n 181002701] and '101 Red' No 2 [c/n 185008305]), and the North Vietnamese Government). Aircraft marked with an asterisk were quasi-civil.

Outwardly the VIP version was usually identical to any ordinary 'V. There was, however, a notable exception. The first IL-18 delivered to Vietnam (BH-195, ex-CCCP-75424, ex-Ghana Airways 9G-AAM, c/n 181003305) was converted from a regular passenger aircraft – and for some obscure reason it had a non-standard window arrangement with one window less on the port side aft of the wings (3+door+2+2 exits+1+3+door+3).

By virtue of their VIP role the IL-18V 'Salons' were often thrust into the limelight by the famous people they carried on various ceremonial occasions. For instance, on 14th April 1961 CCCP-75717 (c/n 180001904), a 235th IAD aircraft, brought Yuriy Alekseyevich Gagarin to Moscow from Kazakhstan where the world's first spaceman had landed after his historic 104-minute flight on 12th April. The aircraft was captained by P M Vorob'yov, while the air detachment was then headed by Boris Pavlovich Boogayev who later became Minister of Civil Aviation (he held this office in 1980-87). On 8th August 1961 the second man in space, Gherman Stepanovich Titov, returned to Moscow aboard IL-18V 'Salon' CCCP-75742 (c/n 181002802) captained by Yevgeniy P Barabash after completing his space mission. On 17th August 1962 cosmonauts Andrian G Nikolayev and Pavel Romanovich Popovich, the first two-man space crew, came back to Moscow after their mission in CCCP-75823 (c/n 182004902), a 235th IAD IL-18V 'Salon' captained by Distinguished Pilot Ye K Kalinovskiy. Similarly, the world's first three-man space crew (Vladimir M Komarov, Konstantin P Feoktistov and Boris B Yegorov) was brought back to Moscow on 19th October 1964 by IL-18V 'Salon' CCCP-75873 (c/n 183006003) captained by S N Kechayev.

IL-18V Polar Version

The need to ensure rapid resupply of and personnel rotation at Soviet Antarctic research stations led the Il'yushin OKB to modify a single IL-18V for the task. Registered CCCP-75743 (c/n 181002901), this one-off aircraft featured additional fuel tanks to increase range. Because of the longer endurance (necessitated by the long overwater legs of the route to the Antarctica) the engines' oil supply was also increased; the new oil tanks were too large to fit under the standard cowlings, and a large bulge had to be made on the port side of each engine nacelle over the oil tank.

Further changes included the installation of additional radio and celestial navigation equipment – a fact revealed by the additional aerials and a star tracker for the astrosextant mounted in line with the second cabin window (fuselage frames 9-10), slightly offset to starboard. An angular fairing of unknown purpose was installed ventrally just ahead of the forward entry door; two tandem sets of dipole aerials for radio altimeters were fitted immediately aft of the nose gear and ahead of the wing trailing edge. The port forward emergency exit had a 'solid' cover minus window.

Originally the aircraft had a standard colour scheme except for the addition of small '*Pol**yar**naya avi**ah**tsiya*' (Polar Aviation) titles on the forward fuselage augmenting the usual Aeroflot titles and winged logo. Later, however, the outer wings, tail surfaces and the top of the fuselage were painted Dayglo red for high definition against snow and ice in the event of a forced landing.

The aircraft flew a successful resupply mission from Moscow to Ice Station Mirnyy between 15th December 1961 and 2nd February 1962; see Chapter 4 for more details.

IL-18V-26A Long-range Polar Supply Aircraft (IL-18D – First Use of Designation)

The success of the Antarctic mission of 1961-62 led the Il'yushin OKB to develop another special version of the *Coot* designated IL-18V-26A. According to some sources, the aircraft was also called IL-18D (***dahl'****niy*, long-range) – a designation reused two years later for another version.

To increase range a large auxiliary fuel tank was installed in the fuselage *above* the wing centre section, blocking up the centre cabin completely and rendering the overwing emergency exits unusable. Thus the aircraft had four separate cabins, Nos 1 and 2 being accessible only via the forward door and Nos 3 and 4 via the rear door. The non-functional emergency exits were closed by windowless covers, reducing the number of windows to 13 on the port side (3+door+2+1+4+door+3) and 14 to starboard (3+2+1+4+1+3). A star tracker was installed in line with the second cabin window but offset to port, not to starboard. Another external identification feature was the bulges on the port side of all four engine nacelles over enlarged oil tanks, just as had been the case with IL-18V CCCP-75743.

Five *Coots* registered CCCP-75844 through -75848 (c/ns 182005304, 182005305 and 182005401 through 182005403) were completed to this standard. Between 20th November 1963 and 11th January 1964 CCCP-75845 participated in the second IL-18 mission to Antarctica.

Subsequently all five IL-18V-26As were converted to standard IL-18Vs by removing the extra tank and reinstating the emergency exits. They served on with various Aeroflot divisions, including the Moscow Territorial CAD/Domodedovo UAD/212th Flight (CCCP-75845), the Uzbek CAD/Tashkent UAD/202nd Flight (CCCP-75846) and the Krasnoyarsk CAD/1st Krasnoyarsk UAD/214th Flight (CCCP-75847), until finally retired as time-expired.

IL-18V Calibrator Version

In 1978 one of the IL-18Vs operated by the East German airline Interflug (DM-STP, c/n 184007401) was transferred from the regular transport department (IF/VF – Interflug Verkehrsflug) to the Inspectorate of Air Traffic Control & Communication (IF/TP-FSNW – Interflug Technische Prüfung – Flugsicherung und Nachrichtenwesen) and converted for the navaids calibration role. The modified aircraft featured two additional 'hockey stick' aerials mounted dorsally just aft of the forward entry door and ventrally just aft of the rear entry door; there was also an additional retractable light for phototheodolite measurements. The chief identification feature, however, was the overall grey colour scheme with a red-tipped radome and a red band around the fuselage near the No 3 baggage compartment door; this promptly earned DM-STP the sobriquet *Graue Maus* (Grey Mouse)!

The aircraft served as a calibrator long enough to see the change of East Germany's registration prefix from DM- to DDR- in 1981 and German reunification in 1990 when it became D-AOAQ. Soon afterwards it was reconverted to standard passenger configuration.

IL-18 Cargo Aircraft (IL-18G?)

Any IL-18 could be adapted for carrying cargo by the simple expedient of removing the passenger seats. Right from the start the seat modules were designed to be attached to the rails in the cabin floor by quick-release fasteners, and removing the seats took no more than 30 or 40 minutes. After that, the cabins could be filled with items which were small enough to go through the entry doors. This is exactly how the IL-18A was operated by the Vnukovo UAD during route proving flights.

Besides the absence of a large cargo door (which was not developed until much later), this simple adaptation of the IL-18 for the cargo role was handicapped by the absence of any kind of cargo handling and restraining devices. Also, the absolutely superfluous cabin partitions, toilets etc remained, eating up part of the cargo space. Some sources quoted the designation IL-18G for such makeshift freighters; the G was not simply the next letter in the Russian alphabetical sequence but denoted *groozo**voy*** (cargo, used attributively). However, this is doubtful because G could be deciphered as *gov**no*** (shit) and such a designation would not have caught on!

Left: **CCCP-75591 (c/n 185008003), an IL-18V 'Salon' VIP aircraft operated by the Soviet Air Force. This aircraft was assigned to the Transbaikalian Defence District and based at Ulan-Ude/Vostochnyy airport.** Sergey and Dmitriy Komissarov archive

Below: **IL-18V CCCP-75743 (c/n 181002901) was specially modified for long-range flights to Antarctica. It is seen here in its original colour scheme – ie, standard IL-18 colours with additional *Polyarnaya aviatsiya* (Polar aviation) titles. Note the bulges on the port side of the engine nacelles over enlarged oil tanks.** Yefim Gordon archive

IL-18T Cargo/Troop Transport/Medevac Aircraft (IL-18AT, IL-18BT, IL-18VT)

When MMZ No 30 had manufactured a substantial number of IL-18s, the OKB devised a way of substantially expanding the applications of these aircraft without resorting to major design changes. The result was a cargo version which could be converted from any IL-18, regardless of the original variant, in such a way as to eliminate the shortcomings of the IL-18G described above.

There was also another aspect to the matter. It is an open secret that in the Soviet Union even the *civilest* aircraft, such as the IL-18, were designed to meet the requirements of the military as well as those of Aeroflot. (As a result, the Soviet airline was a paramilitary surrogate which, along with the general obsession with security, was one of the peculiarities of the Soviet society.) As already mentioned, one of the first production IL-18s was handed over to GK NII VVS for investigating possible military uses; the development of the cargo version offered some opportunities in this respect.

Thus, as early as 18th December 1958 the Soviet Council of Ministers issued a directive ordering the development of a military transport/casualty evacuation version designated IL-18T (***trahns****portnyy* – transport, used attributively). In medevac configuration the aircraft was to carry 69 stretcher patients and two medical attendants, while the paradrop configuration was to take 118 assault troopers and two instructors.

Depending on the original version (IL-18A/IL-18B/IL-18V), the new cargo variant was initially called IL-18AT, IL-18BT or IL-18VT respectively; later the designation was shortened to IL-18T (just like in the CofM directive) for simplicity's sake. Unlike the IL-18G, the conversion involved reinforcing the cabin floor and deleting the cabin partitions, toilets and galley. To distribute the loads more evenly special inverted-U shaped frames were attached to the fuselage structure inside the cabin; these carried an overhead rail running the full length of the cabin on which an electric hoist travelled. The OKB devised special cargo pallets for the IL-18T to ease loading and unloading; when empty they were stored in the former entry vestibules. The pallets were placed near the doors, one by one, and then whisked away to the cabin by the overhead hoist as they filled up; the process was reversed during unloading. This feature saved the ground personnel a lot of legwork when working with small items of goods.

The efficiency of cargo aircraft operations depends heavily on that of cargo handling operations, and that in turn hinges on having the right equipment. Major airports have fork lift trucks and belt conveyors to ensure speedy loading and unloading, but at smaller and ill-equipped locations the process could take hours. Hence the Il'yushin OKB developed a special foldaway belt conveyor loader for the

Later, CCCP-75743 received Dayglo red tail surfaces, outer wings and fuselage roof for high definition against white backgrounds. Note the star tracker on the forward fuselage and the additional dipole aerials under the wing centre section and just aft of the nose gear. Sergey and Dmitriy Komissarov archive

IL-18T; it was used for moving cargo from ground level or a truck bed to the door sill and stowed in the forward underfloor baggage compartment when not in use.

These measures took care of many problems handicapping the so-called IL-18G, but at the expense of the latter's quick-change capability. Worse, the greatest deficiency – the inability to swallow bulky cargoes – was not overcome until many years later when the IL-18 had almost vanished (see IL-18GrM); the IL-18T could only carry small packages (not necessarily containing good things, in spite of the proverb).

The IL-18T was also suitable for troop transportation (but not for paradropping – this capability took several more years to incorporate, see Chapter 5). In this configuration the overhead cargo hoist was removed and longitudinal beams were attached to the vertical load distribution frames. These carried rigid seats installed along the fuselage sides, with a third row on the centreline to give a total of 114. Each seat was provided with an oxygen mask and supply hose.

For the medevac role 63 stretchers in three tiers, three-abreast, were installed on fittings attached to the transverse frames and special uprights in the cargo cabin. The passages between the rows of stretchers permitted medical attendants to tend any patient. There were two seats for the medical attendants at the front and rear of the cabin, and a doctor's table was installed in the middle row near the forward entry door.

State acceptance trials of the IL-18T prototype (identity unknown) began on 29th February 1960; the aircraft was powered by AI-20 Srs 3 (AI-20A) engines. D A Polikarpov was the engineer in charge of the trials.

IL-18V Mid-life Updates

After the advent of the IL-18V further development of the *Coot* proceeded along two lines. One was a persistent effort to improve the V model; this was not limited to dealing with defects as they surfaced but included the introduction of such goodies as anti-collision lights, Doppler speed/drift sensors etc. Some of these changes disrupted the clean lines of the aircraft while others removed existing excrescences, making the airliner more elegant.

The other main thrust was aimed at developing new versions that would allow the IL-18's potential to be used to the full. In so doing the Il'yushin OKB engineers were willing to resort even to significant airframe changes (which are usually avoided insofar as possible).

The powerplant (specifically, the appallingly low engine life and TBO), the troublesome (and initially incomplete) radio navigation suite and the inadequate level of passenger comfort were the *Coot*'s main failings at that time. Passenger comfort in this instance depended on a multitude of factors, the most important one being cabin noise; the immediate objective was to reduce the noise level at least to the 100dB prescribed by the Soviet standards then in force. Another area requiring attention was the air conditioning and pressurisation system: it was imperative to reduce the rate at which cabin pressure changed during climb/descent so that the aircraft would be easier on the passengers' ears. It was also necessary to improve the cabin ventilation rate, reduce the content of harmful substances in the cabin air, include an air humidification feature and provide air conditioning on the ground when the engines were inoperative. Finally, there was the issue of seat design and seating arrangement.

The OKB had varying success in addressing these issues; radical improvements in some areas were interspersed with only modest results in others, and sometimes the engineers' efforts backfired. Of course, the IL-18V was not the only version to benefit from these changes – new features which proved their worth were incorporated on later versions, and some were retrofitted to earlier versions as well. It would be best to describe the features added to the IL-18V during the production period 'in order of appearance'.

As already mentioned, a TG-16 APU was fitted as standard to all aircraft from IL-18V CCCP-75745 (c/n 181003001) onwards. Until then the only electric power source when the engines were shut down (aside from ground power units) consisted of DC batteries installed at the same location – in the unpressurised rear baggage compartment. IL-18As and most IL-18Bs featured a quartet of 15STsS-45 silver-zinc batteries; from the final IL-18B (CCCP-75709, c/n 189001801) onwards the number was increased to six for more reliable engine starting at low ambient temperatures. From Batch 23 onwards these were substituted by 20 12SAM-28 lead-acid batteries which had a lower capacity but were cheaper. The provision of an APU allowed most of them to be deleted, which gave a significant weight saving. All previously built *Coots* were progressively retrofitted with the APU during overhauls.

Of course, the modifications were not limited to simply switching the excess batteries for the APU and cutting a hole for the exhaust ahead of the starboard tailplane. The TG-16 required a separate fuel tank (with a separate filler cap), fire extinguishing, indication and warning systems.

The avionics were updated, too, which also led to some exterior changes. Originally the

IL-18V DM-STP (c/n 184007401) operated by the East German airline Interflug was converted into a unique navaids calibration aircraft for the Inspectorate of Air Traffic Control & Communication in 1978. It is seen here in its overall grey colour scheme which earned it the nickname *Graue Maus* but prior to the addition of red markings to the radome and aft fuselage. Note the L-shaped aerials on top of the fuselage.
Yefim Gordon archive

IL-18 was equipped with two pairs of communications radios – two RSB-5/1230 HF radios and two RSIU-4 VHF radios (one of each model was a backup unit in case the main one failed). All four radios used a common wire aerial stretched between the flightdeck roof and the top of the fin, with a strake aerial running all the way from the flightdeck to the forward emergency exits as a backup.

The RSIU-5 VHF radio became a standard fit from Batch 39 (ie, presumably from IL-18V CCCP-75782) onwards, replacing the earlier RSIU-4; outwardly it was identifiable by an L-shaped aerial mounted ventrally in line with the No 3 cabin windows. However, some earlier *Coots*, including IL-18B c/n 189001502 (CCCP-75701?), were fitted with the new radio before that – probably for test purposes. Later, the main RSB-5/1230 radio was replaced by a more advanced RSB-70 HF radio.

While making his state visit to the USA in 1959, the Soviet leader Nikita Sergeyevich Khruschchov noticed that American airliners were equipped with anti-collision lights – flashing red beacons making the aircraft more conspicuous during taxying and in flight. On returning to Moscow he instructed the heads of Soviet aircraft design bureaux to incorporate this safety feature on Soviet airliners. This was one of the few instances when Khruschchov had a positive influence on the development of aviation; usually his actions in this field were detrimental as a result of his belief in intercontinental ballistic missiles and his conviction that manned combat aircraft had no future. Thus, starting with IL-18V CCCP-75792 (c/n 181004101), all *Coots* left the factory equipped with SMI-2 anti-collision lights atop the fin and under the aft fuselage just ahead of the fin leading edge; all existing IL-18s were progressively upgraded.

The weather radar was one particular problem area. All Soviet first-generation turbojet and turboprop airliners were designed to have radar – a totally new feature for Soviet civil aviation. However, the Tu-104 featured an RBP-1 radar which had been already put through its paces on the Tu-16 bomber (it was actually a bomb-aiming radar, hence the designation RBP for ***rah****diolokatsi****on****nyy bombardi****rov****ochnyy pri****tsel*** – 'radar bomb sight'). Conversely, the IL-18's RPSN-2 *Emb****lem****a* (Emblem)[25] was a purely civilian weather radar – and was brand-new, which inevitably meant lots of teething troubles. The first 16 batches (c/ns 187000101 through 189001605) – ie, all IL-18As and most IL-18Bs, including all export examples – were completed *without radars* because the RPSN-2 was still too immature, even though the delivery papers accompanying the aircraft stated the radar was installed!

Not until the advent of the V model did the factory begin delivering *Coots* with the radar fitted as standard (actually the last six IL-18Bs, c/ns 189001701 through 189001801, were the first to be completed with the radar in place). Even so, the crews found the radar virtually useless at first because the thing didn't work. When IL-18V CCCP-75713 No 1 (c/n 180001804) was submitted for checkout trials in April 1960, the type had been in service for about 18 months and nearly 90 aircraft had been built. Nevertheless, the results proved unsatisfactory; the trials report read:

'The service reliability of the Emblema radar is absolutely inadequate. Radar operation was unstable throughout the trials period; as a rule, all manner of malfunctions occurred 1.5 to 2 hours after take-off and the radar failed. This is despite the fact that the radar was serviced on the ground and operated in flight by the OKB's and manufacturer's personnel, not the flight crew.

'The operation of the radar is deemed unsatisfactory.'

In practice this meant that, like their colleagues flying the piston-engined IL-14 'putt-putts', IL-18 crews were forced to rely on auxiliary radio navigation aids. The cause of the misery was soon traced to the radar set being installed outside the pressure cabin (in the 'thimble' incorporating the nosewheel well); the pressure and temperature fluctuations experienced in flight ruined the delicate electronics. Things improved radically when the 'brains' were relocated to the pressure cabin, starting with IL-18V OK-OAC (c/n 180002101) – the first 'V delivered to Czechoslovakia. IL-18Bs c/ns 189001701 and 189001703 through 189001801, IL-18S c/n 189001702 and IL-18Vs c/ns 189001802 through 180002005 were modified accordingly, and all previously built aircraft were finally retrofitted with radar.

It will be fair to say that by far the greatest part of the IL-18's plentiful electronics served to make the airliner suitable for adverse weather/night operations.

Apart from the temperamental avionics, another crucial factor affecting flight safety and all-weather operation capability was the operation of the de-icing system. Unfortunately this was another can of worms in the IL-18's early days. In 1959, when the service trials were under way at Moscow-Vnukovo, IL-18As CCCP-Л5819 (c/n 187000102) and CCCP-Л5820 (c/n 187000103) were fitted with sensors and cine cameras for assessing the efficiency of the de-icing system in various conditions. The verdict was unambiguous and distressing: *'The de-icing system fitted to the wings and tail surfaces of the IL-18 aircraft* does not *ensure safe operation in severe icing conditions.'* The system's electric heaters were not powerful enough; however, installing more powerful heaters would require major changes to the electric system and installation of more powerful generators (which were presumably unavailable at the time). Hence the unsatisfactory system remained unaltered and the crews were told to avoid conditions when severe icing was possible.

In 1962 an IL-18 captained by Boris A Anopov almost crashed after losing longitudinal control due to stabiliser icing. When the flaps were fully deployed for landing the horizontal tail stalled and the aircraft suddenly dropped its nose, entering a steep dive. Eleven seconds later the pilots managed to recover from the dive at only 400m (1,310ft). After this incident, changes were made to the stabiliser de-icers and maximum flap deflection was reduced from 40° to 30°.

The situation persisted until 1965 when GosNII GA floated an idea which cured the problem completely without requiring any design changes. Quite simply, instead of increasing the power of the heaters the engineers changed the duration of their operating cycles (heating/cooling phases). The idea was tested successfully on IL-18V CCCP-75559 (c/n 184007703) and put into fleet-wide practice; henceforth icing presented no problem as long as the system was serviceable.

The powerplant also gave plenty of causes for complaint – the engine makers were making slow progress in their efforts to refine the AI-20. An official document dated early 1962 read: *'On account of the positive results obtained during bench testing the service life until the first major overhaul of the AI-20 Srs 3 engine is designated as 400 hours. [...] No reclamations are accepted for AI-20 Srs 3 engines concerning combustion chamber burnout after 200 hours' total time and concerning all other malfunctions after 300 hours' total time; the engines are to be considered off-warranty at this point.'* Well, if engine life remained so ridiculously low four years after the AI-20 had entered large-scale production, the state of affairs can be characterised with one word: *lousy.*

It was another year before MMZ No30 started manufacturing IL-18s powered by AI-20K (AI-20 Srs 5) engines with an initial TBO increased to 2,000 hours. The first aircraft to receive the new engines was IL-18V CCCP-75506 (c/n 183006404). Concurrently the *Coot* was equipped with an IV-41 engine vibration monitoring kit (*izmeritel' vibrahtsii*) to enhance operational safety; the system had undergone rigorous testing on IL-18B c/n 18800604 (presumably CCCP-75660).

Despite its many shortcomings (quite apart from the gravest ones described above, there was a list of no fewer than 33 items which needed attention!), the IL-18V entered large-scale production; in fact, it became the most widespread variant of the *Coot*, staying in production far longer than any other version. By the end of 1961 the factory had increased the IL-18's production rate appreciably, turning out six aircraft per month instead of four. By then production of the An-10 and the Tu-104 had come to an end, leaving the IL-18 as the sole 'survivor' among the Soviet Union's first-generation turbine-engined airliners.

This was just the kind of aeroplane Aeroflot needed – more fuel-efficient and less demanding to runway length and surface quality than the Tu-104 but longer-legged than the An-10. And this is exactly why the IL-18V was the centre-stage actor for both the world's largest airline, which Aeroflot was then, and the Soviet Union's East European satellites. (In contrast, the Tu-104 achieved negligible export sales, while the An-10 was not exported at all.)

Meanwhile, after improving the efficiency of cargo operations by developing the IL-18T the Il'yushin OKB decided to tackle this noble task in the field of passenger transportation as well. This was achieved by simply increasing the number of seats to 110 – without stretching the fuselage. After the galley area had been shrunk and the sleeperette seats in the rear cabin replaced with ordinary ones, the only way to cram more passengers into the *Coot* was by further (and much more significantly) downgrading passenger comfort. The increase in seating capacity was bigger this time, but it was achieved solely by packing the luckless passengers together like sardines in a can.

The need to install six-abreast seating instead of five-abreast led to the development of new seats: the single seats used hitherto gave place to triple units. The seat backs were no longer reclining, but the armrests could be raised and the backs of the seats in front folded forward for ease of access to the window seats. The width of the aisle was decreased to 350mm (13¾in) – or just 280mm (11in) if the people in the aisle seats were keeping their elbows on the armrests. Getting past a person standing in the aisle or pushing a catering trolley along the aisle now became a real problem.

The seat pitch, too, was a mere 75cm (29½in); this was called the economy class. There was a joke that a person of normal build and average height had to exhale when getting into his seat – and then hold his breath until the end of the flight because his knees were pressed into his stomach. This arrangement increased the number of seats in the forward and centre cabins to 24 (four rows) and 72 (12 rows) respectively; the rear cabin was thus the most comfortable of all, retaining the 14 tourist-class seats five/four-abreast at 84cm pitch.

As noted earlier, the 'Sardines'R'Us' configuration was tested on IL-18V CCCP-75881 (c/n 183006104) in the spring of 1963 and introduced in Aeroflot service shortly afterwards. The discomfort associated with the high-density layout was tolerable during flights lasting one or two hours – especially since the 110-seater solved many problems during the summer season when thousands of holidaymakers rushed from Moscow, Leningrad and Kiev to resorts on the Crimea Peninsula and in the Caucasus. Later, however, this configuration became standard on longer flights to the Siberia and Central Asia – with the usual justification of making more money.

The larger-than-ever number of passengers exacerbated an old problem, namely the need to provide an acceptable microclimate (ie, improve ventilation and cabin air exchange), introduce air conditioning on the ground with the engines running at low rpm and, of course, reduce cabin noise levels. For three years (during which the *Coot* had been in production by then) the OKB had repeatedly tried to solve the noise problem – with scant success. Using thicker soundproofing mats, increasing the size of the ventilation system's air outlets and deleting the system's bypass valve gave little effect. In fact, the latter measure proved harmful because the air exchange rate became much worse and the carbon monoxide content in the cabin air – which was high as it was – increased further.

In late 1963 the engineers attacked the problem from another angle. Since the propellers were the main source of noise, the OKB attempted to attenuate the noise they generated by synchronising the propellers. Unusually, Il'yushin chose to use the de Havilland XQ-1120 propeller synchronisation system

IL-18V CCCP-75827 (c/n 182005001) taxies in after a flight. The aircraft was constantly refined in the course of production; this example already has anti-collision lights introduced on Batch 41, but the Doppler speed/drift sensor system under the aft fuselage has yet to be installed. Yefim Gordon archive

rather than develop a similar system in-country. The system was fitted to IL-18B CCCP-75681 (c/n 189001101),[26] showing promising results; however, it transpired that major modifications to the AI-20's reduction gearbox and other changes would be required, and the idea was dropped. Sorry, folks, no *Quiet Coot*.

A little earlier IL-18B CCCP-75666 (c/n 188000705) and IL-18V CCCP-75834 (c/n 182005104) underwent testing with a modified ACS which permitted air conditioning on the ground with the engines running. Once again, however, the OKB was not in a hurry to introduce this feature into mass production.

Almost a year after the trials of the 110-seat version another *Coot*, the *etalon* of 1963 – IL-18V CCCP-75523, c/n 183006801) manufactured on 11th November (some documents say 30th December) 1963 – underwent rigorous testing to see if it met the specs. The result was deplorable: apart from the many complaints concerning the aircraft's avionics (primarily poor electromagnetic compatibility of various systems), the test report pointed out that noise levels were practically unchanged as compared to earlier IL-18s. The noise level peaked at 118dB in the rear seat rows of the forward cabin and 112dB in the front rows of the centre cabin; in the forward vestibule the noise was even louder, reaching 119dB!

IL-18I Development Aircraft

21st December 1960 saw the beginning of the manufacturer's flight test of a new version designated IL-18I or IL-18/variant 21A. The suffix letter I was not in the Russian alphabetical sequence, so chances are that it stood for *is**sled**ovatel'skiy* (research, used attributively).

Bearing the out-of-sequence registration CCCP-75888, the IL-18I was reportedly converted from the second prototype of the *Coot* which was originally an IL-18 *sans suffixe*. However, there are reasons to doubt this because the window arrangement was similar to that of the IL-18V; since the latter model was in production by then, it would be far easier to pull an IL-18V off the assembly line for conversion.

This time the redesign was far more extensive. The aircraft could carry 119-125 passengers; the increase in seating capacity was obtained by moving the rear pressure bulkhead 1.64m (5ft 4½in) aft from fuselage frame 56 to frame 62, thereby eliminating the unpressurised rear baggage compartment and the associated door between frames 58-61. (The oft-cited allegation that the IL-18I had a fuselage stretch of 1.64m is misconception; the overall length remained unchanged.) The new interior layout caused a change in cabin window arrangement, with two extra windows on each side (3+door+2+2 exits+1+4+door+5 to port and 3+2+2 exits+1+4+1+5 to starboard).

Maximum range was extended from the IL-18V's 4,700km (2,920 miles) to 6,500km (4,040 miles) thanks to a 28% higher fuel capacity – total fuel was increased from 23,700 litres (5,214 Imp gals) to 30,300 litres (6,666 Imp gals). This necessitated major structural changes in the wings: two more manufacturing joints were added immediately outboard of the inner engines and the flexible fuel cells in the inner wing portions between the inner and outer engines gave place to integral tanks. Additionally, fuel cells were installed in the wing centre section which was integral with the fuselage. The fuel system comprised 26 tanks (four integral tanks and 22 bladder tanks); refuelling was possible both under pressure and by gravity. There was no fuel jettison system.

The IL-18I was the first *Coot* to feature an auxiliary power unit. Unlike late-production IL-18Vs *et seq*, the TG-16 APU was not installed in the rear fuselage; it was mounted under the fuselage on a special frame which swung down on parallel arms before the APU could be started. When retracted the APU was almost invisible – its presence was revealed only by a small flattened teardrop fairing immediately ahead of the wing leading edge.

The maximum take-off weight increased to 64 tons (141,090 lb) and the payload rose from 13.5 to 14 tons (from 29,760 to 30,865 lb).

The State acceptance trials of the IL-18I, which were duly completed on 14th December 1961, included several non-stop long-range flights from Moscow to Irkutsk, along the Irkutsk – Moscow – Leningrad – Moscow route, from Moscow to Tashkent via Vladivostok and Khabarovsk, and from Tashkent to Moscow via Arkhangel'sk. The flights lasted from eight to 12 hours; all systems worked OK. The results were encouraging and, although the IL-18I was destined to remain a one-off, many of its new features were incorporated in the next major production versions superseding the V model – the IL-18E and IL-18D.

IL-18D Medium/Long-haul Airliner (Second use of designation)

In 1964 IL-18V production was drawing to a close as the Il'yushin OKB prepared to launch a new variant which was destined to be the ultimate commercial version of the *Coot*. Continuing the Cyrillic alphabetical sequence (A, B, V, G etc), the aircraft received the designation IL-18D used earlier for the 'limited edition' purpose-built IL-18V-26A. This was logical enough, since the new aircraft was designed as an extended-range version (though not specifically for Antarctic supply missions) and the explanation ***dahl'****niy* was perfectly applicable.

Development of the 'IL-18D Mk II' was long and arduous, as some of the features which went into this model had been under development since 1961. Also, development problems with the IL-18D resulted in another version, which was actually developed later and bore a higher suffix letter, 'jumping the queue' to the production line (as described in the IL-18E section below).

In 1964 Aleksey G Ivchenko's OKB-478 was putting the finishing touches to a new uprated version of the AI-20 turboprop – the AI-20 Srs 6, aka AI-20M (*modifit**see**rovannyy* – modified). Now there were no more obstacles to creating – at long last – a fully-fledged long-haul version of the *Coot* where seating capacity and payload were not compromised by fuel tanks in the middle of the cabin, as had been the case with the IL-18V-26A. Compared with the latter model, the prospects of the 'IL-18D Mk II' looked far more optimistic.

When the AI-20M entered series production at MAP factory No 478 (now called Motor-Sich), it was truly a milestone for Soviet civil and military transport aviation. Until then, the AI-20 had often been the subject of serious criticism – and well-deserved criticism it was. However, the advent of the AI-20M showed the people at OKB-478 had been doing their homework; the new version elicited the most favourable comments from those who had to deal with it.

To begin with, the AI-20M's take-off power was increased from 4,000 to 4,250ehp, which not only made a long-planned gross weight increase possible but enhanced flight safety in the event of an engine failure during take-off. Also, remember that engine power drops in hot-and-high conditions; thus the new engine considerably alleviated the problems associated with serving destinations in the Soviet Union's Central Asian republics and the Caucasus, not to mention such places as Afghanistan and Ethiopia. The power increase was modest, amounting to some 7%, but it was apparent in almost all flight modes. For instance, at maximum power rating at 8,000m (26,250ft) the AI-20M delivered 3,420ehp versus 3,180ehp for the AI-20K; at nominal (ie, maximum continuous) power rating at the same altitude it was 2,980ehp and 2,800ehp respectively. Equally important was the fact that maximum continuous power could now be used for much longer periods, making it possible to use a more efficient flight profile (flying at the aircraft's service ceiling as long as possible to save fuel) or climb to a given flight level more quickly.

However, these are by no means all of the AI-20M's virtues. At all power settings the specific fuel consumption (SFC) was reduced as compared to the AI-20K, the reduction amounting to 7% at take-off power, 5% in cruise mode at 8,000m and 3.5% at maximum continuous power at the same altitude. Moreover, the new model was 40kg (88 lb) lighter: the dry weight of the AI-20M was 1,040kg (2,292 lb) versus 1,080kg (2,380 lb) for the AI-20K! Finally, though figures do appear more convincing than words, it should be noted that the AI-20M was simpler in design and more reliable, while the TBO was increased from an initial 1,000 hours to 2,000 hours in 1966, then to 3,000 hours in 1967 and finally to 4,000 hours. (Still, words *can* be convincing; the K model had the disparaging nickname ***ka****kovskiy* bestowed on it by flight and ground crews – with the obvious implication that the engine was *complete kaka*. Needless to say that the Much-Improved M was greeted with enthusiasm!)

The IL-18D was to incorporate all of the new features tested on IL-18V CCCP-75834 (c/n 182005104). Among other things, the D model was to introduce the long-expected upgrades to the air conditioning system. The forward vestibule was redesigned, allowing the centre cabin to be extended forward; as a result, the IL-18D introduced a new window arrangement with 16 windows to port (3+door+3+2 exits+1+4+door+3) and 17 to starboard (3+3+2 exits+1+4+1+3). The extra window ahead of the overwing emergency exits was thus the IL-18D's external identification feature.

The main changes, however, were not so obvious. In addition to the new engines, the fuel capacity was enlarged considerably – this time without any drastic changes to the airframe design. The wing centre section built integrally with the fuselage housed 6,300 litres (1,386 Imp gals) of extra fuel; unlike the experimental IL-18I, the wing centre section torsion box functioned as an integral tank rather than accommodate a set of bladder tanks. The centre section tank (the No 23 tank) increased overall fuel capacity to 30,000 litres (6,600 Imp gals) and the take-off weight to 64 tons (141,090 lb), enabling the aircraft to carry a 6.5-ton (14,330-lb) payload over a distance of 6,500km (4,040 miles) with maximum fuel, including reserves for one hour. With a maximum payload of 13.5 tons (29,760 lb) the IL-18D had a range of 3,700km (2,300 miles) if the engines operated in cruise mode (84-85% of the maximum continuous power rating) most of the time; this operational mode was easier on the engines, as the turbine temperature was lower.

Construction of the first prototype IL-18D began in mid-1964; to this end the newly completed fuselage of IL-18V c/n 184007803 was pulled off the final assembly line to make the short way to the Il'yushin OKB's experimental shop located right across the field, less than a mile away. As modification work progressed on the fuselage, other airframe subassemblies and components were gradually delivered from MMZ No 30 and the aircraft took shape little by little. Thus both plants – or neither plant – can claim to have to have manufactured the aircraft!

Bearing the out-of-sequence registration CCCP-75581,[27] the prototype entered flight test on 31st July 1964. To speed up completion of the trials a second prototype registered CCCP-75572 (c/n 185008001) was laid down in January 1965, undergoing flight tests between 26th July and 21st September of that year. Even before that, on 29th January 1965, the newly-restored Ministry of Aircraft Industry (renamed back from GKAT) and the Ministry of Civil Aviation (MGA – *Ministerstvo grazhdahnskoy aviahtsii*) had passed a joint order to the effect that the 'modified IL-18D airliner' be put into full-scale production. However, this turned out to be wishful thinking and it was some time before the aircraft could enter production (see IL-18E section later in this chapter).

On 13-25th June 1965 IL-18D CCCP-75581 and the second prototype of the IL-62 *sans suffixe* four-turbofan long-haul airliner (CCCP-06153, c/n 30002; ASCC reporting name *Classic*) had their international debut at the 26th Paris Air Show. The occasion was all the more notable because the Soviet Union had thus re-established its presence at Le Bourget after a long, long break.

On 30th May 1966 the two prototypes of the IL-18D completed their checkout trials after all development problems with the new powerplant had been overcome. The D model finally entered production in July 1966 from Batch 93 onwards, remaining the sole commercial version until production of the civil *Coot* ended (military derivatives of the IL-18 continued in production for many years and are described separately in Chapter 5). It was the second-largest version as far as production volumes are concerned, having a production run of 122.

Three years later came the day which everybody knew would come sooner or later – but

CCCP-75888, the sole example of the IL-18I which was a stepping stone towards the IL-18D. This view shows clearly the three additional windows where the No 3 baggage door should be. Sergey and Dmitriy Komissarov archive

Apart from the extension of the rear cabin aft at the expense of the No 3 baggage compartment and the provision of additional fuel tanks, the IL-18I differed from the 'V in having the TG-16 APU mounted on a retractable frame under the wing leading edge. Here it is shown in retracted position, its presence revealed by a shallow teardrop fairing.
Sergey and Dmitriy Komissarov archive

The IL-18I's APU lowered into position for running. Sergey and Dmitriy Komissarov archive

the knowledge did not make the occasion any more festive for those concerned. On 17th April 1969 MMZ No 30 manufactured the final commercial *Coot* – an IL-18D with the c/n 189011304. (Note: Some sources give the manufacture date as 28th March 1969; thus 17th April may be the rollout date or the first flight date.) Registered 5T-CJL, this aircraft was delivered to the Mauritanian airline Air Mauretanie, subsequently being resold to Vietnam as VN-B198. Thus commercial production of the IL-18 totalled 564 aircraft built by MMZ No 30.[28] These aircraft formed the backbone of Aeroflot and the flag carriers of the East European nations ('the Eastern Bloc'), and the Soviet aircraft industry had good reason to be proud of these winged machines.

On 10-13th February 1980 IL-18D CCCP-74267 (c/n 188011105)[29] flew an Antarctic support mission for the Arctic & Antarctic Research Institute (AANII – ***Ark**tiko-Antark**tich**eskiy na**ooch**no-is**sled**ovatel'skiy insti**toot***), flying from Moscow-Sheremet'yevo to Ice Station Molodyozhnaya along the so-called central route via Odessa, Cairo, Aden and Maputo. Unlike IL-18V CCCP-75743 and the special IL-18V-26As, the aircraft, which wore Aeroflot's smart red/white 1973-standard Polar colours, was almost unmodified, apart from the addition of some navigation equipment (including a star tracker for the astrosextant on the flight-deck roof) – that is, there were no enlarged oil tanks and the engine cowlings were perfectly standard. The two flight crews working in shifts were captained by Yevgeniy P Boonchin and A N Denisov, while overall mission coordination was performed by Boris Dmitriyevich Groobiy. The aircraft covered the 15,992-km (9,933-mile) route in 26 hours' flight time. The return journey followed the same route and took place on 19th-23rd February; all in all, the airliner covered a distance of 45,660km (28,360 miles), staying airborne for 78 hours 54 minutes. For this performance a group of Aeroflot personnel who were involved in preparing and completing the mission received 'Excellent Aeroflot employee' badges on 18th July 1980.

Exactly one year later, on 10th February 1981, the same aircraft took off from Leningrad-Pulkovo, bound for Ice Station Molodyozhnaya on the central route with the 26th Soviet Antarctic Expedition on board. CCCP-74267 arrived at her destination on 12th February, covering a distance of 17,190km (10,677 miles).

IL-18D/VIP Version (IL-18D 'Salon')

Again, a VIP transport version of the IL-18D was developed for government flights and air forces. Once again no separate designation is known, so the VIP version is referred to here as the IL-18D 'Salon'. There were no obvious external recognition features to distinguish such an aircraft from an ordinary 'D.

Twelve IL-18D 'Salons' have been identified. Known Soviet examples were registered CCCP-75412 (actually the first production IL-18D, c/n 186009301), -75498* and -75499*, although three more aircraft registered CCCP-75453, -75454 and -75464 may have been originally built as IL-18D 'Salons', too. Export aircraft include Royal Afghan Air Force T-001, Bulgarian Air Force LZ-BEA*, Chinese People's Liberation Army Air Force '50853 Red' and '50856 Red' (the latter example became B-226), East German Air Force DM-STM*, Romanian Government YR-IMM and YR-IMZ No 2 (c/n 187009802),[30] and Yugoslav Government YU-AIA (later Yugoslav Air Force '7501' and '73201') and YU-AIB. Aircraft marked with an asterisk were quasi-civil.

IL-18D VIP/Communications Relay Aircraft

Long after the 235th OAO (*ot**del'**nyy aviao**tryad*** – Independent Air Detachment), the Soviet government flight based at Moscow-Vnukovo airport, had stopped using the *Coot* as a VIP transport it still operated three IL-18Ds – CCCP-75453, CCCP-75454 and CCCP-75464 (c/ns

Late-production IL-18Ds on the assembly line in 1968; the nearest aircraft features an 'elephant's ear' air intake for the APU on the No 3 baggage compartment door. Note the IL-38 ASW aircraft in the background. Sergey and Dmitriy Komissarov archive

Yugoslav Air Force '73201' represented the IL-18D 'Salon' VIP version. Yefim Gordon archive

187010103, 187010104 and 187010401). They were fitted out as communications relay aircraft supporting HF communication between the Kremlin and suitably equipped VIP aircraft. The comms relay IL-18Ds had the same antenna farm on the forward fuselage as the IL-18V ABCPs described in Chapter 5 (CCCP-75602 and -75606), plus a third identical set of blade aerials located ventrally aft of the wing trailing edge. The mission equipment was compact and the aircraft could still be used as an airliner, carrying officials between Moscow and Leningrad. Some sources refer to this version as the IL-18RT (*retranslyator* – relay station); see also Chapter 5 for this designation.

The three IL-18s continued in service when the 235th IAD became the Russia State Transport Company in September 1993, remaining operational until the late 1990s.

IL-18E Medium/Long-haul Airliner

As already mentioned, the IL-18D was ordered into production on 29th January 1965. However, as a line from an 18th Century Russian soldiers' song goes (in English translation), *'On the map there were no hitches, // Alas! forgotten were the ditches // Which we would have to cross'*. The tests of the AI-20M engine (and consequently the IL-18D) looked set to become a protracted affair, and the *Coot* was overdue for an upgrade in many areas. Hence the Il'yushin OKB and MAP decided to stick to the AI-20K engines and 61.2-ton (134,920-lb) take-off weight while incorporating the improvements already tested on the 122-seat IL-18V CCCP-75834 (c/n 182005104) in the autumn of 1964 – specifically, up to 14th November 1964. In this guise the *Coot* would be produced until the IL-18D was deemed ripe for production.

In order to discern it from the original IL-18V the interim version was designated IL-18E (though IL-18Ye would be a more accurate rendering perhaps); this was logical enough, since the letter E is sixth in the Cyrillic alphabetical sequence (after the D) and the IL-18E indeed appeared after the IL-18D prototypes. It can be argued, however, that the sequence was not observed because CCCP-75834 effectively became the IL-18E prototype after all those modifications!

The IL-18E was unveiled at the 1965 Paris Air Show, albeit not in hardware form yet. The new version started rolling off the assembly line in September 1965. The first production IL-18E was CCCP-75592 (c/n 185008502) which first flew on 30th September and completed check-out trials on 15th December 1965. Twenty-three examples were built before production finally switched to the IL-18D in 1966.

'IL-18M'

It may be mentioned here that in September and October 1965 the East German magazine *Aerosport* mentioned a 'further upgrade' of the *Coot* designated IL-18M. The aircraft allegedly had a reinforced wing centre section/inner wing assembly and landing gear allowing the maximum TOW to be increased by 2,000kg (4,410 lb) and the seating capacity to 126; range with a maximum payload of 14.5 tons (31,970 lb) was stated as 1,600km (990 miles). However, this designation was obviously erroneous, as no mention of it can be found in Soviet/Russian sources.

IL-18E/VIP Version (IL-18E 'Salon')

Once again, a VIP version of the IL-18E existed; as no separate designation is known, it will be referred to as the IL-18E 'Salon'.

Seven aircraft have been identified. Interestingly, unlike the other VIP versions, all IL-18E 'Salons' were export aircraft. Three were operated by the Chinese Government (208, later B-208) and the PLAAF ('50852 Red' and '50855 Red'). The Polish Air Force had two examples serialled '101 Red' No 3 (c/n 185008503/to SP-LSK) and '102 Red' No 2 (c/n 186008905/to SP-LSI), while the Czech Ministry of the Interior's government flight and the Romanian Government operated one each – OK-BYZ and YR-IMZ No 1 (c/n 185008602).

IL-18Gr Cargo Aircraft

Starting in 1978, a number of *Coots* displaced from the passenger routes by the Tu-154 were converted to freighters designated IL-18Gr (*groozovoy* – cargo, used attributively). Outwardly they were no different from any other IL-18 but the interior was stripped out and the cabin floor reinforced. To speed up loading/unloading an electric hoist was provided for moving pallets, travelling along an overhead rail running the full length of the cabin, just as was the case with the IL-18T. However, just like the IL-18T, the IL-18Gr lacked a large cargo door and could only carry items small enough to go through the entry/baggage doors.

The first aircraft to be converted to IL-18Gr configuration was probably IL-18V CCCP-75785 (c/n 181003904). IL-18Gr freighters were operated by various Aeroflot divisions, including the Krasnoyarsk CAD/1st Krasnoyarsk UAD/214th Flight and the Urals CAD/1st Sverdloysk UAD/120th Flight; they typically carried fruit and other perishables from the southern regions of the USSR to Siberia and the Far East. In post-Soviet days many of the surviving *Coots* flew in this configuration.

IL-18 Combi

From the late 1980s onwards many airlines operating the IL-18 modified their *Coots* to a mixed passenger/cargo configuration; the cargo was accommodated in the forward cabin while the passengers sat at the rear where noise levels were lower. Different payload combinations were possible.

IL-18GrM Cargo Aircraft
IL-18V/F (SCD), IL-18D/F (SCD)

Development of this version was brought about first and foremost by the changing economic situation (to be precise, economic crisis) in Eastern Europe and the Soviet Union in the late 1980s. Actually the idea to develop the IL-18 into a fully capable freighter by incorporating a large pressurised cargo door came into being back in the 1960s when the *Coot* was still in production. This modification would enable the IL-18 to swallow a much wider range of loads (which, as it was, was severely limited by the size of the entry and baggage doors); later on, when containerised freight transportation became the big hit, the 'Cargo *Coot*' would have proved suitable for it, too.

Left: **The IL-18E 'Salon' VIP version was built for export only. This is one of the seven aircraft identified to date – OK-BYZ (c/n 186009004) of the Czech Ministry of the Interior's government flight. It was later sold to ČSA Czechoslovak Airlines as OK-VAF and converted to passenger configuration.** Yefim Gordon archive

Below left: **An IL-18Gr freighter about to be loaded by means of a self-propelled belt conveyor loader based on the UAZ-451M 0.8-ton four-wheel drive 'truckster'. The lack of a large cargo door meant the aircraft could only carry small items.** Yefim Gordon archive

For obvious reasons the height of the cargo door was less critical as far as the size of the cargo was concerned. Since the fuselage was of circular cross-section, a door making up one fourth of the fuselage's circumference provided an aperture equal to 70% of the cabin height! The door's width was far more important; it had to be chosen carefully because the wider the door, the more it compromised structural integrity.

A possible alternative was to have the entire forward fuselage (as in the case of the Aero Spacelines Guppy freighter based on the Boeing C-97 Stratofreighter) or rear fuselage swing open, making the entire cabin cross-section available for loading and unloading. Canadair successfully used this feature for the CL-44D, a dedicated freighter derivative of the Bristol Britannia four-turboprop airliner on which the rear fuselage was hinged to starboard. Besides the obvious advantage of allowing the carriage of long items as long as the cargo cabin itself, all cargo handling operations took place well away from the wings and tail surfaces, facilitating the procedure and minimising the risk of ground damage.

Having already gained some experience with side cargo doors on the IL-12D and IL-14G/IL-14T (which, true enough, were unpressurised), the engineers of OKB-240 did not have to break new ground. Back in the early 1970s the Il'yushin OKB came up with a project envisaging installation of a large upward-opening cargo door on the port side of the IL-18's forward fuselage. Yet not a single aircraft was converted in this fashion for many more years.

The reasons were chiefly economic ones – specifically, the economic relations within the Soviet Union and with the 'customer nations' abroad. The chief customer targeted by the project – Aeroflot Soviet Airlines – already had a sizeable fleet of specialised Antonov An-12 *Cub* freighters, and the type was still in production at plant No 84 in Tashkent. Moreover, the Il'yushin OKB's own brainchild, the IL-76 *Candid* freighter with both military and civil applications, was about to enter production. Both Aeroflot and the Soviet Air Force had a lot riding on it, which is why Aeroflot, the largest IL-18 operator in the Soviet Union, was none too enthusiastic about such a conversion. This attitude was shared by some people inside the OKB who feared that a possible 'IL-18C^3' (*Coot* Cargo Conversion!) would spoil the *Candid*'s chances.

Foreign IL-18 operators – first and foremost the East European airlines – were in a different situation, having no specialised cargo aircraft (with a few exceptions). As the newer and more capable Tupolev Tu-134 *Crusty* short-haul airliner, Tu-154 medium-haul airliner and IL-62 long-haul airliner were introduced, displacing the old *Coot* from the more prestigious routes and the ones with heavy traffic, the numerous

Bottom: **Seen here at Moscow/Vnukovo-1 in its days as a passenger aircraft, IL-18V CCCP-75785 (c/n 181003904) was the first *Coot* to be converted to IL-18Gr configuration in 1978.** Sergey Komissarov

IL-18s found themselves 'laid off' in some cases. Therefore the airlines were interested in prolonging the active career of these rugged and dependable workhorses; after conversion to fully capable freighters these aircraft would fit in perfectly again.

Over the years several air carriers, including Balkan Bulgarian Airlines and the East German airline Interflug, addressed the Il'yushin OKB with requests to develop a freighter version of the IL-18 with a side cargo door. However, each time the OKB, which had plenty of other irons in the fire, and Aviaexport (the Soviet agency handling aircraft sales to foreign customers) presented such a huge bill for the order that the customers backed off hastily. Things may have worked out differently if the customers had teamed up and placed a joint order, splitting the costs between them to make the venture affordable. Yet for some reason they did not, and eventually most East European airlines were forced to modify some of their Coots in a similar fashion to the Soviet IL-18Gr.

It was not until the late 1980s that the Il'yushin OKB finally took on this not very prestigious project. The times were changing. Inflation had depreciated the work of the designers considerably and the book of government orders had shrunk to almost nothing; but then, now that the exaggerated security restrictions were gone, the design bureaux were now authorised to deal directly with foreign customers. Thus within just a year (1990-91) the OKB developed a classic side cargo door installation for the IL-18.

The large upward-opening cargo door measuring 3.5 x 1.825m (11ft 5¾ in x 5ft 11⅞in) was located on the port side, replacing the three foremost cabin windows, and was actuated by a single hydraulic ram with a breaker strut. A large rectangular reinforcement plate was riveted onto the port side of the fuselage around the aperture. The forward entry door built into the cargo door was moved forward in comparison with the passenger version. The second and third windows on the starboard side were deleted, probably again for structural strength reasons. Thus the window arrangement for a converted IL-18V looked like this: door (with another door in it)+2+two exits+1+4+door+3 to port and 1+2+two exits+1+4+1+3.

The door featured a perimeter seal, allowing cargo cabin pressurisation to be retained – a major advantage over the An-12 which had an unpressurised freight hold and thus could not carry perishable cargo. The soundproofing mats in the cargo cabin were partially deleted to save weight. For handling containerised or palletised cargo the reinforced cabin floor was equipped with a 'ball mat' near the door and roller conveyors elsewhere, plus cargo tiedown points. Of course, small packages could still be carried in the cabin and the baggage compartments; loading and unloading would be performed using belt conveyors or fork lift trucks. The performance was unchanged as compared to the passenger version.

Surprisingly, the new version did not receive any official designation. Unofficially, however, the *Coot* freighter with a side cargo door was known as the IL-18GrM (*groozo**voy**, modifit**see**rovannyy* – cargo, modified).

The launch order was placed by the German airline BerLine which wanted to convert two of the five IL-18s it had acquired after German reunification and the demise of Interflug. The first of the two, IL-18D D-AOAS (ex-DDR-STM, c/n 188010805), arrived at Zhukovskiy in September 1992; the conversion took place in the hangar of the Il'yushin OKB's flight test facility and was completed within four months. The rear passenger cabin and toilet were retained on this aircraft so that passengers accompanying the cargo (or a relief crew) could be carried. The second aircraft, IL-18V D-AOAP (ex-DDR-STI, c/n 185008404) arrived for conversion in February 1993 and was redelivered in June. Depending on their original version, these two aircraft were unofficially known as the IL-18V/F (SCD) and IL-18D/F (SCD) – ie, freighter, side cargo door.

The Romanian carrier Romavia was next, also ordering the conversion of two IL-18Ds. The first of these, YR-IMZ No 2 (c/n 187009802), a former VIP aircraft, arrived for conversion in December 1994 and the job was completed on 27th March 1995. YR-IMM (c/n 187009904) was scheduled to be next in 1995 but was never converted; a photo taken at Malta-Luqa in June 1998 shows clearly the aircraft still had no large cargo door!

The fate of the second IL-18GrM – the IL-18V/F (SCD) – deserves to be mentioned because it was an unusual and an unlucky one. After changing hands several times D-AOAP moved to the Russian civil register as RA-75554 (inheriting the identity of another IL-18V – but that is another story). On 17th December 1997 the aircraft failed to become airborne at Johannesburg-International due to overloading and was damaged beyond repair after overrunning the runway when the take-off was aborted. Three years later the aircraft was broken up but the cargo door was not lost; it was fitted to IL-18D EX-75466 of Phoenix Airlines (c/n 187010403) and the conversion took place at Sharjah between 1st and 3rd March 2001! As for the other IL-18GrMs, D-AOAS became LZ-AZZ with the Bulgarian airline Air Zory, then CU-T132 with the Cuban carrier Aerocaribbean and finally CU-C132. YR-IMZ was sold to the Sri Lankan airline Expo Aviation as 4R-EXD.

IL-18DORR
Fishery Reconnaissance Aircraft

In the mid-1980s two IL-18Ds, CCCP-75462 (c/n 187010304) and CCCP-74268 (c/n 188011201), were transferred to the Polar Institute of Oceanic Fishery and Oceanography (PINRO – *Pol**yar**nyy insti**toot** mor**sko**vo **ryb**novo khoz**iais**tva i okeano**gra**fii*), most probably from the Moscow Territorial CAD/Domodedovo UAD/212th Flight. After an extensive refit the two *Coots* received the designation IL-18DORR (***dahl'****niy oke**ahn**skiy raz**ved**chik **ry**by* – long-range ocean fishery reconnaissance aircraft). As the designation implies, their mission was to hunt down large shoals of fish for fishing flotillas in international waters.

The mission equipment included spectrometers, multi-mode thermal imaging, photo and video recording equipment and the Iney (Hoar frost) photo telegraph data link system for downloading maps to fishing trawlers. Aside from detecting fish, it could measure the concentration of plankton and algae, the water temperature and detect water pollution. The aircraft could also monitor the numbers of marine mammals and perform ice patrol missions in case of need. The *Coot*'s long range and endurance meant it could replace more than a hundred fishery reconnaissance vessels!

Outwardly the IL-18DORR could be identified only by two small angular fairings mounted dorsally and ventrally in line with the tenth cabin window on each side (frames 34-36), two additional ventral strake aerials between frames 35-42, an extra wire aerial running from the fin leading edge to a strut at frame 23 and six observation blisters in all three cabins (windows Nos 3, 13 and 14 to port/3, 13 and 15 to starboard). Since they would be flying up north, the aircraft wore Aeroflot's red/white polar colour scheme (also used by regular transport aircraft flying in the northern areas of the USSR, not only by Polar Aviation aircraft) for high definition against white backgrounds. To make it patently clear this was not a spyplane a large Ministry of Fisheries badge was painted on the nose – no vain precaution, as it turned out, because the aircraft was sometimes intercepted by NATO fighters over international waters.

The first post-conversion flight of CCCP-75462 took place in early 1985. Operations began in 1986, the aircraft operating over the Sea of Norway and the Barents Sea from its temporary base at Murmansk-Murmashi airport where it first arrived from Moscow-Domodedovo on 25th February. These missions implied flying at 300m (990ft) and 320km/h (198mph), tacking to and fro like an anti-submarine warfare aircraft searching for a sub. Quite often these flights took place at night or at dusk due to the habits of some kinds of fish, added to which, the weather over the Barents Sea was often foul. Missions lasted up to nine hours, including four or five hours on station; all of this placed high demands on the skill of the crew. Up to 26,000 square miles of sea surface could be reconnoitred in a single flight.

Sadly, in the political and economic chaos that followed the demise of the Soviet Union there seemed to be no use for these unique aircraft, so both IL-18DORRs were reconverted to standard IL-18Ds and sold to the Domodedovo Civil Aviation Production Association, subsequently gaining the Russian nationality prefix. RA-75462 was retired in late 1998 and dumped at Moscow-Domodedovo, still in basic polar colours. RA-74268 was more fortunate; after serving with the Domodedovo CAPA in full blue/white colours it was sold to Tyumen' Airlines in 1999.

IL-18D 'Pomor' Fishery Reconnaissance Aircraft

Another IL-18D belonging to GosNII GA (identity unknown) was converted into an ocean fishery reconnaissance aircraft known as the IL-18D *'Pomor'*. (In olden days the Pomors were a fishing people living along the White Sea coast in what is now the Arkhangel'sk Region; the word 'pomor' itself means 'person who lives by the sea'.) As compared to the IL-18DORR it had more capable mission equipment. Unfortunately no other details are known.

IL-18USh Navigator Trainer

A single IL-18V (CCCP-75523, c/n 183006801) was converted into the sole IL-18USh (*oochebno-shtoormanskiy* [*samolyot*] – navigator trainer). Outwardly it differed from the standard *Coot* only in having two dorsal astrosextant blisters (star trackers) located in tandem on the forward fuselage between frames 9-10 and 13-14, slightly offset to port, and two ADF strake aerials in a staggered-tandem arrangement atop the centre fuselage (frames 24-31 and 32-40). The cabin featured 20 trainee workstations with radio and celestial navigation aids.

The aircraft was tested by GosNII GA in early 1972 and found acceptable. The Soviet Air Force, however, rejected the aircraft, probably because the turboprop IL-18 was too slow. Since 1962 it had two versions of the Tu-124 *Cookpot* short-haul twinjet to fill the navigator training role – the Tu-124Sh-1 for long-range bomber crews and the Tu-124Sh-2 for tactical bomber crews; these were later succeeded by the Tu-134Sh-1 and Tu-134Sh-2, military derivatives of the Tu-134A.

Several more *Coots* were reportedly converted to IL-18USh configuration. The prototype was probably reconverted to passenger configuration and transferred to Aeroflot's Moscow Territorial CAD/Domodedovo UAD/212th Flight. It was ultimately struck off charge at Moscow-Domodedovo and scrapped.

IL-24N Ice Reconnaissance Aircraft

In the 1970s Aeroflot's Polar division operated at least two Antonov An-24LR Toros ice reconnaissance aircraft (CCCP-46211 and CCCP-46395) and a single An-24LR Nit' ice reconnaissance aircraft (CCCP-47195). Designed to measure the thickness of ice fields along the Northern Sea Route and indicate the best route for icebreakers leading ship convoys, they were An-24B *Coke* 50-seat regional airliners equipped with the *Toros* (Ice hummock) side-looking aircraft radar (SLAR) or an An-24RV with a Nit'-S1 (Thread-S1) SLAR respectively; LR stood for *ledovyy razvedchik* – ice reconnaissance aircraft.

However, a major weakness of the An-24LR Toros and An-24LR Nit' was their rather limited range and endurance. Hence an ice reconnaissance derivative of the IL-18D was developed as a long-range alternative to Antonov types. The aircraft received a completely separate designation, IL-24N; the designation was reused – the original IL-24 was a 'paper airplane', a four-jet bomber project of the early 1950s.

Left: **CCCP-75523, the sole IL-18USh navigator trainer. Note the tandem star trackers and the staggered-tandem ADF strake aerials on top of the fuselage.** Yefim Gordon archive

Bottom and opposite page top: **CCCP-75449, the first of two IL-24N ice reconnaissance aircraft.** Yefim Gordon archive

Right: **Close-up of the IL-24N's characteristic ventral pod housing the Nit'-S1 side-looking aircraft radar (SLAR) used to measure the thickness of pack ice. The pod itself was identical to that of the IL-20M ELINT aircraft but the contents were different.** Yefim Gordon archive

Below right: **The second IL-24N, CCCP-75466, on the western apron at Moscow/Sheremet'yevo-1, the ramp of the State Civil Aviation Research Institute (GosNII GA) which owned it.** Sergey and Dmitriy Komissarov archive

Outwardly the IL-24N was extremely similar to the IL-20M *Coot-A* reconnaissance aircraft (see Chapter 5), featuring the same huge pod of elliptical cross-section under the forward fuselage. The pod stretched from frame 8 to frame 27; it was attached via a short blended pylon which began at frame 10. Here the similarity ended, however. Despite the identical design, the pod housed a different SLAR – the civilian Nit'-S1 instead of the military Igla-1, hence the N suffix to the designation. The radar set was housed in the forward baggage compartment and the installation required some local structural reinforcement; the operator's workstations were located in the passenger cabin. An extra wire aerial ran from the fin leading edge to a strut at frame 30. All other excrescences characteristic of the IL-20M were missing, there were two small ventral dielectric blisters fore and aft of the SLAR pod (frames 6-7 and 41-42) which the *Coot-A* did not have and the window and baggage door arrangement was no different from the standard IL-18D, save that observation blisters were provided in the rearmost window to port and the last-but-one window to starboard.

Two late-production aircraft – CCCP-75449 (c/n 187010004) and CCCP-75466 (c/n 187010403) – were converted to IL-24N standard in the mid-1980s. For obvious reasons they wore the red/white polar version of Aeroflot's livery. The IL-24Ns were apparently operated by GosNII GA and were frequently seen on the GosNII GA ramp at Moscow/ Sheremet'yevo-1 where they were home-based.

Alas, the IL-24N fell victim to the changing economic situation in Russia in the early 1990s when making money was more important to many people than anything else. After 1992 CCCP-75466 was reconverted to a standard IL-18D; RA-75449 followed suit shortly afterwards. In 1994 both aircraft were sold to the Chelyabinsk-based airline Ramair.

TEST AND RESEARCH AIRCRAFT

The IL-18 was also used a lot for various test and research work. There were several reasons making it eminently suitable for this role. Firstly, the spacious cabin (which, as already mentioned, was pressurised – a major advantage over the An-12, another all-time favourite for conversion into testbeds) could accommodate a lot of test equipment and a large crew of researchers, plus all the necessary amenities for the latter. Secondly, the IL-18 possessed a high payload, long range and endurance and the ability to operate anywhere from the tropics to the polar regions. Last but not least, the aircraft was well mastered by flight and ground crews alike and had a broad maintenance network by the time testbeds began appearing.

Avionics testbeds

About 20 IL-18s were operated by various Ministry of Electronics Industry (MRP – *Minister-stvo* ***rah****dioelek****tron****noy* ***pro****mysh**lennosti*) divisions as avionics testbeds. While not formally Air Force aircraft, they were mostly used to test military avionics.

IL-24N CCCP-75449 in stripped-down condition (but still with SLAR pod) awaits refurbishment at aircraft overhaul plant No 402 at Moscow-Bykovo. Yefim Gordon archive

IL-18A CCCP-75643 was converted into the IL-18SL testbed for the IL-38's Berkoot search radar. It is pictured here at Moscow/Vnukovo-1 on 26th June 1965; the radome aft of the nose gear is clearly visible. Sergey Komissarov

CCCP-75786, the SL-18V testbed for the Tu-160's Obzor-K radar, at its home base in Pushkin as originally flown in pre-1973 colours. Note the large acquisition round for the Kh-45 Molniya air-to-surface missile under the forward fuselage. Yefim Gordon archive

Bottom two pictures: **The SL-18V pictured at Leningrad-Pulkovo in later years. The aircraft has been repainted in 1973-standard livery and carries no acquisition round.** Yefim Gordon archive

a) By June 1965 IL-18A CCCP-75643 (c/n 188000302) was converted into a testbed for the Berkoot (Golden eagle) 360° search radar developed for the IL-38 anti-submarine warfare aircraft (see Chapter 6). The radar was a product of the Leningrad-based NII-131, aka LNPO[31] Leninets (Leninist), which owned the aircraft. This establishment, a division of the State Committee for Electronic Equipment (GKRE – *Gosoo**dar**stvennyy komi**tet** po **rah**dioelek**tron**ike*)[32] was one of the Soviet Union's leading avionics specialists.

The radar was installed in identical fashion to the IL-38 in a large quasi-spherical radome immediately aft of the nose gear unit on a faired mounting ring. An unidentified tubular pod on a short pylon was fitted ventrally just aft of the wings. By then the aircraft had been retrofitted with a TG-16 APU. CCCP-75643 has been referred to as the IL-18SL (for *samo**lyot**-labora**tor**iya* – laboratory aircraft); however, in keeping with Leninets's system of designating its avionics testbeds the aircraft was probably designated SL-18A.

b) In the early 1990s CCCP-75643 became a testbed for the *GLONASS satellite navigation system* (*glo**bahl'**naya navigatsi**on**naya **spoot**nikovaya sis**tem**a*, the Soviet equivalent of GPS). The search radar was removed (with no sign of its ever being there) and large cruciform markings were applied to the nose radome; these were possibly some kind of angle reflectors. By May 1992 the aircraft was finally retired and preserved at the 32nd Kilometre railway station a short way from Pushkin (near St Petersburg), home of LNPO Leninets's flight test centre.

c) IL-18V CCCP-75786 (c/n 181003905) became the SL-18V testbed with LNPO Leninets. The aircraft was used to test the Obzor-K (Perspective-K) navigation/attack radar developed for the Tupolev Tu-160 *Blackjack* bomber. The radar supplanted the RPSN-1 Emblema weather radar and was enclosed by a long conical radome, its large diameter requiring the use of a special adapter with a ventral cutout for the nose gear doors. Small blister fairings were located low on the aft fuselage sides in line with the fin fillet/fin junction.

A special pylon was installed under the fuselage in line with the forward entry door, permitting carriage of fixed missile acquisition rounds for systems integration purposes. For instance, at an early stage of the programme the SL-18V carried a large acquisition round simulating the Kh-45 ***Molniya*** (Lightning) air-to-surface missile. Originally developed by the Pavel Osipovich Sukhoi's OKB-51 as the main weapon for the Sukhoi T-4 bomber (development was later continued by the NPO Raduga (Rainbow, pronounced ***rah**dooga*) weapons design bureau in Doobna north of Moscow), this weapon was a candidate to arm the Tu-160 for a while but was rejected in 1977.

d) At a late stage of its test career the SL-18V (by then reregistered RA-75786) was retrofitted with an APM-60 Orsha magnetic anomaly detector and associated IL-38-style boom removed from the IL-18D-GAL geophysical survey aircraft (CCCP-74267, see page 46). The purpose of the modification is unknown. Thus RA-75786 became the longest IL-18 ever! By 1995 the aircraft was retired in Pushkin; it was still sitting there engineless in August 2001 as 75786 with no nationality prefix.

e) An IL-18D with the out-of-sequence registration CCCP-75713 No 2 (c/n 186009403) was yet another LNPO Leninets avionics testbed designated, logically enough, SL-18D. This aircraft was used in several test programmes and the end result was rather bizarre. The RPSN-1 weather radar gave way to a bulbous nose fairing accommodating a ground mapping radar developed for the Antonov An-124 Ruslan (*Condor*) transport. This was compounded by a Berkoot search radar aft of the nose gear unit; hence a mudguard was fitted to the nosewheels and the nose gear doors bulged accordingly. (Interestingly, the would-be SL-18A had no such mudguard.)

A small canoe fairing with forward-looking infra-red (FLIR) or laser equipment was mounted on the flightdeck roof, with a larger dorsal canoe fairing over the wing centre section and a small ventral canoe fairing ahead of the wing leading edge. The underside aft of the wings bristled with small blade and 'hockey stick' aerials and dielectric blisters. Finally, as on the SL-18V, two small fairings were located low on the aft fuselage sides.

In 1993 the SL-18D was stripped of experimental equipment and leased to Djibouti-based Daallo Airlines as RA-75713 until October 1994. Today it is operated by NPP Mir, the commercial flying division of the Leninets Holding Company; all non-standard fairings except the mounting for the Berkoot radar have been removed and the aircraft wears large 'NPP Mir' and 'Olimpiada 2004 – Sankt-Peterburg' titles to indicate that St Petersburg is a candidate to host the 2004 Olympic Games.

f) IL-18V CCCP-75804 (c/n 182004305) became yet another avionics testbed with LNPO Leninets. Since the V suffix letter was by then allocated to CCCP-75786, this testbed was designated SL-18I or SP-T18I. Again, the aircraft participated in several test programmes, each one leading to more tell-tale

Top: **By 1996 the SL-18V had been retrofitted with an APM-60 MAD removed from IL-18D CCCP-74267, making it the longest *Coot* ever. It is seen here languishing engineless in Pushkin.** Yefim Gordon archive

Centre and below centre: **The SL-18D (CCCP-75713 No 2, c/n 186009403) was another testbed used by LNPO Leninets in various programmes. The aircraft had a ground mapping radar for the An-124, a Berkoot search radar and so on.** Yefim Gordon archive

Bottom: **This in-flight shot of the SL-18D shows well its assorted 'bumps and bulges'.** Yefim Gordon archive

Above and left: **IL-18V CCCP-75804 was one of the most exotic-looking testbeds of LNPO Leninets. Known as the SL-18I or SP-T18I, it featured a combined search radar/SLAR fairing, a dorsal dielectric cupola and a 'Roman nose' housing a development radar.** Yefim Gordon archive

Left: **This view of CCCP-75804 shows the aft end of the SLAR 'sausage' and the superstructure associated with satellite navigation equipment.** Yefim Gordon archive

Below: **The SL-18I sits at Pushkin, flanked by an An-26 and the SL-18D.** Yefim Gordon archive

'bumps and bulges'. Firstly, the SL-18I was fitted with a non-standard nose radar in an elongated and downward-curving radome; the type and application is unknown.

Secondly, once again a Berkoot search radar was installed (with appropriate modifications to the nose gear unit), but this time a large perfectly cylindrical pod with a hemispherical rear end was grafted on to its radome from behind; this housed a Mech (Sword) SLAR. Next, a fairly large dorsal teardrop fairing with a hemispherical cupola on top was mounted in line with the overwing emergency exits; this was presumably associated with satellite navigation. Finally, the aircraft had the small blister fairings on the aft fuselage which were common to almost all of the SL-18 series.

In 1993 the SL-18I was likewise stripped of experimental equipment and leased to Daallo Airlines (initially as 75804, with no nationality prefix, and then as RA-75804). Curiously, all of the non-standard fairings except the dorsal dome were retained, undoubtedly causing many a raised eyebrow wherever the aircraft went; the eye-catching SLAR 'sausage' even caused one Polish writer to mistake 75804 for an IL-20! Today the aircraft wears an Aeroflot-style colour scheme with 'NPP Mir' titles; the ventral radomes have been removed during the latest overhaul but the non-standard nose radome and the mounting for the Berkoot radar still remain.

g) The last of the SL-18 series – in Russian alphabetical order, that is – was the SL-18P converted from the last IL-18E, CCCP-75411 (c/n 186009205). Once again, the aircraft had at least four configurations; unlike the SL-18D and SL-18I, however, it carried equipment associated with only one programme at any one time.

The original configuration dated back to 1966 when the factory-fresh aircraft was converted to test the Taïfoon-M (Typhoon-M) fire control radar developed for the Sukhoi Su-15TM *Flagon-F* interceptor; this may explain the P suffix (for *pere**khvaht**chik* – interceptor). Inevitably, the radar (enclosed by a long ogival radome) was installed on an adapter supplanting the standard RPSN-1 weather radar. A small weather radar 'egg' was mounted on a short pylon on the flightdeck roof.

h, i) The second and third configurations of the SL-18P (in pre-1973 and post-1973 colours respectively) were very similar, being used to test two versions of the PNS-24 Tigr (Tiger) navigation/attack system (*pri**tsel'**no-navigatsi**on**naya sis**tema***) designed for the Su-24 *Fencer* tactical bomber. In both cases the short, almost rectangular-section radome terminated in a pitot – a horizontal H-shaped structure (as fitted to the delta-winged Sukhoi T6-1 prototype) on the second version and the characteristic F-shaped pitot (popularly known as the 'goose' among Su-24 crews) on the third; the radome itself was dark green and white respectively.

An optical sensor offset to starboard and covered by a rotating hemispherical guard was fitted aft of the nose gear on the second version; on the third this was replaced by a different optical system in a small teardrop fairing. Additionally, a short ventral canoe fairing was located just ahead of the wing leading edge *à la* SL-18D. The 'egg' over the flightdeck was removed.

j) The final version of CCCP-75411 was a testbed for the Korshoon (Kite, the bird) 360° search radar developed for the Tu-142M *Bear-F Mod* ASW aircraft. The large teardrop radome was fitted, IL-38 style, aft of the nose gear unit; the RPSN-1 weather radar was reinstated and a mudguard added to the nose gear unit. In this

IL-18E CCCP-75411, the SL-18P testbed, as originally flown with the Su-15TM's Taïfoon-M fire control radar. The shiny factory finish is readily apparent. Note the weather radar 'egg' above the flightdeck. Yefim Gordon archive

The SL-18P later in its career. The Su-15 nose has been replaced with a late version of the Su-24 bomber's PNS-24 Tigr navigation/attack system with two radars in a common radome. Note the ventral sensor fairings. Yefim Gordon archive

guise the aircraft flew until the early 1990s when the radar was removed; the mounting for the radome remained for a while but was deleted during the next overhaul. Currently RA-75411 is operated by NPP Mir. The Korshoon radar then underwent further tests on a converted IL-38 (see Chapter 6).

k) IL-18B CCCP-75703 (c/n 189001505) was converted into an avionics testbed of some sort, probably by Moscow-based NPO Vzlyot (Take-off), another Ministry of Electronics Industry division. The aircraft had a shallow boxy fairing immediately aft of the port wing root (probably a synthetic-aperture radar), with two round antennas mounted in line with it closer to the fuselage centreline. The shape of this fairing was changed later in the aircraft's career. A flat-bottomed bulge aft of the nose gear unit revealed that CCCP-75703 had been fitted with a Berkoot search radar which was subsequently removed and the bottom of the mounting ring closed with sheet metal. A small blade aerial was mounted dorsally ahead of the forward entry door and another ventrally aft of it.

CCCP-75703 was based at LII in Zhukovskiy south of Moscow. Despite its age, the aircraft stayed operational long enough to see the breakup of the Soviet Union, being last noted at Moscow-Bykovo as RA-75703 in June 2000.

l) In 1984 NPO Vzlyot added IL-18V CCCP-75851 (c/n 182005501) to its aircraft fleet; this was a re-export aircraft, having served with Cubana de Aviación since 1964 as CU-T832. CCCP-75851 became a testbed for missile guidance systems. The weather radar was replaced by a cylindrical metal fairing terminating in a cone with a dielectric tip and a pronounced 'chin'; the fairing was made to fit instead of the standard radome without requiring any additional modifications to the airframe.

A large detachable pod with the missile's guidance system (acquisition round) was pylon-mounted ahead of the wings, and small angular fairings housing additional equipment were located aft of the nose gear unit and on the fuselage tailcone. By August 1992 CCCP-75851 was reconverted to standard configuration and is now operated by Elf Air (the commercial division of NPO Vzlyot) at LII as IL-18Gr RA-75851.

m) An obscure avionics testbed presumably operated by NPO Vzlyot was IL-18V CCCP-78732 (ex-CCCP-75794, c/n 181004103). The non-standard registration is noteworthy, since the 787xx series is allocated to Il'yushin IL-76MD *Candid-B* transports and IL-78/IL-78M *Midas* tankers. The aircraft had a small cylindrical pod with a hemispherical front end and a pointed rear end under the forward fuselage, plus a small flat-topped dorsal fairing near the aft entry door. It was last noted at LII as RA-78732 in March 1994 minus the ventral pod and was reportedly operated by Volare, LII's own airline, at the time.

n) A former Soviet Air Force IL-18V with the out-of-sequence registration CCCP-75894 (c/n 182004801)[33] was converted into an avionics testbed by NPO Vzlyot before 1992. The aircraft featured a large canoe fairing under the forward fuselage (with a downward-projecting 'thimble' at the rear) and a smaller ventral 'bathtub' aft of the wings, both incorporating large dielectric panels. It is still based at LII in this condition as RA-75894.

Top and above: **The final test configuration of the SL-18P. The standard RPSN-2 weather radar has been reinstated and a Korshoon search radar for the Tu-142M ASW aircraft fitted aft of the nose gear.** Yefim Gordon archive

Left: **IL-18B CCCP-75703, an avionics testbed of unknown purpose, seen at aircraft overhaul plant No 402 (Moscow-Bykovo) in the late 1980s. Note the antenna array immediately aft of the port wing root (probably a synthetic-aperture radar) and the bulge aft of the nose gear where a Berkoot search radar used to be.** Sergey and Dmitriy Komissarov archive

o) Sometime before 1973 IL-18V CCCP-75811 (c/n 182004504) was converted into the IL-18REO testbed (***rah**dioele**ktron**noye obo**roo**dovaniye* – electronic equipment) by the Moscow Institute of Electronics (MIREA). The nose radome gave place to a long cylindrical metal 'plug' terminating in a dielectric fairing which looked like an outsize match (to use the *mildest* of comparisons). A large cylindrical pod with a hemispherical front end and an ogival rear end (both dielectric) was installed under the forward fuselage. The nature of the equipment tested is unknown.

p) Later, CCCP-75811 was fitted with a Berkoot search radar, a small dielectric blister over the wing centre section and a small ventral canoe fairing aft of the wings. Interestingly, the search radar was located noticeably farther aft than on the SL-18D and SL-18I. In 1996, long after this equipment had been removed, the aircraft was operated by the Il'yushin OKB's own airline, Ilavia, as RA-75811; by August 1999 it had been sold to Elf Air.

q) In 1965-66 (some sources say 1968) the Experimental Machinery Plant headed by Vladimir M Myasishchev (formerly OKB-23) operated an aircraft designated IL-18 'Polosa' (Stripe, pronounced *polo**sah***) or IL-18P.[34] Quite probably this was a modified IL-18V in pre-1973 colours with the non-standard registration CCCP-06180 No 2 (c/n unknown). It had small cheek fairings immediately aft of the radome and small square fairings on the sides of the fin about halfway up; these incorporated slot aerials. Incidentally, the registration had previously been used on the second prototype Mil' Mi-2 *Hoplite* utility helicopter (c/n 0102); it was later 're-reused' for a production Mi-26 *Halo* heavylift helicopter (c/n 226205) owned by the Mil' OKB.

The IL-18P served to test an experimental navigation system permitting automatic flight along a predesignated route and automatic approach/landing. The test programme was performed by pilots Ghennadiy N Volokhov and Aleksandr M Tyuryumin.

r) An IL-18A or IL-18B with the non-standard registration CCCP-48093 No 1 (c/n unknown) was another obscure avionics testbed with a large flat-bottomed fairing under the extreme aft fuselage. It was preserved in a pioneer camp near St Petersburg, in its original pre-1973 colours; the registration was reused in 1989 for an Antonov An-32 *Cline* transport (c/n 0703).

Top and below: **IL-18V CCCP-75851 was used by NPO Vzlyot in the mid-1980s to test missile guidance systems.** Yefim Gordon archive

Centre and bottom: **Close-up of the highly modified nose on CCCP-75851. The adapter carrying the missile seeker head was completely interchangeable with the standard radome.** Yefim Gordon archive

Top left: **Close-up of the fixed acquisition round under the forward fuselage of CCCP-75851.** Yefim Gordon archive

Above and top right: **IL-18V RA-75894, another obscure avionics testbed belonging to NPO Vzlyot, takes off from runway 12 at Zhukovskiy. The large ventral canoe fairings are clearly visible.** Yefim Gordon archive

Left: **IL-18V CCCP-78732 was an avionics testbed of unknown purpose apparently belonging to the Myasishchev OKB. This view shows the small cylindrical pod under the forward fuselage.** Yefim Gordon archive

Above: **CCCP-75811, the IL-18REO avionics testbed, in early configuration. No comments on the shape of the nose please...** Yefim Gordon archive

Right: **Front view of the IL-18REO as originally flown.** Yefim Gordon archive

Above and right: **IL-18REO CCCP-75811 in later guise, probably photographed at Pushkin. Note that the Berkoot search radar is positioned much further aft than on the IL-18SL and SL-18D. The dorsal dielectric blister and ventral 'bathtub' are just visible.** Yefim Gordon archive

Below: **IL-18V CCCP-06180 was probably the IL-18 'Polosa' (IL-18P) used by the Myasishchev Design Bureau to test an experimental navigation system. Note the fairings on the sides of the nose and fin housing slot aerials.** Yefim Gordon archive

Weather Research Aircraft

a) IL-18V 'Meteor'

One of the first *Coots* to be converted for research purposes was IL-18V 'Salon' CCCP-75716 (c/n 180001902), a 235th IAD machine which was the 'presidential' aircraft of Nikita S Khruschchov. By 1963 the aircraft had been transferred to Aeroflot's Polar Directorate, giving way to a newer example.

In 1963 CCCP-75716 was fitted out as a weather research aircraft by MMZ No 30 for the Central Aerologic Observatory (TsAO – *Tsen**trahl'**naya aerolo**gich**eskaya observa**tor**iya*), a division of the Soviet Union's State Committee for Hydrometeorology and Environmental Control (*Goskomghidro**met***). Part of the equipment was installed in a large canoe fairing under the forward fuselage (outwardly identical to that of CCCP-75431, see page 45); another recognition feature was a satellite communications antenna in a small dielectric dome above the wing centre section. The total weight of the equipment suite was about 4 tons (8,820 lb).

At first the aircraft's mission was to corroborate the data supplied by weather research satellites. These were only beginning to gain wide use and had yet to prove their worth. Later, when the weather research satellites had earned their credentials, TsAO started using CCCP-75716 for other tasks; the first of these was monitoring the launches of weather research rockets in the High North. Such flights sometimes involved participation of foreign scientists who were quick to recognise both the considerable capabilities of the aircraft itself and the skill of its crew (high flying skill was a requirement, since the flight profiles were rather unusual). They were in a position to make an opinion, as prior experience they had gained with other weather research aircraft was far less encouraging. The aircraft was also used for investigating jetstreams, turbulence (the clear-air turbulence phenomenon was already recognised as a flight safety threat in those days), storm fronts and the upper reaches of the atmosphere, including the ozone layer.

As the research missions changed, so did the equipment suite. The refitting jobs were performed at various locations – the LII airfield in Zhukovskiy, MMZ No 30's flight test facility in Lookhovitsy and Chkalovskaya airbase. Incidentally, some of the missions took CCCP-75716 over some of the Soviet Union's top secret military installations and sensitive areas where any other aircraft would have been shot down at once. Makes you wonder if some of the aircraft's missions were not exactly of a peaceful nature...

In 1968 IL-18V CCCP-75716 was formally reassigned to the 63rd Flight which was part of the Central Directorate of International Services (TsUMVS – *Tsen**trahl'**noye oopravleniye mezhduna**rod**nykh voz**doosh**nykh so'ob**schchen**iy*) at Moscow-Sheremet'yevo. Its actual owner and operator, however, was GosNII GA and the aircraft was still used for weather research. The canoe pod under the forward fuselage was removed in March 1974, since the programme it was associated with had ended. Since CCCP-75716 could work in conjunction with Meteor weather research satellites, it was unofficially designated IL-18V 'Meteor'. Experience gained with this aircraft allowed the best ways of placing and using various meteorological equipment to be determined and later put to good use on other aircraft.

b) IL-18E 'Meteor'

In 1976 the IL-18V 'Meteor' (CCCP-75716), which was manufactured in 1960, was approaching the limit of its 20-year designated service life and a replacement aircraft had to be procured. It was another year before GosNII GA was able to provide the Central Aerologic Observatory with a newer *Coot*, IL-18E CCCP-75598 (c/n 186008802), which also possessed longer range. Interestingly, the aircraft was released by the factory on 18th July 1967, despite the third digit in the c/n being a 6.

Not wishing to be snared in miles of red tape while obtaining all the necessary clearances for outfitting a new weather research aircraft from scratch, TsAO chose the way of the least resistance, requesting permission to simply transfer the existing equipment from CCCP-75716 to the new aircraft. Permission was quickly granted and the work went ahead. On 24th October 1977 the former IL-18V 'Meteor' was

IL-18E CCCP-75598 was modified by GosNII GA as the IL-18E 'Meteor' weather research aircraft. These views show the many sensors and air probes on the fuselage and the ventral fairing of the BMR-1 weather research radar.
Yefim Gordon archive

This aspect of the IL-18D 'Tsiklon' weather research aircraft gives a good detail view of the instrumented nose probe, the K-11M weather research radar aft of the nose gear, the ventral ASO-2I rainmaking chemical dispensers immediately aft and the plethora of sensors on the forward fuselage. Yefim Gordon archive

transferred to Ul'yanovsk Higher Flying School, the Soviet Union's top-notch civil aviation flying college, where it served on as a trainer until finally retired in 1980.

Designated IL-18E 'Meteor' by analogy with its predecessor, CCCP-75598 featured a specialised BMR-1 weather research radar in a rather large round flat-bottomed radome under the wing centre section. The radar set was located in the forward baggage compartment and accessible via several maintenance hatches from the cabin allowing the radar to be promptly fixed in flight, should it fail. Numerous air sampling traps and sensors were mounted on short struts on the forward and centre fuselage and under the wingtips, including sensors in the foremost and rearmost cabin windows to starboard and air traps in the forward emergency exit (7th window) and 13th window to starboard. Observation blisters were provided in the second and 13th window to port and the third window to starboard. Vertical cameras were installed in the front and rear cabins, while the toilets were transformed into photo processing labs, allowing the films to be developed and photos printed on board.

The conversion job, which was performed by MMZ No 240 at Moscow-Khodynka, was completed in June 1977. For more than 20 years that followed, CCCP-75598 served faithfully in the weather research role, taking part in numerous scientific experiments held both in the Soviet Union and abroad. By 1991, however, it became clear that the IL-18E 'Meteor' was not being used to the full in its intended capacity, so GosNII GA proposed using the aircraft for cargo charter flights in order to generate revenue. Then, faced with the decline in the number of research programmes and skyrocketing aircraft leasing charges, TsAO had to give up using the IL-18E 'Meteor' and the aircraft was modified for other uses (see later).

c) IL-18D 'Tsiklon' (IL-18DTs)

In the mid-1970s the Central Aerologic Observatory commissioned development of a series of weather research aircraft under the common name *Tsiklon* (Cyclone). By then radar technologies of weather and atmospheric research had gained wide use. The use of radar made it possible to study the distribution of clouds and precipitation over wide areas and follow the development of storm nuclei. The techniques employed by the Tsiklon system allowed stand-off research of clouds – ie, without actually entering them, which could change their shape and structure (and could be dangerous for the aircraft itself, considering that severe turbulence is often encountered in clouds, not to mention lightning!). Apart from studying the principal thermodynamic and electric parameters of the atmosphere and cloud formations, the aircraft were designed to perform cloud-seeding missions in order to make rain – eg, when it was necessary to prevent an impending hailstorm which could destroy crops, or to scatter rain clouds which could ruin a public holiday.

One of the seven assorted aircraft in the series was based on the long-range IL-18D – specifically, CCCP-75442 (c/n 187009702) – and designated IL-18D 'Tsiklon', though some sources refer to it as the IL-18DTs.[35] Outwardly it was something of a spotter's delight, sprouting all manner of 'bumps and bulges' so characteristic of research aircraft. A pointed boom 4m (13ft 1½in) long tipped with sensors was installed on the left side of the nose ahead of the flightdeck glazing, increasing overall length to 40m (131ft 2¾in); it was painted in black and white stripes to avoid damage by ground vehicles. A specialised K-11M weather research radar in a deep teardrop radome with flattened sides was installed immediately aft of the nose gear unit (frames 5-11), requiring the nosewheels to be fitted with a mudguard and the nose gear doors bulged accordingly.

In a similar fashion to the SL-18D, a FLIR or laser equipment housing was mounted on the flightdeck roof and two shallow dielectric canoe fairings were fitted above and below the fuselage in line with the wing leading edge. Two elongated square-section fairings with apertures for laser measurement systems took the

The IL-18D 'Tsiklon' (CCCP-75442) in flight, showing off its other sensors on the centre/aft fuselage and wings. Yefim Gordon archive

place of the two rearmost cabin windows on each side. Numerous sensors were mounted on struts just aft of the flightdeck. The lower fuselage incorporated two optically flat camera windows protected by sliding doors for take-off and landing. An extra wire aerial ran from the fin leading edge to a strut above the forward entry door. Finally, two observation blisters were provided in the third and 13th cabin windows on each side, plus two dorsal observation blisters in line with the rear pair of emergency exits and ahead of the rear entry door. The eye-catching *Tsiklon* emblem was painted on the nose to clarify the aircraft's 'storm chaser' role.

The mission equipment comprised a measurement suite, data recording/processing suite and cloud-seeding equipment. The measurement suite included, first of all, a *thermodynamic measurement system* comprising a TsSV-3M-1KM central air data system, an EM TsAO electrometeorograph, a PK G-load measuring kit, an SAMB-70 airborne automatic weather research module, an ASTA-74 airborne automatic thermoanemometer, a DISS-013-134 Doppler speed/drift sensor system, an RV-18Zh radio altimeter, an ISVP airflow structure meter and an SG-1 airborne humidity meter. It recorded the outside air temperature and its fluctuations, the aircraft's speed and heading, the wind speed and direction, airflow pulsations, static and dynamic air pressure, barometric and true altitude, vertical gusts and G loads acting on the aircraft. The second major component was the *cloud and precipitation microstructure measurement system* comprising an RP-73 translucency recorder, an IRCh water/ice particle size meter, an SALYa atmospheric ice particle counter, SEIV-3 and IVO airborne electric cloud water content meters and an AFSO cloud phase analyser. The third component was the *meteorological radar and laser system* – the aforementioned K-11M search radar, a BMR-1A radar for vertical scanning of the atmosphere and an LR-3P laser polarisation meter. The fourth component was the *radiometric system* comprising the RAK radiometric/actinometric complex and the TETA radiometric module. Finally, there was a PNP meter for measuring electric fields and the aircraft's electric charge.

The BARS-1 data recording/processing suite ('bars' means 'snow leopard', but in this case it is an acronym for *borto**va**ya avtoma**tich**eskaya reghist**ree**ruyuschchaya sis**tem**a* – on-board automatic recording system) consisted of a K60-42 magnetic recorder, a PTU-31-1-7 video recording system, a forward-looking AKS-2 cine camera on the starboard side of the nose, an AFA-BAF-21S photo camera and an SYeO common time indication system. The K60-42 automatically recorded signals generated by the thermodynamic and cloud measurement systems for future computer analysis. The cine and photo cameras were used to film the outside conditions, using the observation blisters. A photo/cine film processing lab was provided in the rear cabin.

The cloud-seeding equipment designed to generate rain from cumulus and stratus clouds included three KDS-155 dispensers and three ASO-2I dispensers mounted on the centre fuselage underside (frames 17-23). The most widely used among the latter is silver iodide. The KDS-155 and ASO-2I were adapted from stock chaff/flare dispensers used on Soviet military aircraft for passive electronic and infra-red countermeasures. Instead of bundles of chaff, aluminium-coated glass needles or PPI-26 IRCM flares they fired special PV-26 cartridges with chemicals triggering the formation of ice crystals. The latter would become too heavy to be supported by the air currents inside the cloud and start falling as hailstones; however, these melted and turned into rain before reaching the ground.

The aircraft had a flight crew of five and a 34-man team of researchers who sat behind 19 single and dual equipment consoles installed in all three cabins.

The first post-conversion flight took place on 4th April 1980. The maximum take-off weight was 64 tons (134,480 lb), including a 9.4-ton (20,720-lb) payload; the various external outgrowths reduced the cruising speed to 625-650km/h (388-403mph) and the service ceiling to 8,650m (28,380ft). The aircraft could stay airborne for up to eight hours, with a maximum range of 4,270km (2,650 miles).

The IL-18D 'Tsiklon' had quite an active service career, flying both at home and abroad (among other things, it periodically deployed to Cuba to chase tropical hurricanes). After the demise of the Soviet Union it was reregistered RA-75442. Unfortunately, like some other research aircraft, the IL-18D 'Tsiklon' fell victim to unscrupulous businessmen to whom profits were more important than science – in 1997 it was stripped of all mission equipment, leased to a succession of airlines and finally sold.

Geophysical Survey Aircraft

a) IL-18V CCCP-75431

In late 1972 or early 1973 an IL-18V with the out-of-sequence registration CCCP-75431 (c/n 180002003)[36] was transferred to the Main Geophysical Observatory named after A I Voyeikov (GGO – ***Glahv****naya* ***gheofizich****eskaya observa*-***tor****iya*) and converted to a geophysical survey aircraft. It featured a large canoe fairing under the forward fuselage with four ventral apertures for sensors and cameras; the lower portion could swing open to starboard for maintenance. A large teardrop metal fairing incorporating three dielectric panels was installed ventrally just aft of the wings. The first, third and tenth cabin windows to starboard were blanked off with metal plugs mounting small sensors; another sensor was installed on a short strut above the wing trailing edge. The rear emergency exit to starboard had a non-standard windowless cover.

CCCP-75431 participated in assorted research programmes held in the interests of various ministries. For instance, on 3rd April 1975 the aircraft took off from its home base of Leningrad-Pulkovo, heading for Karaganda. The mission was to survey arable lands in the Karaganda, Tselinograd and Kokchetav Regions of Kazakhstan in preparation for the wheat sowing campaign, determining how much water was needed for irrigation. The IL-18 worked in conjunction with a specially modified Antonov An-24 *Coke* airliner and an An-30 *Clank* photo survey aircraft. Similar land survey missions were flown in the spring of 1977 in such far-apart areas of the Soviet Union as the Krasnoyarsk Region in West Siberia, the Stavropol' Region in southern Russia and the southern regions of the Ukraine, all major providers of grain.

A year earlier CCCP-75431 was involved in the POLEX-Sever-76 ('Polar Experiment – North 1976') together with another 'curious *Coot*', CCCP-75716 (see page 42), operating from Amderma and working together with the research vessel M/V *Professor Wiese*. During 1976 it participated in the SAMEX (*Sovetsko-amerikahnskiy [mikrovolnovyy] eksperiment* – Soviet/US Microwave Experiment), measuring the condition of the sea and atmosphere in two designated areas of the Pacific near the Kamchatka Peninsular and the Kurile Islands. The experiment also involved a modified Convair CV 990 Coronado, the Soviet research vessel M/V *Akademik Korolyov* and the Nimbus-5 and Nimbus-6 satellites.

In due course the aircraft's pre-1973 IL-18 colours gave way to a variation on Aeroflot's 1973-standard livery. The cheatline was red but the red colour on the tail surfaces and outer wings as required by the Polar version of the

IL-18V CCCP-75431, a survey aircraft operated by the Main Geophysical Observatory, flies over the North Sea during the POLEX-Sever-76 experiment. The picture was taken from IL-18V 'Meteor' CCCP-75716 shortly before the latter was reconverted to standard configuration. Yefim Gordon archive

CCCP-75431 in the 1980s following an overhaul and repaint in basic 1973-standard colours, save that the cheatline was red and the Aeroflot titles/logo were missing. The angular object 'on top of the forward fuselage' is actually situated beyond the aircraft, being part of a ground structure! Yefim Gordon archive

Above: **Sometime before 1987 IL-18D CCCP-74267 was converted into the IL-18D-GAL geophysical survey aircraft (aka IL-18D 'Antarktida') equipped with an IL-38 style MAD boom.** Sergey and Dmitriy Komissarov archive

Left: **This close-up shot of the IL-18D-GAL's rear end was taken while the aircraft was performing a scientific mission in Antarctica in the autumn of 1987.** Yefim Gordon archive

Opposite page:

IL-18V CCCP-75423 was modified for environmental monitoring and research by LII and MIREA. This view shows the large SLAR pod under the forward fuselage and the ventral 'bathtub' aft of the wings. Yefim Gordon archive

IL-18V CCCP-75423 passes overhead on final approach to Zhukovskiy after a test mission, showing off the dielectric panel in the 'bathtub' and three of the four optical/laser sensor windows. Yefim Gordon archive

colour scheme was missing, as were the Aeroflot titles and logo. By 1992 CCCP-75431 had been transferred to LII and all survey equipment had been removed, thus turning the aircraft into an IL-18Gr freighter.

b) IL-18D-GAL (IL-18D 'Antarktida')

By 1987 IL-18D CCCP-74267 (c/n 188011105), which by then had been in use for several years for supporting Soviet Antarctic research stations, was modified for geophysical survey in Antarctica. To this end a boom an APM-60 magnetic anomaly detector borrowed from the IL-38 was fitted instead of the usual tailcone; the mission equipment and operators' workstations were installed in the cabin. In some documents the modified aircraft was called IL-18D-GAL (*gheofi**zich**eskaya **a**erolabora**tor**iya* – geophysical flying laboratory), while a newspaper publication referred to it as the IL-18D *'Antark**tid**a'* (Antarctica).

The 'stinger tail' IL-18D took part in the 32nd Soviet Antarctic expedition which started off from Moscow on 26th September 1987. Taking off from Moscow-Sheremet'yevo, the aircraft captained by V Ya Shapkin made the short trip to Leningrad to pick up the team of polar researchers, then flew to Ice Station Molodyozhnaya via Odessa, Cairo, Aden and Maputo. Upon arrival CCCP-74267 made a series of survey flights over the South Pole and the so-called Pole of Inaccessibility – a part of the ice continent which is truly hard to reach.

In 1989 the aircraft was contracted by the Norwegian company Amarok to undertake magnetic monitoring of the sea shelf north of Spitsbergen (Svalbard) Island commissioned by the Conoco, Elf-Aquitaine and Statoil petroleum companies. The work was performed by a team from the Soviet specialist organisation *Sev**morg**heo**log**iya* (the North Sea Geology Trust). Operating from Murmansk-Murmashi and Longyear (the airport of Spitsbergen), CCCP-74267 flew 14 sorties between 20th April and 11th May, inspecting an area of 50,000km^2 (19,305 miles^2). The flights proceeded at 600-900m (1,970-2,950ft) along 23 routes.

In post-Soviet days the IL-18D-GAL was stripped of special equipment and sold to the Domodedovo Civil Aviation Production Association as RA-74267.

c) IL-18E CCCP-75598

In the autumn of 1991 the former IL-18E 'Meteor' weather research aircraft (CCCP-75598) was stripped of its mission equipment and converted into a geophysical survey aircraft with an *izdeliye* A-723 side-looking aircraft radar. The SLAR, which was developed by the Electronics Institute of the Ukrainian SSR's Academy of Sciences, was installed in a large slab-sided fairing aft of the port wing and a ventral fairing in line with it.

Upon completion of this programme GosNII GA reconverted the aircraft to standard configuration with a 72-seat tourist-class layout and leased it to the Moscow-based charter airline IRS-Aero as RA-75598.

75423
CCCP

Left and below: **IL-18A CCCP-Л5821 (ie, SSSR-L5821) was converted by LII as the IL-18LL de-icing systems testbed. Here the test article installed on top of the fuselage is a section of the IL-62 airliner's wing.** Yefim Gordon archive

Environmental Monitoring Aircraft

In the 1980s an IL-18V with the out-of-sequence registration CCCP-75423 (c/n 182005601)[37] was extensively modified for environmental monitoring and research tasks by LII and MIREA. Its main external identification feature was a large cylindrical pod under the forward fuselage housing a SLAR based on the Mech system (the pod was very similar to the one on the SL-18I); this and the unusual 'wannabe-Polar' red/white colour scheme identical to the one worn by CCCP-75431 until 1995 caused some people to mistake it for an IL-24N ice reconnaissance aircraft. At an altitude of 6,000m (19,685ft) the radar could scan a strip of land 20km (12.4 miles) wide.

A flat-bottomed ventral 'bathtub' with a large dielectric panel (similar to that of IL-18V CCCP-75894) was installed aft of the wings. Finally, four angular fairings incorporating optically flat glass windows were provided for operating optical sensors or spectrometers. One such window was located dorsally just aft of the forward entry door, looking up at about 45° to the direction of flight; a second window placed on the port side just a bit further aft was directed about 45° upwards, while the other two windows located fore and aft of the rear entry door looked 45° down.

CCCP-75423 was in the static park at MosAeroShow-92, Russia's first real international airshow, at LII's airfield on 11-16th August 1992. The exhibitors catalogue at the MAKS-93 airshow referred to this aircraft as the IL-18 ZhLIIP (*Zhoo**kov**skoye **lyot**no-ispy**tah**tel'noye is**sled**ovatel'skoye predpro**yah**tiye* – Zhukovskiy Flight Test & Research Enterprise), though this is hardly its proper designation.

In 1993 the aircraft received the Russian prefix; by August 1997 it had been withdrawn from use. Finally, in early 1999 RA-75423 was restored to airworthy status for ferrying to Pushkin; there in the course of a complete refurbishment at the Russian Navy's aircraft overhaul plant No 20 it was stripped of all non-standard appendages and converted to combi configuration for its new owner, IRS-Aero.

IL-18LL De-icing Systems Test Aircraft

In the early 1960s the sixth production IL-18A, CCCP-Л5821 (c/n 188000201), was transferred to LII which converted it for testing the de-icing systems of other aircraft. In this form the aircraft was known as the IL-18LL (*le**ta**yuschchaya lab-ora**tor**iya* – lit. 'flying laboratory').[38]

The changes to the IL-18LL's airframe were quite extensive. A flat-topped superstructure was fitted to the centre fuselage for mounting the test article – a section of airframe incorporating de-icing system elements. To create artificial icing conditions a large circular sprinkler grid with bracing struts at the front was mounted ahead of the wing centre section; the cabin accommodated a water tank and test equipment consoles. Cine cameras installed in angular fairings on top of the outer engine nacelles captured the ice formation on the test article.

In 1965 the IL-18LL was used to verify the de-icing systems developed for the wings of the IL-62 long-haul airliner and engine air intakes of various high-speed combat aircraft. According to test pilot Yakov I Vernikov who captained the aircraft, the added area above the centre of gravity (CG) made the testbed sensitive to crosswinds, complicating flying, especially at low speeds. Towards the end of the decade CCCP-Л5821 was reregistered CCCP-75637 and converted for testing data link system components, yielding its icing test mission to An-12BK CCCP-48974 No 2 (c/n 6344510).[39]

IL-18LL Engine Testbed

Rather confusingly, the IL-18LL designation also applied to a very different testbed, and not a Soviet one at that. In late October 1987 VZLÚ (*Výzkumní a zkušební létecký ústav* – Flight Test and Development Organisation), the Czech equivalent of LII, purchased IL-18V DDR-STC (c/n 180002202) from the East German carrier Interflug. After a minor overhaul the airliner was registered OK-018 on 30th November 1987; the three digits instead of the usual three letters after the registration prefix indicated test or development status.

The aircraft's mission was to test the intended powerplant of the Let L-610 twin-turboprop regional airliner. A Walter (later Motorlet) M-602 turboprop with a take-off rating of 1,824ehp and a cruise rating of 938ehp driving an Avia V-518 five-blade reversible-pitch propeller was installed in place of the No 2 AI-20K. The aircraft retained basic Interflug colours.

The tests went successfully, allowing the L-610 prototype (OK-130, later reregistered OK-TZB) to enter flight test on 28th December 1988. Meanwhile, the IL-18LL was withdrawn from use and scrapped in January 1996 after sitting at Prague-Letňany for several years.

IL-118 Airliner Project

Trying to prolong the active life of the IL-18s remaining in service, in the summer of 1984 the Il'yushin OKB proposed an upgrade eloquently designated IL-118. It involved a rather radical redesign of the *Coot*, the four AI-20 turboprops being substituted with two 10,900-ehp Lotarev D-236T propfan engines driving SV-36 low-noise contra-rotating propellers of 4.2m (13ft 9⅜in) diameter. Developed by the Stoopino Machinery Design Bureau, the SV-36 featured glassfibre blades with a hollow composite spar and integrated electric de-icing threads. The front and rear rows had eight and six blades respectively, running at 1,100 and 1,000rpm respectively; the difference in speed was intended to reduce noise and vibration. The engine was a derivative of the D-136 turboshaft powering the Mi-26 helicopter.

The IL-118 was to offer much better fuel efficiency compared with not only the standard *Coot* but with the jet transports then in service. Yet the project never got beyond the preliminary design stage because all-new designs seemed to offer greater advantages.

Top: **Close-up of the test article and water sprinkler grid on the IL-18LL; note the fairings on the outer engine nacelles housing cine cameras to capture the ice build-up.** Yefim Gordon archive

Centre: **The IL-18LL in a different configuration for testing an air intake de-icing system. Judging by the shape of the test article, the intake assembly appears to be that of the Yakovlev Yak-28 tactical bomber.** Yefim Gordon archive

Below: **Yet another configuration of the IL-18LL with a test article resembling an F-104 style air intake – as used on, for example, the Tupolev Tu-128 heavy interceptor.** Yefim Gordon archive

Chapter Three

IL-18 Anatomy

The following structural description applies to the standard IL-18V as the most widespread version of the *Coot*. Details of other versions are given as appropriate.

Type

Four-engined medium/long-haul airliner. The airframe is of all-metal construction and is mostly made of D16A-T duralumin; some structural elements are made of AK6 an AL9 aluminium alloys, ML5-T4 magnesium alloy and 30KhGSA or 30KhGSNA grade steel. The flight crew comprises the captain, first officer (co-pilot), navigator, flight engineer and radio operator.

Fuselage

Semi-monocoque stressed-skin structure of beam-and-stringer construction with 78 stamped frames which are mostly set at 0.5m (1ft 7⅝in) intervals; the skin thickness varies from 1.5 to 1.8mm (0.06 to 0.07in). The structure is riveted, using rivets with lentil-shaped heads. Fail-safe design features are incorporated in highly stressed areas. For instance, frames 16-20 located in and around the propellers' plane of rotation are interceded by doubler frames at 0.25m (9⅞in) intervals – ie, frames 16, 16A, 17, 17A etc – for higher fatigue resistance. Frames 25, 43 and 72 also have doublers. To ensure adequate pressurisation special elastic tape is placed between the skin and internal structural members (frames and stringers) before they are riveted together.

The cross-section is mostly circular, changing to quasi-oval at the rear; maximum fuselage diameter 3.5m (11ft 5¾in), maximum cross-section area 9.6m^2 (103.2ft^2), fineness ratio 9.85. Duralumin reinforcement plates 1mm (0.039in) thick are attached to the fuselage sides via rubber gaskets in the inboard propellers' plane of rotation between frames 16A-19 (port) and 17-19 (starboard) to protect the skin against pieces of ice breaking away from the propeller blades and pebbles kicked up from the runway by the propellers. The shape of these plates is different, the starboard plate being much taller than the port one.

Structurally the fuselage is divided into three sections. The *forward fuselage* is the flightdeck section (frames 1-3); actually, however, the flightdeck continues aft into the centre fuselage, terminating in a flat rear bulkhead (frame 5). An unpressurised nose fairing is grafted on ahead of the forward pressure dome, partly covering the manufacturing joint with the centre fuselage; its frames are designated by letters in Russian alphabetic sequence (A-B-V-G-D). This fairing accommodates part of the nosewheel well which goes all the way from frame A to the forward fuselage's frame 4, encroaching on the pressure dome; frame A is a bulkhead mounting the weather radar dish and the glassfibre radome hinged to starboard.

The flightdeck glazing features four windscreen panels made of birdproof triplex glass, six side windows and six eyebrow windows with Perspex glazing. The foremost side windows are sliding direct vision windows which

The forward fuselage, showing the flightdeck and the unpressurised nose ahead of the forward pressure dome accommodating the nosewheel well. Yefim Gordon archive

The IL-18's radome swings open to starboard. The black-tipped radome of this machine is unusual. Yefim Gordon archive

can be used for emergency evacuation on the ground.

The *centre fuselage* accommodates the passenger cabins, the Nos 1 and 2 underfloor baggage compartments (frames 9-24 and 34-56) and three avionics/equipment bays (frames 7-9, 24-26 and 26-32), terminating in a rear pressure dome (frame 56). It has a ventral cutout for the wing torsion box carry-through structure. Together with the flightdeck it forms a single pressure cabin; however, the flightdeck bulkhead is designed to withstand the pressure differential if either the flightdeck or the cabin decompresses. The inside of the pressure cabin is covered with glasswool heat- and soundproofing mats.

The cabin floor is made of thick plywood panels attached to transverse beams. Maximum cabin width is 3.232m (10ft 7¼in), floor width 3.077m (10ft 1⅛in) and cabin height 2.0m (6ft 6¾in).

The centre fuselage features two 0.45 x 0.75m (1ft 5¾in x 2ft 5½in) overwing emergency exits of perfectly oval shape on each side between frames 27-28 and 29-30, as well as circular windows of 400mm (1ft 3¾in) diameter. The windows feature double glazing; each of the two panes can withstand the pressure differential at high altitude. The number of windows varies, depending on the version. (The IL-18 *sans suffixe* had 15 cabin windows each side; on the IL-18A/IL-18B the number was originally 15 to port and 17 to starboard but this was later reduced to 14 and 16 respectively. The IL-18V had 15 windows to port and 16 to starboard, while on the IL-18E and IL-18D this was changed to 16 to port and 17 to starboard.)

Two plug-type entry doors measuring 0.758 x 1.4m (2ft 5⅞in x 4ft 7⅛in) are provided on the port side, with rain gutters above them; they are of quasi-oval shape (the aperture is 'flattened' at the bottom to provide an even threshold). On the IL-18 *sans suffixe* and IL-18A/B the doors were located at the extremities of the cabin (frames 9-11 and 49-51) and opened by pushing inwards and sliding towards the nose (forward door) or the tail (rear door). The IL-18V introduced a different arrangement with the doors located closer to the wings (frames 14-16 and 45-47) so that the entry vestibules separated the three cabins; the forward door slid towards the tail when opened and the rear door towards the nose. The baggage compartments are accessed via rectangular doors measuring 0.75 x 1.25m (2ft 5½in x 4ft 1¼in) on the star-

The centre fuselage and wing/fuselage joint of an IL-18V. Yefim Gordon archive

One of the overwing emergency exits located between frames 27-28 and 29-30. These were not easy to spot on aircraft in pre-1973 colours where they were not heavily outlined. Yefim Gordon archive

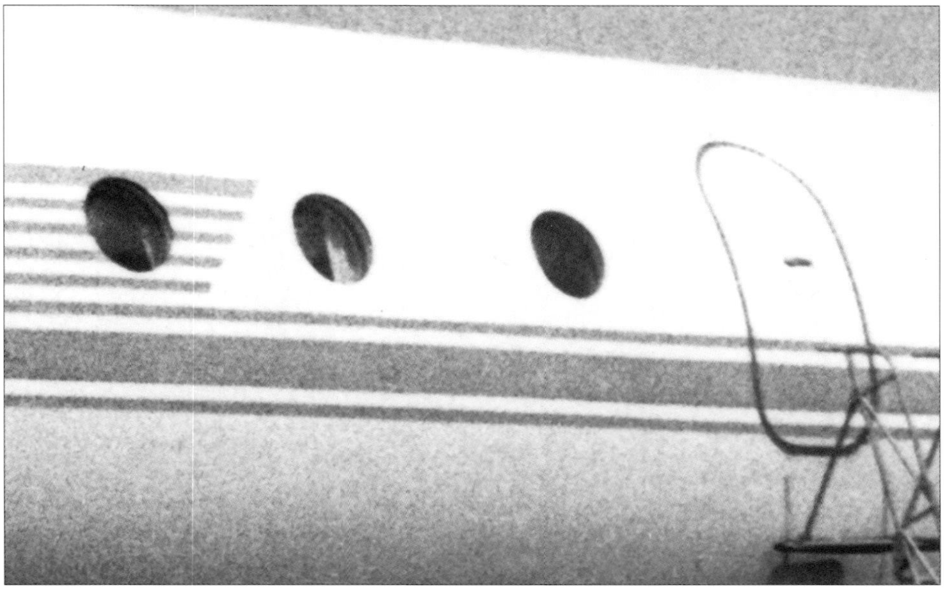

The forward entry door of the IL-18V *et seq* is located between frames 14-16; note the three windows ahead of it. Yefim Gordon archive

board side (frames 13-16 and 42-45) opening inwards and upwards. They feature transverse rubber belt mesh barriers spaced 1m (3ft 3⅜in) apart to restrain the baggage.

The *rear fuselage* is unpressurised. Besides carrying the tail surfaces, it incorporates the No 3 baggage compartment with an upward/outward-opening door measuring 0.9 x 1.28m (2ft 11½in x 4ft 2⅜in) to starboard.

Wings

Cantilever low-wing monoplane of trapezoidal planform. The wings are of all-metal, stressed-skin construction. Dihedral 3°, incidence 3°, camber –1° on the outer wings, aspect ratio 10, taper 3.

Structurally the wings are made up of three pieces: the monobloc centre section built integrally with the inner wings, which carry the engine nacelles, and detachable outer wing panels. The three-spar centre section/inner wings spanning 21.9m (71ft 10¼in) have 49 ribs (of these, the centreline rib 0 and ribs 1-4 on each side are inside the fuselage), 36 stringers and a skin thickness of 3-6mm (0.11-0.23in). The wing spars are attached to centre fuselage mainframes 26, 29 and 32. The inner engine nacelles are mounted in line with rib 10 and the outer ones in line with rib 20. The outer wings are of two-spar construction, featuring 18 ribs, 26 stringers and a skin thickness of up to 2.5mm (0.098in). A TsAGI S-5 airfoil is used on the inner wings between the fuselage and the inboard engines, while the two-spar outer wings utilise the S-3 airfoil, with an intermediate airfoil section in between; thickness/chord ratio 16% at the roots and 13% at the tips.

The inner and outer wings are joined by attachment fittings and splice plates. The leading-edge fairings are detachable; the wing skins incorporate numerous removable panels for access to the control runs, hydraulic and electric system components, fuel tank filler caps and fuel meters. The wing/fuselage joint is covered by a fairing.

The inner wings are equipped with one-piece double-slotted flaps occupying 19.39% of the wing area and 63.8% of the span. The flaps are actuated by an MPZ-9A twin-motor electric drive (*mekha**nizm** pere**sta**nov**ki** za**kryl**kov*) located on the rear spar via drive shafts and combined angle drives/screwjacks (two for each flap); they move on curved tracks housed entirely inside the wings. Flap settings are 15° or 30° for take-off and 40° (later reduced

In contrast, the rear entry door in this view belongs to an IL-18B and is located between frames 49-51 (note the location of the door relative to the fin fillet). Yefim Gordon archive

The outer wing of IL-18B CCCP-75689 (c/n 189001204) preserved in Simferopol'; the vertical rods are supports associated with the display and not part of the aircraft's structure! This view shows the large aileron and the outer portion of the flap. Yefim Gordon archive

to 30°) for landing. For structural integrity reasons indicated airspeed is limited to 320km/h (198mph/173kts) with 15° flap and 300km/h (186mph/162kts) with 30° flap. There are two-section ailerons on the outer wings (see Control system); each aileron section is suspended on three brackets.

Tail Unit

Conventional tail unit of all-metal stressed-skin construction. All tail surfaces feature a modified NACA-00 symmetrical airfoil with a thickness/chord ratio of 12%.

The trapezoidal *vertical tail* consists of a fin with a small curved root fillet beginning at centre fuselage frame 50 and a one-piece rudder (see Control system). Sweepback at quarter-chord 21°34', aspect ratio 2.2, taper 2.88. The fin is a three-spar structure with a detachable leading edge; skin thickness 1.2mm (0.047in). The fin spars are attached to aft fuselage mainframes 61, 65 and 68. There are three rudder mounting brackets, plus upper and lower rudder mounting posts. Rudder skin thickness is 0.8mm (0.03in) at the leading edge and 0.6mm (0.023in) elsewhere.

The cantilever trapezoidal *horizontal tail* of similar three-spar construction consists of two stabilisers with detachable leading edges and one-piece elevators (see Control system). Sweepback at quarter-chord 6°50', no dihedral, incidence –1°; aspect ratio 5.02, taper 2. The stabiliser spars are attached to aft fuselage frames 68, 70 and 72. The skin thickness is 1mm (0.039in) on the stabilisers and 0.8/0.6mm on the elevators. The stabiliser spars are attached to aft fuselage mainframes 68, 70 and 72. Horizontal tail span 11.8m (38ft 8½in); the effective horizontal tail area equals 17.38% of the wing area. There are two elevator mounting brackets on each stabiliser.

Landing Gear

Hydraulically-retractable tricycle type, with free-fall extension in emergency; all three units retract forward. The ground angle is 0°. Maximum landing gear transition speed is 350km/h (217mph).

The levered-suspension steerable nose unit attached to fuselage frame 2 is equipped with twin 700 x 250mm (27.55 x 9.84in) K-275/M non-braking wheels. On some special mission versions these are replaced by larger 800 x 225mm (31.50 x 8.85in) wheels equipped with a mud/snow/slush guard to prevent damage to various fairings associated with the mission equipment.

The main units attached to the middle spar retract into the inboard engine nacelles; they feature four-wheel bogies which are tilted into vertical position (nose up) by separate hydraulic rams/rocking dampers before retraction to lie inverted beneath the engines when retracted. The bogies have 930 x 305mm (36.61 x 12.00in) KT-81/3 wheels equipped with five-disc brakes (KT = *ko*le*so tormoz*no*ye* – brake wheel) and an anti-skid unit.[1] A non-retractable faired tail bumper is installed on the rear fuselage underside between frames 63-66 to protect the fuselage in the event of overrotation on take-off or a tail-down landing.

The nosewheel well is closed by twin lateral doors and two small doors in line with the gear fulcrum; there are two small cutouts immediately aft for the nose gear jury struts which are sealed after retraction. Each main unit has two large main doors, two small clamshell doors in line with the gear fulcrums and two more small door segments for the aft-mounted drag strut. The large wheel well doors open only when the gear is in transit.

Each landing gear strut is provided with an uplock, a downlock and gear door locks. Normally all locks are hydraulically operated. In the event of a hydraulics failure the co-pilot can open the uplocks and gear door locks mechanically, using a lever and a system of cables.

Landing gear position is indicated by pilot lights plus a mechanical indicator (a pop-up rod ahead of the windscreen) for the nose gear unit. A warning horn sounds if the throttles are retarded to flight idle when the gear is up; it is silenced automatically by gear extension.

All landing gear struts have oleo-pneumatic shock absorbers and scissor links. The steerable nose unit controlled by a separate handwheel on the captain's control column can turn ±43° for taxying and is equipped with an RDM-1 steering mechanism/shimmy damper (*raspre-de*li*tel'no-demp*fee*ruyuschchiy mekha*nizm* – distribution/damping mechanism). Tyre pressure is 8 bar (86psi) for the mainwheels and 6 bar (114psi) for the nosewheels.

Powerplant

(IL-18V) Four Ivchenko AI-20 Srs 2 engines or, from Batch 20 onwards (mid-1960), identically rated AI-20A (AI-20 Srs 3 or AI-20 Srs 4) engines or, from Batch 65 onwards (August 1963), AI-20K (AI-20 Srs 5) engines which also power the IL-18E. Take-off power 4,000ehp, max power at 8,000m (26,250ft) 3,180ehp; nominal (max continuous) power 2,800ehp at 8,000m.

The IL-18B was powered by identically rated AI-20 Srs 1 or, from Batch 14 onwards (August 1959), AI-20 Srs 2 engines. The IL-18D is powered by AI-20M (AI-20 Srs 6) engines with a take-off rating of 4,250ehp, a maximum power rating at 8,000m (26,250ft) of 3,420ehp at 8,000m and a nominal power rating of 2,980ehp at the same altitude.

The engine was manufactured by the Zaporozhye Engine Factory (ZMZ) in 1958-63; in 1966 production switched to the Perm' Engine Production Association (PPOM).

The AI-20 is a single-shaft turboprop with an annular air intake, a 10-stage axial compressor, an annular combustion chamber, a three-stage uncooled turbine and a fixed-area jetpipe with a conical centrebody. Bleed valves are provided aft of the fifth and eighth compressor

The rear fuselage and tail unit of IL-18A CCCP-75643 (c/n 188000301) and IL-18V CCCP-75554 (c/n 184007501). Yefim Gordon archive

stages. Power is transmitted via a planetary gearbox with a reduction ratio of 0.08732; the gearbox incorporates a torque meter. Engine pressure ratio 9.2, mass flow at take-off rating 20.7kg/sec (45.6 lb/sec), turbine temperature 1,173°K. Engine speed is 10,400±200rpm at ground idle and 12,300±90rpm at all operational settings.

Acceleration time from ground idle to take-off power does not exceed 20 seconds. Maximum operating time at take-off rating 15 minutes. Specific fuel consumption (AI-20M) 0.243kg·hp/hr (0.53 lb·hp/hr) at take-off rating and 0.197kg·hp/hr (0.43 lb·hp/hr) in cruise mode.

Length overall 3,097mm (10ft 1⅜in), width 842mm (2ft 9⅛in), height 1,180mm (3ft 10½in); dry weight 1,080kg (2,380 lb) for the AI-20K or 1,040kg (2,292 lb) for the AI-20M. The AI-20M has a 24,000-hour service life and a 7,000-hour time between overhauls.

Construction is of steel and magnesium alloy. The spool rotates in three bearings: a roller bearing in the air intake assembly (with an extension shaft to the reduction gear), a ball thrust bearing and a roller bearing in the combustion chamber casing. The air intake assembly has inner and outer cones connected by six radial struts and is de-iced by engine bleed air. The combustion chamber has ten burner cones, with igniters and pilot burners at the top. The outer casing is split horizontally for access to the burner cones.

Two accessory gearboxes (dorsal and ventral) are provided, the accessories proper being mounted on the forward casing. The upper gearbox mounts the starter-generators, an R-68D constant-speed governer and a ground anti-windmilling lock, while the lower gearbox drives the oil pumps, the Model 707-I fuel delivery pump, a KTA-5F fuel control unit (*ko**mahnd**no-**top**livnyy agre**gaht*** – FCU), an SGO-8 or SGO-12 AC generator and, in the case of the inboard engines, an NP-25-5 hydraulic pump.

The pressure-feed lubrication system uses a 75/25% mixture of MK-8 grade oil and MS-20 or MK-22 grade oil. Each engine has a 56.5-litre (12.43-Imp gal) oil tank installed on the left side between the rods of the engine bearer truss, a GMN-20 main oil pump, twin MF-20 mesh-type oil filters, a Model 875 oil cooler and an oil pump.

The AI-20 is started by twin STG-12TMO-1000 starter-generators using DC power from the APU or a ground power source; time from initiation to ground idling rpm is 70 seconds.

The main gear units feature four-wheel bogies with KT-81/3 brake wheels. The aft-mounted strut on the right is for the downlock.
Yefim Gordon archive

The main gear units retract forward into the inner engine nacelles. Yefim Gordon archive

SPN-4 igniters are provided for engine starting. Operational ambient temperature limits are –60°/+50° C (–76°/+122° F). An IV-41 sensor system monitors the engines' vibration levels (IV = *izmeritel' vibrahtsii*). The engines are controlled by means of a centrally mounted bank of throttles and a system of cables which operates the FCUs and the fuel shutoff cocks.

The engines are mounted in individual nacelles attached to the upper surface of the inner wings. They are carried on V-shaped truss-type bearers via vibration dampers; the engine attachment lugs are mounted on the forward and centre casings. Each nacelle comprises a fixed rear portion attached to the wing, extending slightly beyond the wing trailing edge, and a cowling which consists of a one-piece annular forward fairing, two upward-hinged side panels and, in the case of the inboard nacelles, a lower panel. On the inboard nacelles the forward fairings incorporate ventral oil coolers with air intakes and airflow adjustment flaps; on the outer nacelles the oil coolers are located in the fixed rear portion just ahead of the wing leading edge.

The rear portions of the nacelles house long, gently curving jetpipes made of 1Kh18N9T grade heat-resistant steel and feature dorsal generator cooling air intakes and removable access panels. Each jetpipe is carried on four attachment points. The engines are separated from the jetpipe bays and, in the case of the inner engines, from the mainwheel wells by titanium firewalls. There are gaps between the jetpipes and the wing upper surface allowing cooling air to pass; additionally, the wing skin below the jetpipes is protected by air-cooled titanium heat shields.

The engines drive Stoopino Machinery Design Bureau AV-68I Srs 02 or AV-68I Srs 03 four-blade variable-pitch constant-speed propellers turning clockwise when seen from the front; diameter 4.5m (14ft 9⅛in), weight 370kg (815 lb). The propellers are equipped with spinners.

Power is adjusted by altering the propeller pitch (blade settings vary from 0° for start-up on the ground through 83°30' in fully feathered position; the coarsest pitch used in flight is 55°, and the reverse thrust setting for slowing the aircraft after touchdown is 12°). Blade pitch is adjusted hydraulically. To ensure flight safety in the event of an engine failure or in-flight shutdown there are several automatic and manual propeller feathering systems featuring hydraulic, electric or pneumatic actuation. The propeller blades feature electric de-icer cuffs.

A Kazan' Machinery Design Bureau TG-16M APU is installed in the No 3 baggage compartment for self-contained engine starting and ground power supply. The APU consists of a 100-shp GTD-16 gas turbine and a GS-24A starter-generator driven via reduction gear; it is enclosed by a protective cowling and draws air from the compartment itself. The exhaust is located on the starboard side between frames 64-65 and surrounded by a heat-resistant steel plate). Maximum continuous power is 81.6hp and rotor speed is 24,000rpm. The APU has a self-contained fuel system with a 50-litre (11-Imp gal) fuel tank installed high on the port side of the fuselage between frames 58-60.

Control System

Conventional mechanical dual control system with push-pull rods made of duralumin pipes, control cranks and levers (everywhere except for the aileron control circuit where cables and rollers are used to transmit inputs from the control wheels to the push-pull rods). Cables and rollers are also used to actuate the elevator trim tabs. The control rods and cables are provided with bronze/rubber pressure seals where they exit the pressure cabin. Gust locks are provided to prevent damage to the system by high winds while the aircraft is parked.

The system includes an AP-6E autopilot. The autopilot servos are connected to the control runs in parallel by means of cables and may be disengaged pyrotechnically at the push of a button if they jam. The servos feature overriding clutches, allowing the pilots to take corrective action when the autopilot is engaged.

Roll control is provided by two-section ailerons on the outer wings occupying 6.5% of the wing area and 35.3% of the span. The ailerons have both aerodynamic and mass balancing; maximum deflection is ±20°. The inboard section of the starboard aileron features a trim tab actuated by an MP-100MT electric drive; maximum deflection is ±4°.

Pitch control is provided by one-piece elevators with aerodynamic and mass balancing which account for 40% of the effective horizontal tail area. Maximum deflection is +15°/–24°30'. Each elevator is equipped with a manually controlled trim tab deflected +13°/–7°30'.

Directional control is provided by a one-piece rudder. The rudder is likewise aerodynamically balanced and mass-balanced and features a spring-loaded servo tab at the root, with a trim tab actuated by an MP-100MT electric drive above it, which make up 6% and 4.6% of the rudder area respectively. Maximum deflection is ±25° for the rudder, ±9° for the trim tab and ±15° for the servo tab.

The port inner engine nacelle (with Ivchenko AI-20A turboprop) and AV-68I Srs 02 variable-pitch propeller of a Batch 29 IL-18V. Note the oil cooler intake below the spinner and ventrally mounted airflow adjustment flap.
Yefim Gordon archive

The port outer engine nacelle and AV-68I Srs 02 variable-pitch propeller of a Batch 29 IL-18V. On the outer engines the oil coolers are located much further aft. Note the de-icer strips on the propeller blade leading edges.
Yefim Gordon archive

The IL-18's engine cowlings are designed to provide good access to the engines and their accessories. The sloping brace of the engine bearer is just visible. Yefim Gordon archive

Fuel System

Twenty-two fuel tanks with a total capacity of 23,700 litres (5,214 Imp gals), which equals a fuel load of 18.6 tons (41,000 lb). Bag-type tanks (fuel cells) made of kerosene-proof rubber are housed in special containers in the inner wings, except the wing centre section (the portion of the monobloc which is inside the fuselage). The fuel cell containers have removable ventral access panels. Tanks Nos 1-3 on each side are located between the engine nacelles (ribs 11-15), Nos 4-5 in line with the outer engines (ribs 15-20), Nos 6-7 outboard of these (ribs 20-24), Nos 9-10 between the fuselage and the inner engines (ribs 4-9) and Nos 11 in line with the latter (ribs 9-11). The outer wing torsion boxes serve as integral tanks (the No 8 tanks) holding 3,300 litres (726 Imp gals) each. Normally the port and starboard halves of the fuel system are isolated, but the system features a cross-feed valve enabling each engine to draw fuel from any group of tanks.

Tanks Nos 1-8 form the main group (the so-called Group 1) holding 16,400 litres (3,608 Imp gals); these are used for routes not exceeding 3,000km (1,860 miles). Group 2 (tanks Nos 9-11) holding 7,300 litres (1,606 Imp gals) is filled for long-range flights only, and the fuel in these tanks is used up first.

The IL-18D features a 6,300-litre (1,386-Imp gal) integral tank in the wing centre section torsion box increasing overall fuel capacity to 30,000 litres (6,600 Imp gals) and the fuel load to 23.5 tons (51,900 lb).

A vent system is provided for pressurising the fuel tanks in flight, with six ram air intakes under each wing.

The engines on each wing are fed by paired PNV-2 delivery pumps in the service tanks (Nos 1); if one goes down, the other pump can still provide adequate delivery. Tanks Nos 6 and 9 feature identical PNV-2 fuel transfer pumps. Each engine has a Type 661 high-pressure delivery pump, a TF-6 crude fuel filter and a Model 12TF-15 final fuel filter. Fuel system operation is automatic, but the transfer pumps can also be activated manually.

The IL-18 has two-point pressure refuelling, with a standard refuelling connector on the right-hand side of each mainwheel well near the main gear fulcrums (the large main gear doors have to be opened for access). The fuel pressure is 1.5-3kg/cm² (21.42-42.85psi) and simultaneous refuelling by two fuel bowsers takes 16-20 minutes. Refuelling by gravity is also possible via three filler caps on each wing (tanks Nos 10, 6 and 8). Defuelling is done via four valves but approximately 200 litres (44 Imp gals) of trapped fuel remains in the system; there is no provision for fuel jettisoning in flight. Fuel grades used are Russian T-1, TS-1 or T-2 jet fuel, Western Jet A-1, DERD.2494 and DERD.2498 (NATO F35 and F43) or equivalent.

Electrics

The electric system serves for engine starting and operates the avionics, equipment and part of the de-icing system, fuel system components etc. The greater part of the system uses a single-wire layout.

Main DC power (27V) is supplied by eight 12-kilowatt engine-driven STG-12TMO-1000 starter-generators. If two generators fail or an engine is shut down, the remaining six generators are sufficient to serve all DC-powered equipment. Backup DC power is provided by three 12SAM-28 lead-acid batteries housed in the electrics bay under the wing torsion box carry-through structure between frames 26-28. Alternatively, some aircraft (notably VIP aircraft) have 20 12SAM-28 batteries in the No 3 baggage compartment for higher engine starting reliability.

115V/400Hz single-phase AC for the propeller and windscreen de-icers is supplied by four 8-kilowatt SGO-8 or 12-kilowatt engine-driven SGO-12 synchronous generators (three main units and one back-up) and a PO-1500 AC converter (*preobrazo**vah**tel' odno**fahz**nyy*) at frame 6 in the forward avionics bay. 36V/400Hz three-phase AC for the radar, compass system, autopilot, Put'-4M navigation system, NAS-1B autonomous navigation system and artificial horizon is supplied by two (main and backup) 1-kilowatt PT-1000TsS AC converters (*preobrazo**vah**tel' tryokh**fahz**nyy*) at frame 8 in the forward avionics bay.

DC power is distributed via four distribution panels in the inner engine nacelles (on the right-hand side of each mainwheel well between frames 2-5), outer engine nacelles (frame 4), a rear distribution panel (between frames 65-67 to port) and circuit breaker panels in the flightdeck and galley. The buses of the inboard distribution panels supply DC power to all equipment in the fuselage via two cables each to maximise reliability. AC power is distributed via a distribution panel in the ventral electrics bay and a circuit breaker panel in the flightdeck.

On the ground electric power is supplied by the batteries and the APU which drives a TS-24A starter-generator. Four ground power receptacles (three for DC power to starboard and one for AC power to port) are provided between frames 25-26.

Hydraulics

The hydraulic system operates the landing gear, nosewheel steering mechanism, wheel brakes, propeller feathering actuators and windscreen wipers. It features two NP-25-5 pumps with a 40 litre/min (8.8 gal/min) delivery rate driven by the inboard engines, a 48-litre (10.56-Imp gal) hydraulic reservoir, a drain tank (both installed on the port side near frame 8), two hydraulic accumulators for the general system in the mainwheel wells (at nacelle frames 3-4) and two separate hydraulic accumulators for the wheel brakes (normal braking system) in the nosewheel well between frames B and V.

The system uses AMG-10 oil-type hydraulic fluid (*aviatsi**on**noye **mah**slo ghidrav**lich**eskoye*); total system capacity 75 litres (16.5 Imp gals), nominal pressure 210kg/cm^2 (3,000psi). Three filters are provided they may be removed for cleaning without requiring the entire system to be emptied. For greater reliability all hydraulic lines are made of stainless steel.

Nitrogen System

The nitrogen system serves for emergency wheel braking and emergency propeller feathering/engine shutdown. It features a 12-litre (2.64-Imp gal) bottle pressurised at 150kg/cm^2 (2,140psi) for the brakes in the nosewheel well at frame B and a 3-litre (0.66-Imp gal) bottle pressurised at 65kg/cm^2 ($\approx$ 930psi) for the propeller feathering circuit in the No 1 baggage compartment.

De-icing System

The wing and tail unit leading edges, propeller blades and spinners, pitot heads and flightdeck windscreen panels have electric de-icing; the windscreen temperature is automatically maintained at +35°C (+95°F). Most of the electric de-icers operate on DC power, with the exception of the heated windscreen and propeller blade/spinner de-icers which use 115V AC. The engine air intakes, inlet guide vanes and oil coolers are de-iced by engine bleed air. An SO-4A icing detector (*signali**zah**tor oblede**neni**ya*) is provided. The system retains sufficient performance even with two engines shut down.

Oxygen System

The oxygen system normally serves the co-pilot only; the other crew members use it only in an emergency. Portable oxygen bottles with KP-21 breathing apparatus (*kislo**rod**nyy pri**bor***) are provided in the event any of the passengers should feel unwell.

Fire Suppression System

Six 8-litre (1.76-Imp gal) OS-8 fire extinguisher bottles (three on frame 1 of each inboard engine nacelle which is the firewall); on aircraft manufactured from 1963 onwards they were augmented by ten 2-litre (0.44-Imp gal) OS-2-IL fire extinguisher bottles. All stationary fire extinguishers (OS = *ognetoo**shi**tel' statsio**nar**nyy*) are charged with '3.5' grade extinguishing agent. The system is designed to fight fires in the engine nacelles, inside the engines proper and in the No 3 baggage compartment/APU bay. The part of the system which serves the engine nacelles has a two-stage operating algorithm; the first shot is triggered automatically by flame sensors, the second one is fired manually at the discretion of the crew. Each nacelle features three sprinkler rings distributing the extinguishing agent inside the engine bay and jetpipe bay; they are mounted on frames 1, 3 and the detachable nose fairing. Impact sensors are installed on the underside of the inboard engine nacelles to trigger all fire extinguishers automatically in a wheels-up landing.

The aircraft is equipped with an SSP-2A fire warning system (*sis**tem**a signali**zah**tsiï po**zhah**ra*) providing audio and visual warnings. Portable CO_2 fire extinguishers (five on the IL-18A/B and four on the IL-18V) are provided for fighting fires in the cabin and flightdeck.

Air Conditioning and Pressurisation System

The IL-18 features a ventilation-type pressure cabin with a volume of 237.5m^3 (8,387ft^3) pressurised by engine bleed air to a pressure differential of 0.5kg/cm^2 (7.14psi). Sea level pressure is automatically maintained up to 5,200m (17,060ft) in order to keep the passengers comfortable. At 8,000m (26,250ft) the pressurisation system maintains a cabin pressure equal to 1,500m (4,900ft) above sea level; at 10,000m (32,800ft) the cabin pressure equals 2,400m (7,870ft) ASL.

The air is cooled by heat exchangers located in the wing roots and humidified before being distributed through the pressure cabin. The system can be adjusted to automatically maintain the cabin temperature anywhere between +5° and +20°C (+41° and +68°F). The cabin air is completely exchanged up to 30 times per hour, with a delivery rate of 2,600-4,000kg/h (5,730-8,820 lb/h). Excess air is spilled to the atmosphere via a valve located on the starboard side of the forward fuselage; on the IL-18A/B it was located symmetrically to port.

The crucial components of the pressurisation/air conditioning system are duplicated for greater reliability. Should the automatic controls of the system fail, the parameters may be set manually.

A mobile air conditioning unit may be connected to the aircraft on the ground. The IL-18E/IL-18D have an upgraded ACS which allows the cabins to be conditioned on the ground with the engines at ground idle.

Avionics and Equipment

The IL-18 is fully equipped for all-weather day/night operation, including automatic flight assisted by an autopilot.

Navigation and piloting equipment: The navigation suite includes an RPSN-2AMG or RPSN-2N Emblema weather radar with a secondary traffic collision avoidance system (TCAS) function, an NAS-1B autonomous navigation system, a Put'-4M (Way-4M) navigation system, a KS-6G compass system, a DAK-DB remote celestial compass (*distantsi**on**nyy **a**stro**kom**pas*), an RSBN-2S Svod (Dome) short-range radio navigation (SHORAN) system with flush antennas built into the fin, an SP-50 Materik instrument landing system, an RV-UM radio altimeter with dipole aerials under the stabilisers, an NI-50BM-1 navigation display, main and backup ARK-11 automatic direction finders (each with a buried loop aerial and a ventral strake aerial under the wing centre section) etc.

The radar set of the RPSN-2 is located at frame 4 on the port side; there are two radar displays (for the pilots and the navigator). The NAS-1B autonomous navigation system comprises a DISS-1 Doppler speed/drift sensor system (***dop**lerovskiy izme**rit**el' **sko**rosti i **sno**sa*) with a flat rectangular antenna under the rear fuselage between frames 58-61 and an ANU-1 autonomous navigation computer. The SP-50 includes a KRP-F localiser receiver, a GRP-2 glideslope beacon receiver (both served by antennas hidden inside the radome), an MRP-56P marker beacon receiver with an aerial under the wing centre section, and a module for working with Western ILS beacons.

Communications equipment: One RSB-5/1230 communications radio, two RSIU-5 (R-802G) command link radios and a 1-RSB-70 backup communications/command link radio. The RSB-5/1230 and 1-RSB-70 radios were served by a wire aerial running from frame 4 to the fin top and a dorsal strake aerial running from frame 4 to frame 27 (the latter was later removed and replaced by a blade aerial on the forward fuselage). The main RSIU-5 radio has a ventral 'hockey stick' aerial between frames 11-12, while the backup RSIU-5 uses an aerial built into the dielectric fin cap.

An SPU-7 intercom (*samo**lyot**noye peregov**or**noye oo**stroy**stvo*) and an SGU-15 public address system (*samo**lyot**noye **grom**kogov-*

*o**rya**schcheye oo**stroy**stvo*) with 15 loudspeakers in the cabins and two in the flightdeck are provided for communication between crew members and crew-to-passengers communication. The handset of the PA system is located in the rear vestibule to avoid adding engine roar to the message insofar as possible.

IFF system: SRO-2 or SRO-2M Khrom (Chromium; NATO *Odd Rods*) IFF transponder (*samo**lyot**nyy **rah**diolokats**ion**nyy ot**vet**chik* – 'aircraft-mounted radar responder'). The characteristic triple IFF aerials serving one of the wavebands are located ahead of the flightdeck glazing (frames A-B) and under the aft fuselage (frames 71-72); four other IFF antennas looking like penny-sized white buttons are located on the sides of the nose (just aft of the radome) and the tailcone. Some military versions are equipped with the later SRO-1P Parol' (Password) IFF transponder with equally characteristic triangular aerials.

The aircraft also features an ATC transponder with aerials inside the radome, on the tailcone and in the fin (a flush antenna located above the SHORAN antennas).

Data recording equipment: MSRP-12-96 primary flight data recorder (FDR) (later replaced by an MSRP-64B FDR), K-3-63 backup FDR and MS-61B cockpit voice recorder (CVR).

Exterior lighting: Port (red) and starboard (green) navigation lights, two at each wingtip (SM-21M lights at the leading edge and BANO-45 lights at the trailing edge); KhS-57 white tail navigation light on the tailcone. Retractable FRS-200 landing/taxi lights on the sides of unpressurised part of the nose (frames G-D) and under the wingtips. Red SMI-2 anti-collision strobe lights under the rear fuselage (between frames 56-57) and at the top of the fin; the lights flash sequentially at one-second intervals.

Three EKSP-39 electric flare launchers (*elek**tricheskaya** kas**set**a sig**nahl'**nykh pa**tro**nov*) are fitted low on the starboard side of the nose between frames D and 2. Each launcher fires four 39-mm (1½-in) signal flares (red, green, yellow and white).

Interior lighting: The flightdeck features a PS-45 overhead light, ARUFOSh-45 ultra-violet lights for the instrument panels to make the dials glow in the dark, KLSRK-45 flightdeck lights and (at the navigator's and radio operator's stations) SLShN-48 movable table lights. The cabin features 23 overhead lights over the aisle and individual reading lights over the seats, plus 'No smoking/Fasten seat belts' illuminated signs. Four PS-45 lights in the Nos 1 and 2 baggage compartments, plus three in the No 3 baggage compartment. PS-45 lights in al three wheel wells for inspection purposes.

Accommodation

The flightdeck is configured for a crew of five, with the navigator's workstation on the left and the radio operator's station on the right. The flight engineer sits on a fold-away seat between and behind the pilots.

The IL-18V can be configured with various seating arrangements from 84 to 127 passengers featuring first class (F), tourist class (CY) and economy class (Y) seating. Provision is made for rapidly changing the seating arrangement of aircraft in regular airline service to suit seasonal fluctuations in demand. Possible layouts include:

- a 73-seat mixed-class layout (CY20+CY45+F8) featuring five-abreast seating (2+3 with an aisle offset to port) at 90cm (35⅜in) pitch in the forward cabin and 102cm (40in) pitch in the centre cabin and two four-abreast rows of sleeperette seats at 112cm (44in) pitch in the rear cabin;
- a similar 78-seat mixed-class layout (CY20+CY50+F8) with 90cm seat pitch in the centre cabin;
- an 84-seat tourist-class layout with 20+50+14 seats at 90cm pitch in all three cabins, five-abreast except for the rearmost row which is four-abreast;
- an 89-seat tourist-class layout with 20+55+14 seats, the extra row in the centre cabin being introduced by reducing the pitch to 84cm (33in);
- a 105-seat all-economy layout with 24+65+16 seats at 84cm pitch, the extra capacity being obtained by shrinking the galley and reconfiguring the rear cabin has two rows six-abreast and one row five-abreast;
- a 110-seat mixed-class layout (Y24+Y72+CY14), the first and second cabins having six-abreast seating at 90cm and 75cm (29½in) pitch respectively while the rear cabin has two rows five-abreast and one row four-abreast at 90cm pitch;
- a so-called 'winterised' 110-seat layout (Y24+Y72+CY14) featuring larger coat closets and a larger galley at the expense of reducing the seat pitch to 87cm (34¼in) in the first and third cabins and a miserable 75cm (29½in) in the second cabin;
- a 111-seat all-economy layout with 24+71+16 seats similar to the 105-seat version, with a twelfth row added in the centre cabin;
- a 127-seat mixed-class layout (Y24+Y89+CY14) with reduced coat closets and a small galley, the first and second cabins having six-abreast seating at 90cm and 75cm pitch respectively while the rear cabin has two rows five-abreast and one row four-abreast at 87cm pitch.

With five-abreast tourist class seating the seats are 45cm (17¾in) wide (measured between the armrests) and the aisle width is likewise 45cm. With six-abreast economy class seating it is 41.5cm (16⅜in) and 35cm (13¾in) respectively.

All IL-18Vs *et seq* in regular airline configuration are equipped with three toilets (two on the starboard side of the forward vestibule and one to starboard just aft of the rear cabin). Most layouts feature three coat closets.

In VIP configuration (IL-18V 'Salon') the aircraft provides accommodation for 35-50 passengers. The 35-seat IL-18V 'Salon' features two cabins for the retinue seating 20 and 14 respectively at 90cm pitch, divided by the forward vestibule featuring a toilet and a coat closet to starboard and a refrigerator and a portable oxygen bottle rack to port. Further aft is a galley of reduced size; next is a VIP lounge (the so-called 'main passenger's cabin') equipped with a sofa, swivelling armchairs, tables, a sofa and a book closet. The rear vestibule with a coat closet separates this from a bedroom and toilet for the VIP at the extreme rear. Another configuration seated 34. There was a single row of four sleeperette seats immediately aft of the flightdeck, followed by an eight-seat compartment for a relief crew. Separated from these by the forward vestibule and the galley was a 14-seat tourist class cabin for the retinue (two rows of five seat and one row of four). Then came two VIP cabins, each with two armchairs facing each other (with a table in between) and a four-seat sofa across the aisle; two seats on the starboard side opposite the rear entry door (presumably for the bodyguards); and finally a bedroom featuring two more armchairs and a table.

The earlier IL-18B featured a different 89-seat tourist class layout with 19 seats at 78cm (30¾in) pitch in the forward cabin (frames 6-14) and a 70-seat main cabin (frames 21-49), the seat pitch being 84cm in the first six rows and 87cm in the next eight rows. The first cabin had three rows of three seats on the starboard side, with a toilet opposite (ie, immediately ahead of the forward entry door) and two rows five-abreast. The space between the two cabins was occupied by coat closets (port and starboard), fire extinguisher racks and the galley; the rear vestibule featured a sofa to starboard and a coat closet to port, with two more toilets further aft. There was also an 80-seat tourist class configuration (15+65) with two seats to starboard in the foremost row, three seats to starboard in the next row and five-abreast seating everywhere else.

The IL-18D was offered in:

- 65-seat (CY14+CY43+F8)
- 90-seat (CY20+CY55+CY15)
- 100-seat (CY20+CY65+CY15)
- 110-seat (Y24+Y71+CY15) or
- 122-seat (Y24+Y83+CY15) configuration.

The cabins feature lightweight seats; the seat backs recline at the push of a button in an armrest. The seats are provided with seat belts and detachable meal trays installed on the armrests. On early-model paired seat modules the centre armrest could be detached, transforming the seats into a sofa. The seat upholstery can be removed for dry cleaning.

The cabin walls, ceiling and overhead luggage racks are upholstered in a special material called Pavinol (a non-combustible fabric coated in flexible vinyl plastic). The cabin floor has lightweight carpet upholstery.

The overhead luggage racks incorporate passenger service units (PSUs) featuring individual ventilation nozzles, reading lights and a flight attendant call button.

The galley is equipped with three UEK-2 water boilers and three EDSh-110 electric food heaters, a sink with hot and cold water and storage lockers for food and drinks.

IL-18 Main Specifications

	IL-18A	IL-18B	IL-18V	IL-18E	IL-18D
Length overall, m (ft)	35.7m (117' 1½")	35.9m (117' 9⅜")	35.9m (117' 9⅜")	35.9m (117' 9⅜")	35.9m (117' 9⅜")
Wing span, m (ft)	37.4m (122' 8½")	37.4m (122' 8½")	37.4m (122' 8½")	37.4m (122' 8½")	37.4m (122' 8½")
Stabiliser span, m (ft)	11.8m (38' 8½")	11.8m (38' 8½")	11.8m (38' 8½")	11.8m (38' 8½")	11.8m (38' 8½")
Landing gear wheelbase, m (ft)	12.755m (41' 10⅛")	12.755m (41' 10⅛")	12.755m (41' 10⅛")	12.755m (41' 10⅛")	12.755m (41' 10⅛")
Landing gear track, m (ft)	9.0m (29' 6⅜")	9.0m (29' 6⅜")	9.0m (29' 6⅜")	9.0m (29' 6⅜")	9.0m (29' 6⅜")
Height on ground, m (ft)	10.165m (33' 4¼")	10.165m (33' 4¼")	10.165m (33' 4¼")	10.165m (33' 4¼")	10.165m (33' 4¼")
Blade tip clearance, m (ft) *	0.778 / 1.078m (2' 6⅝in / 3' 6½")	0.778 / 1.078m (2' 6⅝in / 3' 6½")	0.778 / 1.078m (2' 6⅝in / 3' 6½")	0.778 / 1.078m (2' 6⅝in / 3' 6½")	0.778 / 1.078m (2' 6⅝in / 3' 6½")
Wing area, m^2 (ft^2)	140.0 (1,505.3)	140.0 (1,505.3)	140.0 (1,505.3)	140.0 (1,505.3)	140.0 (1,505.3)
Flap area, m^2 (ft^2)	27.15 (291.9)	27.15 (291.9)	27.15 (291.9)	27.15 (291.9)	27.15 (291.9)
Aileron area, m^2 (ft^2)	18.22 (195.9)	18.22 (195.9)	18.22 (195.9)	18.22 (195.9)	18.22 (195.9)
Vertical tail area, m^2 (ft^2)	17.93 (192.79)	17.93 (192.79)	17.93 (192.79)	17.93 (192.79)	17.93 (192.79)
Rudder area, m^2 (ft^2)	6.83 (73.44)	6.83 (73.44)	6.83 (73.44)	6.83 (73.44)	6.83 (73.44)
Horizontal tail effective area, m^2 (ft^2)	24.32 (261.5)	24.32 (261.5)	24.32 (261.5)	24.32 (261.5)	24.32 (261.5)
Elevator area, m^2 (ft^2)	9.75 (104.83)	9.75 (104.83)	9.75 (104.83)	9.75 (104.83)	9.75 (104.83)
Overall pressure cabin volume, m^3 (ft^3)	237.5 (8,387)	237.5 (8,387)	237.5 (8,387)	237.5 (8,387)	237.5 (8,387)
No 1 baggage compartment volume, m^3 (ft^3)	13.32 (470.39)	3.32 (470.39)	3.32 (470.39)	3.32 (470.39)	3.32 (470.39)
No 2 baggage compartment volume, m^3 (ft^3)	13.68 (483.1)	13.68 (483.1)	13.68 (483.1)	13.68 (483.1)	13.68 (483.1)
No 3 baggage compartment volume, m^3 (ft^3)	7.06 (249.32)	7.06 (249.32) / 2.3 (81.22) †	7.06 (249.32)/ 2.3 (81.22) †	2.3 (81.22)	2.3 (81.22)
Empty operating weight, kg (lb)	29,450-30,579 (64,925-67,414)	n.a.	31,500 (69,440)/ ‡ 34,500 (76,060) **	34,630 (76,340)	34,000 (74,955)/ 35,000 (77,160) **
Maximum taxy weight, kg (lb)	n.a.	n.a.	61,700 (136,020)	n.a.	n.a.
Maximum take-off weight, kg (lb)	58,000-59,350 (127,865-130,840)	61,200 (134,920)	61,200 (134,920)	61,200 (134,920)	64,000 (141,090)
Maximum landing weight, kg (lb)	n.a.	51,200 (112,870)	51,200 (112,870)	51,200 (112,870)	n.a.
Maximum payload, kg (lb)	12,000 (26,455)	14,000 (30,865)	13,500 (29,760) §	13,500 (29,760)	13,500 (29,760)
Payload in maximum-range flight, kg (lb)	n.a.	n.a.	9,400 (20,720) §	n.a.	6,500 (14,330)
Wing loading at max TOW, kg/m^2 (lb/ft^2)	n.a.	n.a.	437.0 (89.59)	n.a.	n.a.
Power loading at max TOW, kg/hp (lb/hp)	3.625-3.70 (7.99-8.17)	3.83 (8.44)	3.83 (8.44)	3.83 (8.44)	3.76 (8.3)
Permissible CG range (loaded), % mean aerodynamic chord	n.a.	n.a.	16-23	n.a.	n.a.
Maximum Mach number	n.a.	n.a.	0.65	n.a.	n.a.
Economic cruising speed, km/h (mph / kts) at 8,000m (26,246ft)	n.a.	n.a.	600 (372 / 324)	625 (388 / 337)	625 (388 / 337)
Top speed, km/h (mph; kts)	n.a.	n.a.	650 (403; 351)	675 (419; 364)	675 (419; 364)
Never-exceed speed (indicated), km/h (mph; kts) in level flight / in a shallow dive	n.a.	n.a.	510 (316; 275) / 612 (380; 330)	n.a.	n.a.
Service ceiling, m (ft)	n.a.	n.a.	12,500 (41,010)	12,500 (41,010)	11,800 (38,710)
Maximum range, km (miles)	5,400 (3,350)	5,400 (3,350)	5,400 (3,350)	5,400 (3,350)	6,500 (4,040) / 7,100 (4,410) **
Range with max payload, km (miles)	3,300 (2,050)	3,300 (2,050)	3,000 (1,860) / 3,300 (2,050) **	3,200 (1,990) / 3,300 (2,050) **	3,700 (2,300) / 4,300 (2,670) **
Take-off run, m (ft)	n.a.	n.a.	1,100 (3,610)	1,100 (3,610)	1,350 (4,430) ‖
Landing run, m (ft)	n.a.	n.a.	n.a.	n.a.	850 (2,790)

* Inboard/outboard engines; † Usable volume (ie, excluding the volume eaten up by the APU) if an APU is installed; ‡ With APU but with the seats removed; § APU-equipped aircraft only; for aircraft lacking an APU the maximum payload was 14,000kg (30,860 lb) and the payload in maximum range configuration was 9,900km (21,825 lb). The maximum payload could be carried on routes not exceeding 2,750km (1,700 miles); ¶ This includes seats, galley equipment, carpets, water for the toilets, oxygen bottles, a ladder and 12 signal flares; ‖ At maximum TOW; ** Different sources give different figures.

Chapter Four

The Airline Workhorse IL-18 in Civilian Service

As already noted, the IL-18's service career with Aeroflot started in January 1958 when the Vnukovo UAD of the Moscow Territorial CAD – the airline's best subdivision which was traditionally entrusted with introducing new hardware – began route proving flights with a handful of IL-18As. For safety reasons the *Coots* carried freight and mail until enough experience had been amassed to allow them to be cleared for passenger flights.

Revenue services finally started on 20th April 1959 in a spectacular fashion. With a little poetic licence, it may be said that in the case of the IL-18 the starter's gun was a double-barrelled shotgun: two scheduled passenger flights were performed on the opening day. First, Vnukovo UAD/65th Flight IL-18B CCCP-75674 (c/n 188000903) captained by Boris A Lakhtin, Hero of the Soviet Union, took off from Moscow/Vnukovo-1, bound for Adler, a popular Black Sea resort. Next, an IL-18B registered CCCP-75672 (c/n 188000901) and captained by A Averkin flew from Moscow-Domodedovo to Alma-Ata, the capital of the Kazakh Soviet Socialist Republic. Though also assigned to the Moscow Territorial CAD, this aircraft apparently belonged to the Domodedovo UAD/212th Flight.

Domestic flights were soon followed by international services. The first scheduled flight abroad by an IL-18 took place on 5th January 1960 when a Central Directorate of International Services (TsUMVS – *Tsen**trahl'**noye oopravl**en**iye mezhduna**rod**nykh voz**doosh**nykh so'ob**schchen**iy*)/63rd Flight aircraft captained by I I Frolov flew from Moscow-Sheremet'yevo to Sofia-Vrazhdebna. On 31st January 1962 another TsUMVS *Coot* captained by Yuriy V Ovsyannikov inaugurated the Moscow-Djakarta service, a route more than 10,000km (6,200 miles) long. On 18th June 1962 an Aeroflot IL-18 captained by V Ye Larionov made a route proving flight that took it nearly 12,000km (7,450 miles) from Moscow to Accra (Ghana) via Rabat (Morocco), Bamako (Mali) and Conakry (Guinea). The type also has the distinction of being the first Soviet aircraft to land at Mexico City airport which is located in a mountainous area high above sea level.

The *Coot* turned into one of the Soviet Union's major export successes among civil aircraft. Starting in November 1959, the Soviet agency Aviaexport delivered new or used IL-18s to 20 nations located as far apart as China, Cuba and Ghana.

The IL-18 became the most widespread Soviet airliner in its weight and capacity category, with a production run of 565 (that is, not counting the specialised military versions) – a figure which was subsequently bettered by the Tu-154, but that was much later. Of these, about 420 units were delivered to Aeroflot alone (again not counting the quasi-civil examples in Soviet Air Force service and industry-operated examples which, like every other civil-registered aircraft in the Soviet Union, wore Aeroflot markings).

Speaking of registrations, under the current five-digit registration system IL-18s were allocated reggies in the 742xx and 75xxx blocks. Known Soviet/CIS IL-18s and derivatives were registered CCCP-74250 through -74270, CCCP-74295, CCCP-74296, CCCP-74298, CCCP-74299, UN 75001 through UN 75005, UN 75111, CCCP-75400 through -75420, CCCP-75422 through -75426, EX-75427, CCCP-75428 through -75475, UR-75475, CCCP-75476 through -75483, CCCP-75489, CCCP-75495 through -75554, RA-75554, CCCP-75555 through -75599, CCCP-75601, CCCP-75602, CCCP-75604, CCCP-75606, CCCP-75614, CCCP-75634, CCCP-75637,

Civil Aviation Directorate	United Air Detachment & constituent Flight	Home base
Armenian CAD	Yerevan UAD/279th Flight	Yerevan-Zvartnots
Azerbaijan CAD	Baku UAD/107th Flight	Baku-Bina
Central Directorate of International Services (TsUMVS)	63rd Flight	Moscow-Sheremet'yevo
Far Eastern CAD	1st Khabarovsk UAD	Khabarovsk
Kazakh CAD	Alma-Ata UAD (218th Flight?)	Alma-Ata
	Karaganda UAD (154th Flight?)	Karaganda
Kirghiz CAD	Frunze UAD (250th Flight?)	Frunze-Manas
Krasnoyarsk CAD	1st Krasnoyarsk UAD/214th Flight	Krasnoyarsk-Yemel'yanovo
Latvian CAD*	Riga UAD (280th Flight?)	Riga-Spilve
Leningrad CAD	1st Leningrad UAD (67th Flight?)	Leningrad-Pulkovo
Magadan CAD	1st Magadan UAD (185th Flight?)	Magadan-Sokol
Tajik CAD	Dushanbe UAD	Dushanbe
Training Establishments Directorate (UUZ)	Ul'yanovsk Higher Civil Aviation Flying School	Ul'yanovsk
State Civil Aviation Research Institute (GosNII GA)		Moscow-Sheremet'yevo
Moscow Territorial CAD	Domodedovo CAPA/212th Flight	Moscow-Domodedovo
	Vnukovo CAPA/65th Flight	Moscow/Vnukovo-1
Northern Territorial CAD†		
Polar Aviation Directorate†		
Turkmen CAD	Ashkhabad UAD (369th Flight?)	Ashkhabad
Ukrainian CAD	L'vov UAD	L'vov
Urals CAD	Chelyabinsk UAD (124th Flight?)	Chelyabinsk
	1st Sverdlovsk UAD/120th Flight	Sverdlovsk-Kol'tsovo
Uzbek CAD	Tashkent UAD/202nd Flight‡	Tahkent-Yoozhnyy
West Siberian CAD	Tolmachovo UAD (384th Flight?)	Novosibirsk-Tolmachovo
235th Independent Air Detachment		Moscow/Vnukovo-2

* The Latvian and Moldavian divisions were called Republican Civil Aviation Production Associations (RCAPA) in the late 1970s but were later renamed Civil Aviation Directorates.

† Liquidated by 1980 in the course of reorganisation.

‡ In some documents the 202nd Flight is stated as part of the Far Eastern CAD/1st Khabarovsk UFD.

Busy scene at Moscow/Vnukovo-1 as a late-standard IL-18B with one window aft of the forward entry door taxies in, with An-10s, Tu-104As and more IL-18s in the background. This colour scheme was standard for Aeroflot's *Coots* until the mid-1970s.
Sergey and Dmitriy Komissarov archive

An IL-18V in early configuration (with dorsal strake aerial) during a turnaround at Dushanbe airport, Tajikistan. Note the typeface of the Aeroflot titles; this was the standard graphic presentation for the type (compare with the titles on the prototype in Chapter 1).
Sergey and Dmitriy Komissarov archive

Passengers board an IL-18B via a standard Soviet self-propelled gangway of the type which can still be found at CIS airports.
Yefim Gordon archive

Ил-18

ДУШАНБЕ
АЭРОФЛОТ
Ил-18

ЭРОФЛОТ
Ил-18

CCCP-75639, CCCP-75641 through -75648, CCCP-75650, CCCP-75651, CCCP-75653 through -75666, CCCP-75668 through -75676 No1, CCCP-75676 No2, CCCP-75677 through -75691, CCCP-75694 through -75711, 75711 (No2), CCCP-75712, CCCP-75713* through -75794, CCCP-75796 through -75850, UR-75850, CCCP-75851 through -75876, CCCP-75879 through -75888, CCCP-75894 through -75903*, CCCP-75904, CCCP-75905, EX-75905, CCCP-75906, CCCP-75908 through -75923, RA-75923 (No 2) and CCCP-75924 through -75929. Some registrations have been used twice; if both aircraft originally wore the CCCP- prefix the registration is marked by an asterisk. The registrations CCCP-75640, -75649 and -75693 fit nicely into the sequence but have not been reported to date, whereas CCCP-75652, -75667 and -75692 were probably allocated but not taken up. Examples registered CCCP-74287, CCCP-74288 and CCCP-74291 have also been reported, but these are probably mis-sightings for CCCP-74267, CCCP-74268 and CCCP-74261 respectively.

Non-standard registrations include CCCP-04330, CCCP-04350, CCCP-04356, CCCP-04770, CCCP-06180 No2, CCCP-33569 No1, CCCP-48093 No1, RA-54460 (possibly a callsign only), 74626, CCCP-78732, EX-011, EX-105, EX-201, EX-405, EX-601, EX-904 and EX-7504. (All registrations are given here with their original country prefix, since many aircraft later moved from one CIS republic to another. No attempt is made here to list the letter-only registrations of some CIS states.)

In Soviet times the IL-18 saw service with civil aviation divisions in ten of the Soviet Union's fifteen republics. Known operators are detailed in the table on page 60.

(Note: During the reorganisation of the former Soviet Union's civil aviation which began in 1992, the former Civil Aviation Directorates were rebranded Regional Air Transport Directorates (RUVT – *Reghio**nahl'**noye oopravl**en**iye vozdooshnovo trahnsporta*). Some of them changed their names in so doing; for instance, the former Leningrad CAD is now the North-Western RATD.)

In comparison to the piston-engined Li-2P, IL-12 *sans suffixe*/IL-12B and IL-14P/IL-14M the turboprop *Coot* offered much more comfortable transportation in 50% less time. Comfort was enhanced by the generous seat pitch (in the early days, although this would change for the worse as Aeroflot sought to cram steadily more passengers into the aircraft) and the tasteful cabin trim. On the minus side, the passengers had to endure vibration and a wearying deep-throated roar which peaked at a deafening 118dB in the propellers' plane of rotation. It was especially bad on the IL-18A/IL-18B, and the stewardesses were the worst off, as the galley was located in the noisiest area; after a few years' work the poor girls were guaranteed to get health problems because of the vibrations! The situation improved slightly on the IL-18V and subsequent versions where the interior layout was changed to get the people away from the worst noise.

Another major drawback was the lack of self-contained air conditioning on the ground, which meant the IL-18 could not compete with jet-powered Western airliners in hot climates. The addition of an APU from Batch 30 onwards did not save the day because the TG-16 was exactly what the Russian designation implied – a turbine-powered generator with no air bleed for the air conditioning system. It was not until the advent of the IL-18D that this problem was partially addressed by providing air conditioning on the ground with the engines running at ground idle.

The passengers may have had mixed feelings about the IL-18, but the flight crews had few doubts. The *Coot* was popular with the pilots for its easy handling and ability to tolerate piloting errors. The aircraft was equipped with an automatic approach system and could make a go-around with one or even both engines on one side inoperative – although the latter case was a dangerous situation demanding considerable flying skills. The navigators liked their workstation which was better laid out in comparison with most other Soviet airliners of the late 1950s and early 1960s. On these aircraft (the An-10, Tu-104, Tu-114, Tu-124 and Tu-134) the navigator sat ahead of and below the pilots in the extreme nose which was extensively glazed; in the case of the An-10 this feature was dictated by the requirement to make the aircraft easily convertible for military transport needs, while on the Tupolev aircraft the situation was kind of reversed because the glazed nose was a consequence of the airliners' bomber ancestry. Now the IL-18, as already mentioned, was designed from the outset as a purely civil aircraft and there was no need to put the navigator in a glazed 'bomber nose' which gave him no real advantages at night and in poor weather. Hence a more rational flightdeck layout could be chosen, decreasing flightdeck volume appreciably.

Speaking of crews, conversion training of both Soviet and foreign flight crews and ground personnel operating the IL-18 took place at the Ul'yanovsk Higher Civil Aviation Flying School (UVAU GA – *Ool'**yah**novskoye **vys**sheye **lyot**noye oo**chil**ischche grazh**dahn**skoy avi**ah**tsii*) which was within the framework of the Training Establishments Directorate (UUZ – *Oopravl**en**iye oo**cheb**nykh zave**den**iy*). This establishment, which handled all Soviet civil types larger than the 50-seat Antonov An-24 *Coke* twin-turboprop regional airliner, was known as the COMECON Civil Aviation Centre (*Tsentr GA SEV*) in the 1970s and 1980s but then reverted to its original name. (The COMECON, or Council for Mutual Economic Assistance (SEV – *So**vet** ekono**mich**eskoy vzaimo**pom**oschchi*), was the Eastern Bloc's counterpart of the EEC.) As a rule, foreign IL-18 crews completed the entire training course before any aircraft were actually delivered so that they could fly their airline's first *Coots* home themselves – a process which meant a lot to the airmen, boosting their morale considerably.

As one might expect of a new airliner, the early years of IL-18 operations were beset by technical problems, including some quite serious ones. The worst of these was the low reliability of the AI-20 engines (especially the K model) and AV-68 propellers (to be precise, the propeller feathering mechanism which left a lot to be desired). On several occasions severe engine fires or the impossibility to feather a dead engine's propeller resulted in fatal crashes. However, the OKB and the factory did their best to rectify the airliner's faults. For instance, titanium firewalls and stainless steel fuel lines in the engine nacelles (replacing the

Early IL-18s had no APU, so it was standard operational procedure to start up the engines from a ground power source to avoid sapping the DC batteries if possible. Note the ground power cable connected to the socket on the centre fuselage. Yefim Gordon archive

A quartet of immaculate IL-18Vs at Moscow-Sheremet'yevo (the terminal which is now Sheremet'yevo-1!) in the mid-1960s. A PAZ-652 bus has just brought a load of passengers to CCCP-75772 (c/n 181003601), probably a Leningrad Civil Aviation Directorate/1st Leningrad United Air Detachment/67th Flight machine. Sergey and Dmitriy Komissarov archive

IL-18V CCCP-75529 (c/n 184006902), a Moscow Territorial CAD/Domodedovo UAD/212th Flight aircraft, shares the apron at its home base with Tu-104B CCCP-42442 in the 1960s. The glazed pavilion in the background is for the suburban train arrivals. Yefim Gordon archive

earlier duralumin pipelines) were introduced after the crash of IL-18S CCCP-75705 on 17th August 1960; all previously built aircraft were modified accordingly. In due course, the powerplant and avionics worked up an adequate reliability level and the IL-18 got a well-earned reputation as 'the most dependable airliner' – especially in comparison with the Tu-104 and An-10.

Teething troubles notwithstanding, IL-18 utilisation steadily increased until it was Aeroflot's principal medium/long-haul airliner. In 1970 the type shouldered 39.8% of the passenger air traffic in the Soviet Union, and this situation continued until 1975. Later, as the Tu-154 fleet grew and took the passengers away from the noisy *Coots*, the IL-18s were modified to serve on as freighters.

The IL-18 was not only the most widespread but also the most cost-effective Soviet airliner in its class. In 1973, when Aeroflot had already started revenue services with the Tu-154 – the type intended to replace the *Coot* – the IL-18 had the lowest operating costs among comparable Soviet aircraft at 560 roubles per flight hour versus 660 SOR/hr for the Tu-134, 810 SOR/hr for the Tu-104 and a staggering 1,340 SOR/hr for the initial-production Tu-154 *sans suffixe*. The ultimate IL-18D was the most efficient of them all, the improved AI-20M engines and increased fuel capacity giving it a range of 4,500-6,500km (2,795-4,040 miles) depending on the payload; as it was, the *Coot* ranked first among Soviet airliners in the same class as far as non-stop range was concerned. In average payload the IL-18 was outperformed only by the Tu-104, and then only by a small margin.

Simultaneously with the start of deliveries to Aeroflot the Soviet Air Force and the Naval Air Arm also started taking delivery of their first IL-18s. These were almost invariably VIP aircraft, often with additional secure communications equipment. In the Air Force most *Coots* were operated by the 8th ADON (***a****viadi****vee****ziya* ***o****so****b****ovo nazna****chen****iya* – Special Mission Air Division) at Chkalovskaya AB near Moscow, although some aircraft were stationed elsewhere; for example, IL-18V CCCP-75591 (c/n 185008003) was assigned to the Trans-baikalian Defence District and based at Ulan-Ude/Vostochnyy airport.

Likewise, the *Coot* served with the Soviet federal government flight from the type's early days in service. The unit was originally called

Top and second down: **This Soviet Air Force IL-18S ('001 Red', c/n 188000801) was used by the then Defence Minister Marshal Andrey A Grechko. Note the colour scheme which differs from the pre-1973 civil livery. The serial '001 Red' later passed to a Tu-104V VIP jet, while this *Coot* became CCCP-33569 and then CCCP-75479.** Yefim Gordon archive

Third down: **IL-18V CCCP-75515 (c/n 183006603?) sported this experimental eye-catching livery which earned it the nickname *Fazahn* (Pheasant). Note the small 'Soviet Airlines' titles (in English!) in white superimposed on the cheatline near the forward entry door.** Yefim Gordon archive

Bottom: **Ground personnel loads crates of fruit from a GAZ-53A 3.5-ton lorry into an IL-18Gr freighter wearing 1973-standard colours at Leninabad airport, Tajikistan. The loader vehicle is a heavily modified UAZ-451M with the right-hand half of the cab cut away.** Sergey and Dmitriy Komissarov archive

MAGON (*Mos**kov**skaya **avia**groop**pa o**sob**ovo naz**na**cheniya*** – Moscow Special Mission Air Group) and based at Sheremet'yevo (which was not yet a civil airport in those days). Later the government flight moved to Moscow-Vnukovo, which is rather closer to the city than Sheremet'yevo, and became the 235th Independent Air Detachment. Of course, the aircraft destined for the government flight were manufactured with extra special care and quality. Nevertheless, after an average three to four years in service the VIP *Coots* were converted to passenger configuration and passed on to ordinary Aeroflot units, yielding their place to newer examples.

Until the mid-1970s Aeroflot's colour schemes were as disparate as the aircraft it operated. Each type had its own livery – or, in the case of such aircraft as the An-10 and the An-24, several liveries; even the graphic presentation of the Aeroflot titles varied. Nearly all IL-18s for the home market were completed in a colour scheme featuring a light blue cheatline with 'feathers' at the front running below the windows, a white tail with a Soviet flag, and natural metal wings and undersurfaces. One IL-18V registered CCCP-75515 (c/n 183006603?) stood out from the crowd, featuring an experimental colour scheme with a heavy bright blue cheatline, an all-blue tail and Aeroflot titles in a different type than hitherto. This eye-catching non-standard paint job earned this particular aircraft the nickname *F**a**zahn*** (Pheasant).

Then the airline's management rightly decided that having such a motley collection of colour schemes was not a good idea and that a common fleetwide standard should be developed. Such a standard was duly developed and endorsed by MAP and the Ministry of Civil Aviation (MGA – *Mini**ster**stvo grazh**dahn**skoy avi**ah**tsii*) in March 1973, but it was a while before the now-familiar blue/white livery with bold Aeroflot titles began appearing on actual aircraft. Since IL-18 production had ended by then, passenger *Coots* were progressively repainted in the new colours as they underwent a major overhaul which required the entire airframe to be stripped of paint anyway; the new-build quasi-civil IL-22 airborne command posts were the only exception.

Speaking of which, it was common practice in the Soviet Union that a single aircraft overhaul plant (ARZ – ***aviaremontnyy zavod***) would repair all aircraft of a given type, regardless of where they were based (including export aircraft); if the aircraft was produced in large numbers, several plants would be assigned. Thus, ARZ No 402 at Moscow-Bykovo[1] handled all IL-18s until Aeroflot phased out the type. Later, almost all IL-18 refurbishment work passed to the Russian Navy's ARZ No 20 in Pushkin near Leningrad (now renamed back to St Petersburg) which already handled military derivatives of the *Coot*. The plant at Moscow-Bykovo switched to the IL-76 transport and the Yakovlev Yak-42 *Clobber* short/medium-haul airliner, but IL-18s were occasionally overhauled there even in mid-2001. Ivchenko AI-20 engines are currently repaired by the Russian Air Force's ARZ No 123 in Staraya Roossa near Velikiy Novgorod.

Unfortunately, there is hardly an aircraft type in the world which has avoided accident attrition altogether; even the best and most popular aircraft sometimes crash, and the IL-18 is no exception. Admittedly some of the accidents were caused by design flaws, especially the powerplant problems of the early days. However, as the reader will see from Appendix 2, IL-18 accident attrition was caused by the telltale human factor as much as by hardware failures and the aircraft itself was often blameless.

Into the White Wilderness

The IL-18 also made an invaluable contribution to the exploration of Polar regions and the development of the Soviet Union's sparsely populated northern and north-eastern regions. These were demanding missions which put both aircraft and people to the test – and they stood up to the challenge with dignity.

Aeroflot's Polar Aviation Directorate (UPA – *Oopravleniye Polyarnoy aviahtsii*) received its first IL-18s in 1959; in keeping with the then-current practice they were mostly registered in the CCCP-04xxx block which was a mixed bag with all sorts of aircraft.

The IL-18's first Polar mission to gain wide publicity was in 1961-62 when the need arose to rapidly deliver a fresh team of researchers at the Soviet Antarctic research station Mirnyy.[2] The mission looked set to be an extremely difficult one. En route the weather would change from the frost and snow of winter Russia to the sweltering heat of the tropics and then back to biting frost, and air humidity would change dramatically, too. A substantial part of the route lay over water or stretches of land devoid of landmarks and ground navigation aids. Only a highly skilled crew could cope with the task – quite apart from the fact that the aircraft itself needed to be modified for the ultra-long-range flight.

On 15th December 1961 IL-18V CCCP-75743 departed from Moscow to Mirnyy in company with the unique An-12TP-2 transport/geophysical survey aircraft (CCCP 04366; the registration was painted on with no dash). The mission was led by Mikhail I Shevelyov (Hero of the Soviet Union), head of the Polar Aviation Directorate. The IL-18 was flown by captain A S Polyakov – a pilot with a wealth of experience of flying in polar regions, first officer M P Stoopishin (HSU), navigator M Dolmatov, flight engineer N Pishkov, mechanic P Goncharov and radio operator V Men'shikov.

The route took the two aircraft across three continents (Eurasia, Australia and Antarctica) and two oceans, including such places as the Bay of Bengal, Sumatra, Java, the Tasmanian Sea, New Zealand and the Ross Sea. In ten days the IL-18 and An-12 covered a distance of 26,423km (16,411 miles)[3] with an average speed of 580km/h (360mph), staying airborne for 44 hours 34 minutes in all. Refuelling stops were made in Tashkent, Delhi, Rangoon, Jakarta, Darwin, Sydney, Christchurch and at the US Antarctic research station McMurdo. The last-but-one leg of the journey (from Christchurch to McMurdo) turned out to be the hardest because the American radio officer would not answer the radio calls of the Soviet crews who needed meteorological and navigation information. Still, despite lacking the all-important weather report, the crews of both aircraft maintained the planned route without deviating. Having nothing but their weather radars and astrosextants to rely on, they circumnavigated tall mountains obscured by clouds, including the 3,743-m (12,280-ft) volcano Mount Erebus, and made a precision approach to McMurdo's airstrip.

The Soviet turboprops spent a full month on the glacial continent; during this time the *Cub* was re-equipped with non-retractable skis,

An IL-18B with the non-standard 'Polar' registration CCCP-04330 being readied for flight on the south apron at Moscow-Sheremet'yevo (the site of today's Sheremet'yevo-2 terminal) in June 1965. Note the small 'Polar Aviation' titles augmenting the Aeroflot titles; the latter were absent in 1962 and the 'Polar Aviation' titles were much bigger. Sergey Komissarov

The specially modified Polar Aviation IL-18V CCCP-75743 (c/n 181002901) was the first *Coot* to perform an ultra-long-range mission to Antarctica. Here it is seen in the late red-tailed colour scheme in which it flew from Moscow to Ice Station Mirnyy and back between 15th December 1961 and 2nd February 1962 together with the An-12TP-2. The wing of the latter aircraft (CCCP 04366) is visible in the foreground. Yefim Gordon archive

IL-18D CCCP-74267 (c/n 187011105) at the Soviet Ice Station Molodyozhnaya in Antarctica in February 1980. Yefim Gordon archive

making several survey flights over unexplored areas of Antarctica to measure the Earth's magnetic field. During the Antarctic sojourn CCCP-75743 also made several flights in support of the expedition, including an unscheduled trip from Mirnyy to McMurdo to evacuate a mechanic from a neighbouring Australian research station who fell ill unexpectedly. On 25th January – 2nd February 1962 both aircraft returned to Moscow along the same route, covering a total distance of 52,800km (32,795 miles) and spending a total of 90 hours in the air during the expedition.

The success of the mission proved that aircraft could resupply the Antarctic research stations much quicker and at less cost as compared to the traditional delivery by ship. All participants in the expedition were awarded government orders; additionally, A S Polyakov and the An-12 captain B S Osipov received the prestigious Hero of Socialist Labour title.

Between 20th November 1963 and 11th January 1964 IL-18V-26A CCCP-75845 participated in the second IL-18 mission to Antarctica together with the same IL-18V CCCP-75743 (again captained by A S Polyakov). The IL-18V-26A was captained by M P Stoopishin who had flown an IL-2 attack aircraft during the Great Patriotic War. Again the flight followed the so-called eastern route – from Moscow to Ice Station Mirnyy via Tashkent, Delhi, Rangoon, Jakarta, Darwin, Sydney, Christchurch and McMurdo and back again. This time the way to Antarctica took just six days, the *Coots* landing on the snow strip at Mirnyy on 25th November. From then on, IL-18s performed regular resupply and occasional research flights to Antarctica until 1986 when the mission was taken over by the more capable IL-76TD freighter.

The factors making the IL-18 indispensible on the northern routes included its long range (ensuring that the aircraft could reach an alternate airfield if the destination airfield was closed by bad weather – a frequent occurrence in Russia's northern areas) and good field performance. Although it does not appear particularly rugged at first glance, the *Coot*'s landing gear allowed it to operate from semi-prepared runways, including temporary ice strips. For instance, IL-18s regularly used *ad hoc* runways laid out on the ice of the frozen Kolyma River near the town of Cherskiy right until the early 1980s when Aeroflot terminated the Moscow-Cherskiy service. As late as 1987 Aeroflot IL-18s flew scheduled flights from Moscow to Chokurdakh (via Igarka and Tiksi) and Tiksi (via Ukhta and Khatanga). Other destinations beyond the Arctic Circle served by the type were Amderma, Anadyr', Batagai, Magadan, Mys Schmidta (Cape Schmidt), Noril'sk and Pevek.

Unusual Missions

Since Aeroflot's airliner and civil transport fleet constituted an immediately available military reserve, Aeroflot IL-18s sometimes had to fulfil missions outside their usual line of work, operating for the Ministry of Defence. (True, in many nations airliners are operated for the military in case of need; eg, in the USA this is practiced under a federal programme known as CRAF – Civil Reserve Air Fleet. However, in most nations this happens only in times of war. As for the USSR, given the militarisation of the Soviet economy and mentality, it appeared more like Aeroflot was operating its airliners by kind permission of the military!) At times these missions took them a long way from home. In the summer of 1960, for instance, Aeroflot IL-18s were heavily involved in the United Nations airlift to Congo, carrying supplies and troops.

IL-18s pressed into military service regularly operated into Kabul during the Afghan War. Apart from carrying personnel and evacuating casualties, they had to fulfil the dreary mission of carrying government issue zinc coffins with the bodies of Soviet servicemen killed in action. For some obscure reason aircraft carrying dead buddies back home, regardless of type, were known in Soviet Army slang as ***chornyy tyul'pahn*** (Black Tulip). Less macabre missions included personnel rotation and evacuation of seriously ill or injured servicemen from remote garrisons to military hospitals in Moscow and other major cities; the latter function was performed by IL-18s commandeered from the Urals CAD and reconfigured for the MEDEVAC role under mobilisation plans. Among other things, troop rotation missions were flown to Eastern Europe where Soviet troops were deployed; eg, in the Group of Soviet Forces in Germany (called the Western Group of Forces in 1989-94) such flights operated into Sperenberg AB near Berlin.

The IL-18 was often used for humanitarian airlift missions. The *Coot*'s high payload, long range, good 'hot-and-high' performance and – traditionally for Soviet aircraft – the ability to operate from poor-quality runways (such as are found in third world countries) made it ideally suited for this role. For instance, in June 1960 an Aeroflot IL-18 performed an earthquake relief mission to Chile, bringing food, clothing and medicines. The IL-18s of the East German airline Interflug also performed such missions from time to time, sporting large red crosses on a white background on the fuselage.

Not all such missions were necessitated by natural disasters. In 1989, shortly before the dissolution of the Soviet Union, an ethnic conflict erupted in Abkhazia, part of the Georgian Republic, when nationalist forces on both sides took advantage of the slackened control from Moscow to unleash war. Urgent measures had to be taken; among other things, on 20-26th July 1989 IL-18V CCCP-75834, an Il'yushin OKB aircraft captained by S S Balooyev, made several flights to Sochi/Adler, Sukhumi-Babushara and Tbilisi-International to evacuate passengers stranded at these airports.

There were cases, however, when IL-18s were involved in anti-humanitarian airlift missions, namely weapons smuggling – sometimes against the will of the crew. In post-Soviet times the LNPO Leninets avionics concern started making money by using some of its avionics testbeds for commercial flights (apparently the sensitive equipment and associated test equipment racks had been removed by then). Thus on 24th June 1992 the one-of-a-kind SL-18I (CCCP-75804) captained by V V Sorogovets took off from Leningrad-Pulkovo, heading south for Baku-Bina where it was to uplift a cargo of TV sets for delivery to Nakhichevan', then pick up another cargo bound for Volgograd and thence head back home. After a late night arrival the crew spent the night in the aircraft. At dawn, however, they were greeted by some very criminal-looking locals who started loading the aircraft with machine-guns, grenade launchers and all manner of ammunition which they hadn't even taken the trouble to package properly, never mind conceal. When the justifiably alarmed airmen protested they had been contracted to carry TVs, not run guns, they were given a choice: mind your own business or die. Reluctantly the crew had to take their deadly cargo to Nakhichevan' where the aircraft was loaded with wooden crates – supposedly containing apricots.

Records

Even before it started carrying passengers, the IL-18 demonstrated its capabilities for the first time – but far from the last time – by establishing a series of world altitude records for turboprop aircraft (Class C, Group II, subclass C-1). On 14th November 1958 the third production IL-18A (CCCP-Л5820, c/n 187000103) captained by Il'yushin OKB chief test pilot Vladimir K Kokkinaki climbed to 13,154m (43,156ft) with a 10-ton (22,045-lb) payload. The following day the same aircraft flown by the same crew set another record, reaching 12,471m (40,915ft) with a 15-ton (33,070-lb) payload. One more world record followed on 17th November when the IL-18 attained 13,274m (43,550ft) with a 5-ton (11,020-lb) payload.

On 19th August 1959 IL-18A CCCP-Л5820 gave another show of force by setting a string of world speed records with different payloads. Flying along a 2,000-km (1,242-mile) closed circuit with a payload of 1, 2, 5, 10 and 15 tons (2,205; 4,410; 11,022; 22,045; 33,070 lb), it clocked a speed of 719.496km/h (446 892mph; 388.916kts).

Two more world altitude records were set on 25th November when the same aircraft reached 12,000m (39,370ft) with a 20,114-kg (44,343-lb) payload and 12,118m (39,757ft) with a 20-ton (44,090-lb) payload. The crew consisted of captain Vladimir K Kokkinaki, first officer Eduard I Kuznetsov, navigator V F Voskresenskiy, flight engineer P K Kokkinaki and radio operator I S Siliminov. Finally, on 2nd February 1960 CCCP-Л5820 added yet another world record to its already impressive score, reaching an average speed of 693.547km/h (430.774mph; 374.89kts) with a payload of 1/2/5/10 tons on a 5,000-km (3,100-mile) closed circuit which took the airliner from Moscow to Sevastopol', then to Sverdlovsk and back to Moscow. The crew was the same as in the November 1959 record-breaking flights.

On 7th October 1967 an IL-18D (identity unknown) flew non-stop from Moscow to Magadan-Sokol, covering 6,270km (3,894 miles) in 10 hours 17 minutes while cruising at 10,500m (34,450ft). For the first time in Soviet aviation history such a long-range non-stop flight was performed by a female crew consisting of captain Lyubov' M Oolanova (Hero of Socialist Labour), first officer Valentina Slobodskaya, navigator Ghertrooda Olekhnovich, flight engineer Valeriya Popova and radio operator Maria Adamenko. A week later the same crew set another range record, flying in a straight line from Simferopol' via Moscow to Magadan to achieve an impressive 7,661.949km (4,758.97 miles).

On 20th October 1967 an IL-18 – again with an all-woman crew captained by Lyubov' M Oolanova – set a Class C-1 altitude record, reaching 13,513m (44,333ft) in a zoom climb. On 6th May 1968 an IL-18 captained by B Konstantinov clocked 727.84km/h (452.07mph; 393.42kts) in a straight line on a 15/25-km (9.3/15,5-mile) measurement stage, averaging 706km/h (438.5mph; 381.62kts) over a 100-km (62.1-mile) closed circuit at the same time. Not to be outdone, a crew captained by L M Oolanova established a female speed record on 12th June 1969, averaging 701.068km/h (435.445mph; 378.95kts) over a 5,018-km (3,117-mile) closed circuit (Moscow – Sevastopol' – Sverdlovsk – Moscow). The following day the same crew set a female record of sustained altitude, reaching 12,900m (42,320ft). Finally, on 12th June 1969 the same crew excelled yet again, setting a closed-circuit range record by flying non-stop from Moscow to Omsk, then to Khatanga and thence back to Moscow to cover a distance of 8,023.153km (4,983.32 miles). No absolute record previously existed in this category.

On another notable occasion, IL-18D YR-IMM (c/n 187009904) owned by the Romanian government flight and captained by I Alexandru made a trip around the world in 80 hours with 12 stops en route. The journey was completed on 7th January 1970.

Wrecks and Relics

Sure enough, aircraft don't last forever, but some IL-18s were destined to find 'life after death'. Among other things, the first production IL-18A (CCCP-75634, c/n 187000101) became a ground instructional airframe at the Kiev Institute of Civil Aviation Engineers (KIIGA – ***Kiy**evskiy **i**nsti**toot** inzhe**ne**rov **g**razh**dahn**skoy **a**vi**ah**tsii*) at Kiev-Zhulyany airport. IL-18A CCCP-75642 (c/n 188000301) was an instructional airframe at the technical school in the Ukrainian city of Krivoy Rog, while the fuselage of IL-18A CCCP-75646 (c/n 188000305) was used for instructional purposes at Moscow/Sheremet'yevo-1 until 1995. MALEV Hungarian Airlines IL-18V HA-MOG (c/n 184007103) became a ground instructional airframe at the airline's educational centre at Budapest-Férihegy.

Other retired examples were less lucky. For instance, ČSA Czechoslovak Airlines IL-18V OK-OAC (c/n 180002101) was destroyed during tests of the Semtex plastic explosive on 19-20th November 1984. This is not 'life after death' but sooner 'death after death'!

Most aircraft find their way to museums sooner or later. The Central Russian Air Force Museum in Monino boasts quite a few civil aircraft, including IL-18V CCCP-75737 (c/n 181002702). Unfortunately the aircraft is now in a rather sorry state – over the years the rain and sun have obliterated the blue cheat line completely, and the registration has vanished as a result. The Civil Air Fleet Museum in Ul'yanovsk has IL-18D CCCP-74250 (c/n 187010504) – again in very weathered condition after many years of open storage, although the aircraft was pristine when it arrived. IL-18V CCCP-75525 (c/n 183006803) withdrawn from use at Leningrad-Pulkovo in 1976 was reportedly donated to a museum in the town of Pereyaslav-Khmel'nitskiy in September 1979.[4] Additionally, many Soviet airports boasted airliners preserved on a plinth in front of the terminal building, and mercifully most of these 'gate guards' are still extant. For instance, IL-18V CCCP-75554 (c/n 184007501) – apparently a TsUMVS/63rd Flight aircraft – was preserved in front of the domestic terminal at Moscow-Sheremet'yevo (Sheremet'yevo-1) on 23rd April 1979 to mark the 20th anniversary of the type's service with Aeroflot. An IL-18V with the out-of-sequence registration CCCP-75428 (ex-Air Guinée 3X-GAC, c/n 181003704) was preserved in the airport of Petropavlovsk-Kamchatskiy following retirement in 1980. IL-18V CCCP-75518 (c/n 183006701), a Moscow Territorial CAD/Domodedovo CAPA/212th Flight aircraft which was retired in October 1980, was saved from the breaker's torch to become a 'gate guard' at Vologda-Grishino airport on 28th June 1984; the aircraft was ferried there by a crew under Il'yushin OKB test pilot Aleksandr M Tyuryumin. Another Aeroflot IL-18V (identity unknown) was preserved at Baku-Bina airport by 1985.

Similarly, several IL-18s were among the aircraft preserved in various Soviet cities – mostly to attract people to recreational spots; most of these aircraft, however, were far less lucky. For example, IL-18A CCCP-75643 (c/n 188000302), a former avionics testbed, was preserved near a railway station in Pushkin town by 1992 but broken up in the summer of 1998. Another IL-18A with the out-of-sequence registration CCCP-75748 (c/n 188000105) which had seen service with the Moscow Territorial CAD was preserved in a park in Sverdlovsk (now Yekaterinburg) by September 1977. IL-18B CCCP-75658 (c/n 188000602) became a landmark in Zaporozhye, the Ukraine; a sister aircraft registered CCCP-75659 (c/n 188000603) was preserved at an unknown location by 1995. One more IL-18B (CCCP-75689, c/n 189001204), a Kazakh CAD aircraft, took up residence in a park in the Ukrainian city of Simferopol' and was still around in 1999. IL-18A or 'B CCCP-48093 No 1 was preserved in a pioneer camp near St Petersburg.

IL-18V CCCP-75738 (c/n 181002703) in pre-1973 colours was preserved in a city park at an unknown location. Regrettably, local vandals set fire to the aircraft one night and there was nothing left to do but to scrap the charred remains. IL-18V CCCP-75751 (c/n 181003005) had been placed in a children's playground in Adler by mid-1984 but reportedly scrapped on site in July 1999. IL-18V CCCP-75781 (c/n 181003805) was preserved in a residential area not far from Brest airport. IL-18V CCCP-75844 (c/n 182005304) was preserved in the city centre of Nikolayev, the Ukraine.

Other examples were preserved outside the (former) Soviet Union. These include IL-18B OK-NAA (c/n 189001604) at the Military Museum at Prague-Kbely, IL-18D OK-WAJ (c/n 187010101) in the town of Bakov, in the Czech Republic. In Germany they have IL-18Vs DDR-STA (c/n 180001905) at *Aeropark Leipzig*, DDR-STB (c/n 180002001) at *Luftfahrthistorische Sammlung Finow* (which was sold to *Aeropark Diepensee* and still later to *Automobilmuseum Leipzig-Plagwitz*), DDR-STE (c/n 182005101) in the memorial complex at Borkheide, DDR-STG

IL-18V CCCP-75554 (c/n 184007501) was preserved at Moscow/Sheremet'yevo-1 on 23rd April 1979 to mark the 20th anniversary of the type's service with Aeroflot. Yefim Gordon

Alas, there are plenty of good-for-nothing people who don't give a damn about aviation or history but are ready to destroy anything – as the burnt-out remains of this IL-18 formerly preserved in a park at an unknown location testify.
Yefim Gordon archive

(c/n 182004402) in Erfurt, DDR-STH (c/n 184007305) in Augsburg (later sold to *Sammlung Leo & Peter Junior* in Hermeskeil) and OK-PAI (c/n 181003105) at the *Auto und Technik Museum* in Sinsheim. In Hungary IL-18V HA-MOA (c/n 180001903) is preserved at the Budapest-Férihegy Airport Museum, while sister ship HA-MOE (c/n 182005505) is in the Hungarian Air Force Museum in Szolnok.

In Poland IL-18V SP-LSD (c/n 184007102) was consecutively preserved near Podłaski, near Wikloń where it served as a restaurant, and near the village of Kośćielec close to Częstochowa. IL-18V SP-LSE (c/n 180002504) was originally preserved in Warsaw's Ursynów quarter, then moved near Poniatowski Bridge where it was burned by vandals on 26th September 1995. IL-18V SP-LSH (c/n 181002701) was first preserved as a restaurant in Radomsko. It moved to Strumień in the early 1990s.

In Romania, IL-18V YR-IMC (c/n 182004802) was preserved in Banaşti city 90km (55.9 miles) north of Bucharest. The Chinese People's Liberation Army Air Force Museum at Datang Shan AB in Changping near Beijing boasts three IL-18Ds – '208' (formerly B-224 No 2, c/n 187009703), B-226 (c/n 187009902) and B-230 No 2 (c/n 184007605), while IL-18E B-228 (c/n 185008702) is on show at the China Agricultural Museum in northern Beijing and sister ship B-216 (c/n 185008704) is preserved in a park in Chengdu. In Bulgaria IL-18V LZ-BEV (c/n 185008201) was preserved in a children's playground in Sofia, while IL-18D LZ-BEO (c/n 186009602) became a cafe 75km (46.5 miles) from Sofia on the highway from Bourgas to Zlatitsa.

Other *Coots* which have been found new lives as eateries include IL-18Vs OK-PAE (c/n 181002902) used as a restaurant near Seč, Czech Republic, and OK-PAG (c/n 181004201) used as a restaurant near Slusovice, Czech Republic. (As an aside, it is really sad to see an aircraft 'preserved' in this fashion. It brings to memory a song by the popular Soviet/Russian rock group *Mashin*a ***vrem****eni* (Time Machine) about a once proud but now decrepit ship used as a bar at a waterfront, where anybody with a few coins in his pocket can come aboard and stuff a cigarette butt into a cannon. The closing lines go like this (no attempts at a poetic translation): *I was there, too, and as I was eyeballing the crowd, // I realised with sadness one thing: // It is better to go down at the right time // Than to end up as a museum like this.'*)

Summing up, it can be said that the IL-18's sturdy design, performance and good operating economics (even by modern standards) account for its rare longevity. The *Coot* was, and still is, a true workhorse designed to do its job anywhere and in any environment. In the first 20 years after the type's service introduction the IL-18 fleet logged a total of more than 12,000,000 flight hours, making nearly 5.5 million flights and carrying 235 million passengers. The IL-18 was Aeroflot's only type to achieve individual aircraft total times of 35,000 hours in passenger configuration and 40,000 hours in cargo configuration.

Noisy though it may be, even in the late 1990s the IL-18 still found use as a passenger aircraft. The Domodedovo Civil Aviation Production Association operated its last two remaining 100-seat IL-18Ds (RA-74267 and RA-74268) on the Moscow-Baku and Moscow-Gyandzha services, competing with Azeri carriers AHY Azerbaijan Airlines (operating Tu-154B-2s and Boeing 727-235s to Moscow) and Turan Air (operating Tu-154Ms). Load factors were as high as 85.3% in 1996. Interestingly, the IL-18D took three hours to get from Moscow-Domodedovo to Baku-Bina versus 2 hours 20 minutes for the 168-seat IL-62M, but used up 6 tons (13,230 lb) of Jet A-1 for the flight versus 16 tons (35,270 lb)!

On 23rd November 2002 the Russian Civil Aviation Authority (GSGA – *Gosoo****dar****stvennaya* ***sloozh****ba grazh****dahn****skoy avi****aht****sii*, State Civil Aviation Service) issued an airworthiness directive banning further use of the IL-18 as a passenger aircraft and relegating all surviving civil *Coots* to the cargo role. This was probably not so much out of concern over geriatric airliners but rather a result of various breaches by unscrupulous operators (such as forged paperwork relating to aircraft parts etc) unearthed by the GSGA. Still, even as a cargo aircraft, the IL-18 is assured to be around for many more years yet, as is evidenced by the daily scene at Sharjah, Dubai and other major air cargo staging areas.

An Airliner in Uniform
Dedicated Military Versions

Predictably, the *Coot* quickly found military uses as well. Apart from the obvious VIP transport role and the military airlift reserve function fulfilled by Aeroflot's regular fleet, the IL-18 evolved into several specialised versions for the Soviet Air Force and the Soviet Naval Air Arm. These are dealt with in this chapter.

IL-18TD Troopship

As early as 18th December 1958 the Council of Ministers had issued a directive ordering the development of the IL-18T (***trahns**portnyy* – transport, used attributively) military transport version. In medevac configuration the aircraft was to carry 69 casualties on stretchers, while the paradrop configuration was to take 118 assault troopers and two instructors.

However, for some reason the idea did not materialise until ten years later. In 1968 a single IL-18D (CCCP-74296, c/n 188010603) was converted experimentally into a troop carrier aircraft designated IL-18TD (***trahn**sportno-de**sahn**tnyy* – transport/paradrop, used attributively). The cabin featured tip-up seats along the walls and steel cables for hooking up the paratroopers' static lines. Yet the Soviet Air Force decided it did not want a slow turboprop capable of paradropping only personnel. The project was abandoned and CCCP-74296 was reconverted to a standard IL-18D; it was later operated by Tret'yakovo Aircompany in cargo configuration as RA-74296.

IL-18B Airborne Command Post/ Staff Transport Version

Shortly after completing State acceptance trials at GK NII VVS the aforementioned IL-18B CCCP-75666 (c/n 188000705) was delivered to the Soviet Air Force and outfitted as an airborne command post (ABCP) or staff transport. The new role was revealed by several additional blade aerials associated with secure HF communications system.

IL-18V Airborne Command Post/ Staff Transport Version

Three Soviet Air Force IL-18Vs – CCCP-75516 (c/n 183006604), CCCP-75602 (c/n 182004203) and CCCP-75606 (c/n 182004405) – were converted into ABCPs. The aircraft sported two identical sets of additional blade aerials (one small aerial followed by two large ones) located symmetrically above and below the forward fuselage just aft of the nose gear. No separate designation is known for this version.

The aircraft were based at Chkalovskaya AB east of Moscow, serving with the 8th Special Mission Air Division; later, they were operated by the 223rd Flight Detachment, an 'airline' under Russian Air Force management.

IL-18D Airborne Command Post/ Staff Transport Version

Two Soviet Air Force IL-18Ds – CCCP-75478 (c/n 189011302) and CCCP-75496 (c/n 185011303) – were fitted with a different HF communications gear suite, which also qualifies them as ABCPs. In reality, however, they were primarily systems and avionics testbeds for the purpose-built IL-22 *Coot-B* described later. These aircraft featured a small strake with communications antennas extending forward from the fin fillet and were equipped with a powerful TA-6 APU installed in similar fashion to the IL-22 instead of the standard TG-16; the latter was due to the fact that the APU doubled as an extra generator powering the mission equipment. Again, both aircraft served with the 8th ADON; later they had the mission equipment removed (the TA-6 APU was retained) and now belong to the 223rd Flight Detachment.

Another IL-18D (CCCP-75498, c/n 187009804) in early Aeroflot colours and equipped with a standard TG-16 APU had a set of blade aerials similar to the IL-18Vs described above and a non-standard window arrangement: 3+door+3+2 emergency exits+1+4+door+3 on the port side and 3+3+2 exits (the rear one had a 'solid' cover minus window!)+1+3+1+3 to starboard. Later the aerials were removed and the aircraft was repainted in 1973-standard colours. It also went on to serve with the 8th ADON/223rd Flight Detachment as RA-75498.

IL-18RT Telemetry Relay Aircraft

Two quasi-civil IL-18Vs – CCCP-75528 (c/n 183006901) and CCCP-75840 (c/n 182005301) – were converted into a telemetry relay version designated IL-18RT (*retrans**lya**tor* – relay installation) by LNPO Leninets at the flight test facility in Pushkin. These aircraft were used for supporting the trials of rockets and unmanned aerial vehicles. The date of the conversion is unknown but CCCP-75840 had been in service with the AVMF since 10th February 1965, having been transferred from the Moscow Territorial CAD/Vnukovo UAD/65th Flight.

The IL-18RT could be identified by a thimble-shaped dielectric fairing on a special adapter supplanting the standard tailcone and prominent cigar-shaped fairings at the ends of the horizontal tail; these housed receiver and transmitter antennas. Additionally, CCCP-75840 featured a small elongated fairing low on the port side of the forward fuselage near the nose gear unit and small blister fairings *à la* the SL-18 avionics testbed series on the aft fuselage sides. According to some sources, the IL-18RTs supported the Soviet space programme together with the later IL-20RTs (see later in this chapter).

The first IL-18RT was re-converted to standard configuration, serving with the Pacific Fleet Air Arm. The second aircraft was later reregistered RA-75840, serving with the 240th GvOSAP (*Gvar**deys**kiy ot**del'**nyy **smesh**annyy avia**polk*** – Guards Independent Composite Air Regiment) which was part of the Russian Naval Aviation Combat & Conversion Training Centre at Ostrov AB near Pskov. It was mostly operated not in its intended role but as a proficiency trainer for the Centre's IL-38 crews so as not to waste the *Mays*' service life.

In 1998 RA-75840 was sold to the charter carrier IRS-Aero and the military avionics were removed in the course of refurbishment at ARZ No 20 which was completed on 30th September; the airline took delivery of the aircraft on 2nd October. The non-standard antennas vanished but the adapter for the tail radome remained, the aperture being closed off by a blunt conical fairing. The resulting fat tailcone clearly identified RA-75840 as a former special mission aircraft.

IL-20M *Coot-A* ELINT aircraft (*izdeliye* 17)

This, chronologically the second dedicated military derivative of the IL-18 (after the IL-38), was actually the third aircraft to bear the service designation IL-20. The first IL-20 was a piston-engined ground attack aircraft of 1947 which proved disappointing and did not progress beyond the sole prototype stage. Later the designation was applied to a handful of demilitarised IL-28 *Beagle* tactical bombers which were used by the Soviet airline Aeroflot as mailplanes in the mid-1950s, carrying newspaper matrices to the eastern parts of the USSR.

The IL-20 started life on 23rd March 1965 when the Defence Industry Commission of the CofM Presidium ruled that OKB-240 should develop an electronic intelligence (ELINT) aircraft based on the IL-18. The choice of the Il'yushin airliner as the starting point was dictated by much the same reasons as in the case of the civil special mission versions – the IL-18DORR fishery reconnaissance aircraft and IL-24N ice reconnaissance aircraft.

The requirements applying to an ELINT platform are rather different from those applying to a 'traditional' photo reconnaissance (PHOTINT) aircraft. An electronic snooper generally does not have to overfly the target or get close to it. Thus there is little danger of being shot down, hence high speed/altitude performance and defensive armament are not prime requirements. What matters more is range and especially endurance, as well as sufficient internal volume to accommodate lots of mission equipment and a large team of equipment operators. That said, it is clear that the Soviet Air Force could hardly pick a better platform in the late 1960s – especially from a cost/effect standpoint.

The result was an airborne reconnaissance system with the in-house product code *izdeliye* 17 and the unclassified service designation IL-20M (*modifi***tsee***rovannyy* – modified); for some obscure reason there was never an 'IL-20 *sans suffixe*' version of the *Coot*. Apart from some minor changes associated with the aircraft's military role, the airframe was identical to that of the IL-18D which was the current (and final) production version of the civil *Coot* from 1966 onwards. However, an ELINT aircraft inevitably sprouts non-standard aerials and other 'bumps and bulges', often quite big ones. In the case of the IL-20M the most obvious recognition feature was a massive elliptical-section pod under the forward fuselage. This housed the antenna arrays of the Igla-1 (Needle-1; pronounced *ig***lah**) side-looking airborne radar[1] – which, incidentally, was the Soviet Union's first phased-array radar.

The SLAR pod was mounted on a short pylon between fuselage frames 8 and 27; almost the entire bottom and sides were dielectric and could swing open for maintenance. The nosewheels were fitted with a large mudguard to protect the pod from flying stones and the nose gear doors were suitably bulged. The radar set was located in the former forward luggage compartment of the IL-18. The Igla-1 SLAR generated a radar image of the terrain lying at right angles to the direction of flight as far as the radio horizon and could pinpoint the location of such objects as bridges, dams, road junctions etc.

The IL-20M's imaging capabilities were not limited to SLAR. Two oblong, shallow slab-sided fairings were located on the forward fuselage sides under the forward cabin windows; their front portions incorporated ports for A-87P long-range operation (LOROP) oblique cameras which featured optically flat glass protected by sliding doors which remained closed most of the time. The cameras themselves were mounted in the forward cabin; hence the foremost cabin window on each side was omitted. The front end of the starboard fairing also incorporated a ram air intake for the SLAR's heat exchanger.

Yet the aircraft's primary mission was signals intelligence (SIGINT), notably detection of enemy radars across the border or in the frontline area, and the IL-20M was adequately equipped for this role. The greater part of the abovementioned cheek fairings was dielectric, accommodating the antenna array of the Romb-4 (Rhomboid-4, or Diamond-4) general-purpose SIGINT system; the latter determined the enemy radar's bearing and operating frequency. Apart from that, the aircraft featured a Kvadraht-2 (Square-2) detail SIGINT system with three square-shaped dielectric panels low on the fuselage sides aft of the wings (this detected bearing, operating frequency, pulse rate frequency, relative pulse duration, emission power etc) and a Vishnya (Cherry) communications intelligence system. The latter's twin trapezoidal aerials mounted dorsally on the forward fuselage were another trademark feature of the IL-20M. Four small blade aerials were located under the aft fuselage. There were also three ventral blister radomes aft of the wings pertaining to an unidentified system.

The aircraft had a crew of 13: captain, first officer, flight engineer, navigator, radio operator and eight reconnaissance systems operators ('crows', as they are known in USAF slang). Of these, only the camera operator sat in the forward cabin; the other seven workstations were in the back. The Vishnya COMINT system was worked by a specially trained operator who, besides a good knowledge of the appropriate foreign language, had to be well versed in military radio slang.

Multiple reconnaissance systems were used simultaneously to make sure accurate intelligence was obtained. For instance, the radar imagery generated by the SLAR could be checked against detailed photographs provided by the cameras, while the information supplied by the Kvadraht-2 system could be matched with a radar map of the area.

The installation of the many mission equipment items necessitated some local reinforcement of the fuselage. Due to the aircraft's military role and the greatly reduced number of occupants most of the IL-18's cabin equipment

An IL-20M cruising over international waters as seen from a shadowing Western fighter. This view shows clearly the size of the SLAR's dielectric panel. *Jane's All the World's Aircraft*

Close-up of the pod housing the Igla SLAR and the starboard cheek fairing. Note the closed camera window ahead of the large dielectric panel for the Romb-4 general-purpose SIGINT system and the cooling air intake found on the starboard side only. via Yefim Gordon

Two views of the *Coot-A*'s forward fuselage, showing most of the type's 'bumps and bulges'. The dorsal trapezoidal aerials are for the Vishnya COMINT system. via Yefim Gordon

(toilets, wardrobe, galley etc) was deleted. Changes were made to the oxygen system; all crew seats were designed to accommodate a parachute, and an escape chute and hatch were provided on the starboard side of the aft fuselage where the rear baggage door used to be (a vestige of the door still remained).

An IL-18D airframe was set aside at MMZ No 30 for conversion into the IL-20M prototype, and conversion work began on 6th May 1967. Almost a full year later the aircraft made its first flight from Moscow-Khodynka on 25th March 1968; the test crew was captained by Stanislav G Bliznyuk, while V M Volod'ko was the engineer in charge of the flight tests.

In July 1968 the prototype was very nearly lost during high-alpha trials. The basic IL-18 had passed such trials in its time but renewed trials were found necessary because the IL-20M's many equipment fairings altered the aircraft's aerodynamics considerably. Again the crew was headed by Bliznyuk; as a point of interest, on the following day the same crew was to go to Feodosiya for a month's testing of the IL-38. At extreme angles of attack the IL-20M got into a super-stalled position; losing altitude, it entered thick overcast and the crew became disoriented. 'Looks like we're f***ing finished', someone in the crew said.

Still, Bliznyuk did not lose his presence of mind and, using his experience with fighters and a good deal of courage, managed to recover from this seemingly hopeless situation. For saving the aircraft and crew he received his first – and most cherished – Order of the Red Star.

On 4-10th April 1970 the IL-20M participated in the Dvina military exercise; this was probably the type's service debut. Like the IL-38, the ELINT version had overt military markings but was usually painted a rather yucky swampy green overall (grey-painted examples were rare). The paint had a strong tendency to fade, giving the IL-20M a rather untidy appearance, especially because weathered areas were often painted up without bothering about an exact colour match. Curiously, many IL-20Ms carried no tactical code.

Low-rate production continued until 1976, totalling about 20 aircraft. IL-20 batch numbers started at 114, continuing the numeration of the civil *Coot* which was no longer in production (the last IL-18 off the line was the fourth aircraft in batch 113 built in 1969). Hence the construction numbers followed the system used for the IL-18, except that the first two digits were 17, not 18. For instance, an example coded '21 Red' is c/n 173011504, ie, the fourth *izdeliye* 17 aircraft in Batch 115 built in 1973. When applied visibly – and often it isn't! – the c/n was also located identically (on the fin and under the wing leading edge at the roots).

The West did not become aware of the type's existence until 1978 when an IL-20M was intercepted by NATO fighters over the Baltic Sea. After that the type was code-named *Coot-A*. The actual designation took even longer to become known, and some Western sources erroneously called the aircraft IL-20DSR (***dahl'****niy* ***strategich****eskiy* ***razved****chik* – long-range strategic reconnaissance aircraft).

The IL-20Ms flew very special and hush-hush missions, operating from bases along the Soviet borders (near Khabarovsk in the Far East, near Tbilisi in the South etc). They were also deployed in the Soviet Union's Warsaw Pact allies in order to get closer to the 'potential adversary'; for instance, since the mid-1970s two *Coot-As* were seconded to the Group of Soviet Forces in Germany (renamed Western Group of Forces in 1989).

Due to the highly classified nature of their work the IL-20Ms were not listed on the inventory of regular reconnaissance regiments, reporting directly to the HQ of the Defence District or air army they were working for. The two examples stationed in East Germany were operated by an independent reconnaissance squadron at Sperenberg AB near Berlin which was part of the 16th VA (*voz****doosh****naya* ***ar****miya* – air army); this unit has been reported as the 390th OAO (*ot****del'****nyy aviatsi****on****nyy ot****ryad*** – Independent Air Detachment). In 1978 the *Coot-As* were briefly deployed to Oranienburg AB. Remarkably, they remained at Sperenberg long after German reunification on 3rd October 1990; the last IL-20M departed for Levashovo AB near St Petersburg on 21st June 1994.

The IL-20M can be regarded as the Soviet answer to the Boeing RC-135 series. Still, the *Coot-A* could not equal its American counterpart in speed and range, having no aerial refuelling capability; also, the RC-135 had more capable mission avionics which could be readily configured for a specific mission, as the large cargo door inherited from the C/KC-135 allowed equipment modules to be quickly replaced.

IL-20M *Coot-A* Upgrades and Conversions

By mid-1999 GK NII VVS had outfitted one of the IL-20Ms ('90 Red', c/n 173011501) with an upgraded mission equipment suite. The modified aircraft lacked the rear trapezoidal aerial of the Vishnya COMINT system, and the two rearmost ventral blister radomes were also removed. Even more unusually for a *Coot-A*, it carried a civil registration, RA-75923 – *despite the fact that an IL-22M-11 airborne command post with the same registration (c/n 187010305) already existed* (see page 76)! The reason for this duplication may be that the IL-22 in question was already retired by then and, while the aircraft itself still existed, the registration may have been cancelled years before and then reused.

The existence of the new variant came to light on 15th August 1999 when Chkalovskaya AB was open to the general public on Aviation Day (the second Sunday in August). Though not part of the official static display, IL-20M RA-75923 No 2 was parked on the 8th ADON flight line which the visitors were able to examine, *while RA-75923 No 1 was standing engineless on the grass no more than 100m (330ft) away*! However, in mid-2002 the dead IL-22M-11 was finally broken up, leaving IL-20M c/n 173011501 as the only RA-75923 in existence. Thus order was restored at last. By then the quasi-civil IL-20M had been further modified by fitting a largish square-shaped ventral 'bathtub' (almost certainly dielectric) where the two rearmost ventral blister radomes used to be.

For some reason two IL-20Ms manufactured in 1972 were stripped of all mission equipment and associated fairings and converted to passenger configuration (or perhaps even *built* as such to fill a special order due to the unavailability of new-production IL-18s). The first aircraft (reportedly ex-'07 Red', c/n 172011401) was sold to the Ukrainian carrier Lviv Airlines in 1999 and registered UR-BXD. It wore 'IL-18D' nose titles but the two windows in the forward cabin instead of three and the vestigial No 2 baggage door (not to mention the c/n) gave it away.

The other aircraft (c/n 172011402) was entered into the Soviet civil register as CCCP-75903, albeit an IL-22 *Coot-B* ABCP (c/n 0393610235) with the same registration already existed! (The registration date has been quoted as 2nd February 1988, which automatically makes it the second aircraft to be registered CCCP-75903 because IL-22 production had ended by then!) On the other hand, IL-20 CCCP-75903 No 2 was the first to receive the Russian prefix in 1994, and it was another four years before IL-22 CCCP-75903 No 1 became RA-75903 No 2.

IL-18SIP (IL-18RTL) and IL-20RT Telemetry/Communications Relay Aircraft (*izdeliye* 17)

The IL-18 – or, to be precise, the IL-20 – even participated in the Soviet space programme. When space vehicles are launched the operation of their systems is monitored by means of telemetry. At some stages of the flight, however, it is not possible to collect telemetry data using ground stations or space tracker ships; a suitably equipped aircraft becomes the only option. Hence on 28th September 1970 the Soviet Council of Ministers issued a directive concerning the development of the IL-20RT space tracker/telemetry and communications relay aircraft based on the *Coot-A*; once again RT stood for *retranslyator* – relay installation. The first aircraft was to be completed by August 1972.

Outwardly the IL-20RT had nothing in common with the similarly designated ELINT aircraft as far as 'bumps and bulges' are concerned. To test the space tracker's mission avionics LII converted one of the IL-18As, CCCP-75647 (c/n 188000401). Reregistered CCCP-27220 and designated IL-18SIP (*samolyotnyy izmeritel'nyy poonkt* – airborne measuring station) or IL-18RTL, this aircraft was instantly recognisable by the huge slab-sided dorsal canoe fairing with three access hatches on each side. A flattened teardrop fairing incorporating a square dielectric panel was mounted on each side of the fin about halfway up, and a fairly large thimble radome with a blade aerial above the root supplanted the standard tailcone. Several forward-pointing probe aerials were located on the flightdeck roof, and an antenna array of complex shape was installed ahead of the rear entry door.

Originally the sole IL-18SIP retained pre-1973 IL-18 colours, save for the fact that the Soviet flag was placed unusually low because of the fin antenna fairings occupying its usual place; the dorsal 'hump' was painted black.

The IL-18SIP airborne measurement station (CCCP-27220), the prototype of the IL-20RT, as originally flown in pre-1973 colours. Note the low-set Soviet flag on the tail. Yefim Gordon archive

CCCP-27220 at a later date following repaint in 1973-standard Aeroflot colours. The door and window arrangement typical of the IL-18A/B is readily apparent. Yefim Gordon archive

Top and centre: **CCCP-75480, the first production IL-20RT (c/n 173011405). The colour scheme and door/window arrangement are quite different from the IL-18SIP.** Yefim Gordon archive

Later the aircraft gained 1973-standard Aeroflot livery, except for the unusually small titles because the canoe fairing (which was now white) left no room to apply standard-size titles.

The IL-18SIP was retained by LII and used during tests of air-launched cruise missiles. In the late 1980s, however, the aircraft had reached the limit of its useful life and had to be retired. In 1992-93 the hulk of CCCP-27220 could still be seen on LII's dump in the wooded area between Zhukovskiy's old and new runways but was finally scrapped by August 1995.

The production aircraft were based on new-build IL-20 airframes and hence had IL-20-type construction numbers commencing 17. Apart from the different door and window arrangement *à la* IL-20M, the IL-20RT differed from the experimental IL-18SIP in lacking the fin fairings and the flightdeck roof probes. The nose gear doors were bulged (this was the current production standard) but no mudguard was fitted because there was no longer a ventral radome to be protected from flying stones.

The IL-20RT's mission avionics included the RTS-9 (***rah**diotelemetri**tricheskaya stah**ntsiya* – radio telemetry station) and BRS-4 telemetry receivers, the SYeV-12 time synchronisation system (*si**stem**a yedi**no**vo **vremeni***) and additional voice link and telegraph communications equipment; their antennas were housed in the dorsal and tail fairings. The avionics suite was worked by ten operators.

Four production IL-20RT aircraft – CCCP-75480 through -75483 (c/ns 173011405, 173011503, 173011505 and 173011601) – were built in 1973. They wore a decidedly non-standard colour scheme: the cheatline ran below the windows rather than across them, the registration was applied in black on white rather then white on blue, and the Aeroflot titles and logos were very small (the same size as on the IL-18SIP following repaint) but were placed above the forward cabin windows rather that above the wings.

The production aircraft were based at Yoobileynyy airfield in Leninsk (currently Tyuratam), Kazakhstan, which served the Baikonur space centre; they were used during trials of rockets and satellites. Later, when five new 'aircraft 976' radar picket/telemetry and communications relay aircraft based on the IL-76MD *Candid-B*[2] became available in 1986-87, the IL-20RT space trackers were stripped of all mission equipment and used as transports and trainers; one of the four aircraft, CCCP-75481, ended up in the AVMF Combat & Conversion Training Centre at Ostrov AB. The aircraft still had its trademark 'hump' and tail fairing – and, oddly enough, retained the old Soviet prefix and flag.

IL-22 *Coot-B* airborne Command Post (IL-18D-36 'Bizon', *izdeliye* 36)

This, one of the most enigmatic versions of the IL-18D, was developed by the Myasishchev Experimental Machinery Plant (EMZ), not by Il'yushin. The reason was that OKB-240 was not in a position to handle the project, having its hands full refining the IL-38 and the then-new IL-62 airliner and designing the IL-76 transport.

The choice of the IL-18 as the basis for a new airborne command post was dictated by several reasons. High endurance and long range (4,000-5,000km; 2,485-3,100 miles), high reliability and the ability to operate from relatively short runways away from the home base for extended periods were prime requirements; high speed, on the other hand, was not. This narrowed the choice of platform to turboprop aircraft with proven airframes and powerplants (the IL-62 still had to make its mark and required long runways).

The An-12 transport was unsuitable because of its unpressurised freight hold. Since aircraft factory No 39 in Irkutsk and factory No 64 in Voronezh had stopped building the type, a fully pressurised version of the *Cub* could only be manufactured at the Tashkent aircraft factory No 84 which had more important tasks, gearing up to build the An-22 *Cock* 'big lifter'. The pressurised An-10 airliner from which the An-12 was derived had inadequate range and suffered from design faults which eventually caused it to be retired from Aeroflot service in May 1972. The IL-18, on the other hand, was a widespread and reliable aircraft operated by Aeroflot and the VVS alike and outperformed all competing turboprop designs.

Development of a new ABCP derivative of the *Coot* was initiated by Communist Party Central Committee/Council of Ministers joint directive No 603-215 of 7th August 1968 and by

Bottom: **A typical IL-22 *sans suffixe* (IL-18D-36) in an earthen revetment at Pushkin airfield near St Petersburg. This view shows the various blade aerials and fin top 'cigar', as well as the full-length fairing of the IL-18D-36. All IL-22s had full Aeroflot colours and 'IL-18' nose titles.** Yefim Gordon archive

Above and opposite page: **An IL-22 *sans suffixe* (IL-18D-36) basks in the winter sun at Pushkin, awaiting an overhaul at ARZ No 20. This example is interesting in that it has a dark-coloured ventral canoe fairing.** Yefim Gordon archive

Left: **Detail of the IL-22's fin tip pod from a stripped-down example at Pushkin.** Yefim Gordon archive

MAP order No 312 of 6th September. EMZ was assigned responsibility for this programme and the aircraft was allocated the service designation IL-22.

Originally the IL-22 programme was known in-house as Project 36 and bore the codename *Bizon* (Bison); hence the initial version of the IL-22 received the manufacturer's designation IL-18D-36 'Bizon' or *izdeliye* 36. Ivan Ivanovich Razhev was appointed chief project engineer; Fyodor N Zhookov was also actively involved in the IL-22 programme, along with I P Vorob'yov, V S Boyko, V P Goosev, V N Yepishkin, A A Koorazhkin and S V Belobrov.

The IL-22 (aka IL-18D-36 'Bizon'), or *izdeliye* 36, was immediately recognisable by the characteristic cigar-shaped antenna fairing atop the fin which required relocation of the upper anti-collision beacon to the centre fuselage. Another trademark feature was the shallow ventral fairing of semi-circular cross section running almost the full length of the fuselage; this was dielectric and housed more antennas.

Two small blade aerials were mounted dorsally above the three forward cabin windows, followed by a slightly larger blade aerial just aft of the forward entry door and then a *huge* blade aerial in line with the overwing emergency exits. Another small blade aerial was located low on the starboard side of the nose beside the ventral canoe fairing and three more (one plus two abreast) aft of the said fairing.

The fuselage incorporated some structural changes. The window arrangement differed slightly from the IL-18D (3+door+2+two emergency exits+1+5+door+2 to port and 3+2+two exits+1+4+2+2 to starboard). An escape chute and hatch were installed on the starboard side of the aft fuselage, supplanting the rear baggage door *à la* IL-20M. However, this was not enough, since there would be a high-ranking officer and his staff on board in addition to the crew and the single chute could become a bottleneck. Hence a large slipstream deflector powered by two faired hydraulic rams was mounted ahead of the rear entry door, making it possible to bail out that way. The nose gear unit featured a mudguard to protect the ventral canoe fairing (and suitably bulged gear doors), but many IL-22s had it removed in service.

The aircraft featured a 'war room' and was fitted with various communications equipment (including Satcom), as well as scrambling/descrambling equipment, data storage systems and an ESM suite. The mission equipment (mostly located in the former baggage compartments) used a lot of power and the engine-driven generators could no longer cope with the load, so the IL-18's TG-16 APU was replaced by a more powerful TA-6 unit (the same one as used on the IL-76, IL-62 and Tu-154). The APU breathed through two small intakes located above one another on the port side near the fin fillet and closed by square doors with external stiffening ribs; the exhaust was located under the port stabiliser (rather than on the starboard side ahead of the horizontal tail) and surrounded by a large heat-resistant steel plate. The new APU gave a bonus, allowing the equipment to be operated and the cabin to be air-conditioned on the ground.

Two standard IL-18s (identities unknown) were converted to IL-22 prototypes by Myasishchev's experimental shop at Zhukovskiy in 1970, entering State acceptance trials later that year. The trials went successfully and the aircraft entered limited production at MMZ No 30. The IL-22s were built 'green' and flown to Zhukovskiy where they were fitted out with mission equipment at the Myasishchev facility.

Despite their military role, all IL-22s were quasi-civil and wore full 1973-standard Aeroflot colours (and 'IL-18' nose titles!). A large batch of registrations was set aside for the IL-22s, though the registrations did not always run in sequence. Known IL-18D-36s were registered CCCP-75895 through -75907. Some aircraft had the tips of the propeller spinners painted black or blue.[3]

Interestingly, the IL-22 has a separate construction number system. For instance, CCCP-75903 No 1 is c/n 0393610235. The first three digits are a code denoting the manufacturer – the Moscow Aircraft Production Association named after Pyotr V Dement'yev; the c/ns of Moscow-built MiG-23 *Flogger* fighters and early-production MiG-29 *Fulcrum-As* also begin with 039. The next two digits are a product code (*izdeliye* 36 or IL-18D-36). The remaining five digits do not signify *anything at all*; they are often referred to as the 'famous last five', and the first two and last three of them accrue independently. This is a computer-generated number designed to avoid revealing the batch number/number of the aircraft in the batch (and hence how many have been built). This means that new-build IL-22s also have *fuselage numbers* indicating the batch number and the number of the aircraft in the batch; so far, however, no c/n to f/n tie-ups have been made. The location of the c/n is identical to that of the IL-18.

On 14th May 1994 the IL-22 had its public debut when CCCP-75903 was displayed statically during an open doors day at its home base, Kubinka AB west of Moscow. Later the general public was also able to see the IL-22 at close range during further events at Kubinka and the Aviation Day displays at Chkalovskaya AB. Understandably enough, no visitors were allowed inside, apart from kids who had no idea what they were looking at! After the IL-22 had been discovered by the West it received the NATO reporting name *Coot-B*.

IL-22M *Coot-B* Airborne Command Post (IL-22M-11 'Zebra', *izdeliye* 40)

A second version of the *Coot-B* appeared later, receiving the manufacturer's designation *izdeliye* 40 and the service designation IL-22M-11 'Zebra' or simply IL-22M.[4] There are some indications that it was developed in 1977 under Project 11-U concerning ABCPs with all-new mission equipment, hence the -11 suffix to the designation (the U probably stood for *oopravleniye* – control, referring to command, control and communications centres).

Outwardly the IL-22M-11 differed from the IL-18D-36 primarily in having a much shorter ventral fairing which began just aft of the first cabin window (rather than immediately aft of the nose gear) and terminated at the wing leading edge. The dorsal antenna farm now comprised two small blade aerials, two slightly larger ones in line with the second cabin window and the forward entry door, one small blade aerial, one L-shaped aerial and the familiar huge blade aerial. The ventral fairing was flanked by one small blade aerial to port (at the front) and two medium blade aerials plus one huge one to starboard (at the rear), while the existing blade aerials aft of the wings were complemented by one L-shaped aerial and two small blade aerials on the centreline.

The window arrangement again differed from the standard airliner (3+door+3+two emergency exits+1+4+door+3 to port and 3+two exits+1+4+1+3 to starboard). The slipstream deflector near the aft entry door was deleted, as the chances of bailing out successfully through the door were considered too small.

IL-22M-11 registrations continued the series started by the IL-22s *sans suffixe*; known examples were registered CCCP-75908 through -75929 and again some registrations were out of sequence. Thus the total number of *Coot-Bs* in existence was around 35.

Again, most IL-22M-11s were new-build aircraft and used the same damn fool counter-espionage c/n system. For instance, CCCP-75912 was c/n 0394011097: the fourth and fifth digits indicate *izdeliye* 40. In 1983, shortly after the new variant entered production, the factory code was changed to 296 to further confuse would-be spies. Thus, late-production 'Zebras' have c/ns commencing 296 (eg, CCCP-75925, c/n 2964017557).[5]

At least six 'Zebras', however, were converted from standard IL-18Ds, and these are *not* prototypes. The IL-18 production line at MMZ No 30 had to be ultimately closed after 1983 and the jigs thrown away because the fac-

CCCP-75900, a stripped-down early-production IL-22 *sans suffixe*, shares the ramp at ARZ No 20 with an IL-38 and a Kamov Ka-27 ASW helicopter. The *Coot-B* apparently has to sit in the open in this condition because the hangars are occupied by other aircraft being worked on.
Yefim Gordon archive

tory had to fill a growing demand for the MiG-23 while preparing to build its successor, the MiG-29. However, the Air Force's demand for IL-22s proved unexpectedly high. Since no new-build airframes could be obtained any more, several low-time IL-18Ds were 'requisitioned' from Aeroflot and transferred to the VVS for conversion to IL-22M-11s.

After conversion such aircraft were issued new registrations in the CCCP-759xx block to fit the general pattern; most retained their original c/ns but in some cases the first four digits were replaced with 29640, resulting in a strange hybrid c/n. Known examples are CCCP-75917 (ex-IL-18D CCCP-75451 No 2, c/n 187010105), CCCP-75919 c/n 2964009805 (ex-IL-18D CCCP-75780 No 2, c/n 187009805),[6] CCCP-75923 (ex-IL-18D CCCP-75463, c/n 187010305), CCCP-75926 (ex-IL-18D CCCP-74270, c/n 188011203), CCCP-75928 (ex-IL-18D CCCP-75448, c/n 187010003) and CCCP-75929 (ex-IL-18D CCCP-74251, c/n 187010505). Curiously, the c/n of CCCP-75926 had been painted on in error as 2964011203 at some point but then was changed to the correct one; close inspection of the tail revealed 2964 under the paint on which the figures 188 were superimposed!

IL-22s operated from Chkalovskaya AB (8th ADON), Kubinka AB and Ostaf'yevo AB near Moscow, Dyaghilevo AB near Ryazan', Irkutsk and other bases. For instance, IL-18D-36 CCCP-75896 was based at Vinnitsa in the Ukraine. IL-22M-11 CCCP-75916 was operated by the 50th OSAP at Machulischchi AB near Minsk. A sister aircraft (CCCP-75918) belonged to the 243rd OSAP stationed at Sknilov AB near L'vov; a third IL-22M-11 (CCCP-75929) was operated by a unit stationed near Kishinyov and a fourth (CCCP-75915) somewhere in Kazakhstan. They were taken over by the Belorussian, Ukrainian, Moldovan and Kazakh air forces respectively after the collapse of the USSR.

IL-22s have been seconded to the Group of Soviet Forces in Germany (Sperenberg AB) and the Northern Group of Forces (Poland, Legnica AB). The type has also been reported as based at Pushkin near St Petersburg but this is misconception, as the aircraft were simply on overhaul at the local aircraft repair plant No 20.

Little is known of the *Coot-B*'s actual operational use. For instance, on 17th September 1999 two pairs of Russian Air Force/326th Heavy Bomber Division Tu-95MS-6 *Bear-Hs* took off from Anadyr' and Tiksi to take part in an exercise. The bombers operated in conjunction with an IL-22 ABCP. Apart from their main role, the *Coot-Bs* are sometimes used as VIP transports; as such they participated in the Afghan War of 1979-89 and in the First Chechen War of 1994-96.

IL-22/IL-22M Mid-life Updates

During their service career some aircraft have received new equipment fits, as evidenced by new antenna arrangements. For example, IL-18D-36 CCCP-75906 has had the ventral fairing removed altogether. IL-22M-11s RA-75920 and -75924 have much-reduced dorsal antenna farms comprising two small blade aerials, one slightly bigger one and one L-shaped aerial.

The most impressive upgrade, however, is the latest configuration of IL-22M-11 RA-75909 (c/n 0394011092) which appeared in the spring of 1998. The top of the fuselage ahead of the wings bristles with two small and four large unswept blade aerials, and two huge 'hockey stick' aerials with thin forward-pointing probes are mounted ahead and aft of the overwing emergency exits. The short ventral fairing is replaced by two strake aerials of roughly the same length; there are also two small unswept blade aerials low on the port side of the nose and a small swept blade aerial, a large unswept blade aerial and a 'hockey stick' aerial (identical to the dorsal ones) on the starboard side. At least two more *Coot-Bs*, CCCP-/RA-75908 (c/n 0394011091) and RA-75902 (c/n 0393610226), were later upgraded to this standard, and quite possibly this sub-variant has an as-yet undisclosed separate designation.[7]

Five IL-22s were stripped of mission equipment and used as airliners (and are mostly listed as IL-18Ds in fleet lists!). In so doing most of the aerials associated with the HF comms suite were removed, but the characteristic fin top pod and the unusual APU placement remained. Thus, Ukrainian Air Force IL-18D-36 CCCP-75896 was sold to the Angolan airline ALADA as D2-FFR. One more, RA-75903, became a VIP transport with the Russian Air Force's 16th Air Army; it was displayed as such at Kubinka AB on 8th August 2002 on occasion of the 16th VA's golden jubilee.

IL-22M-11 ER-75929 was sold to Vichi Airlines (the commercial division of the Moldovan Air Force), flying alternately in all-cargo or 105-seat all-economy configuration, and was occasionally leased to the Romanian airline Acvila Air. Another example, YL-LAO (ex-Belorussian Air Force CCCP-75916), has been operated by the Riga-based private airline Concors since June 1998. The aircraft has a mixed-class arrangement, including a luxurious VIP cabin at the rear. Both aircraft have had most of the IL-22's aerials removed, retaining only the fintip 'cigar'; yet this, along with the non-standard APU and the absence of windows on the starboard side of the nose, indicates their origin all too clearly! A third IL-22M-11, UN-75915, became a VIP aircraft. Unfortunately it was written off in a ground collision with a runaway An-12 in January 1995 only a few days after it had been refurbished and converted.

The second version of the *Coot-B*, the IL-22M (aka IL-22M-11 'Zebra'), shows off the short ventral fairing and different antenna farm. Yefim Gordon archive

CCCP-75909 (c/n 0394011092), the second production IL-22M, at Pushkin. This aircraft was later converted into a new version of the IL-22M whose designation remains unknown. Sergey and Dmitriy Komissarov archive

A few *Coot-Bs* found themselves outside Russia when the Soviet Union collapsed. This is 75918, a Ukrainian Air Force IL-22M-11, at Sknilov AB near L'vov. The grey object above the flightdeck is not a non-standard aerial but a tail rotor blade of a helicopter standing beyond! Alfred Matusevich

The Sub Hunter – A Breed Apart

IL-38 *May* Maritime Patrol/ASW Aircraft (*izdeliye* 8)

In the late 1950s new classes of offensive weapons started taking shape in the USA and the Soviet Union; these later formed the so-called nuclear triad comprising ground-launched intercontinental ballistic missiles (ICBMs), strategic bombers and nuclear-powered missile submarines (SSBNs). The Americans were quick to realise the advantages conferred by arming a submarine with ICBMs. Submarines could travel quickly and stealthily to the required areas of the world ocean to launch missile strikes against the Soviet Union and possibly its Warsaw Pact allies. While the first sea-launched ballistic missiles (SLBMs) required the sub to surface, more advanced models capable of underwater launch were soon developed, adding to the submarine's stealth and minimising the danger of being destroyed.

This, in turn, sparked the development of appropriate defensive weapons. For instance, the advent of submarines armed with SLBMs led to new developments in anti-submarine warfare (ASW) systems. Maritime patrol and ASW aircraft became one of the primary means of countering the submarine threat both in the USA and the USSR.

The USA quickly started the Polaris programme which was to bolster the US Navy with 41 SSBNs, as well as appropriate land support systems and, at a later stage, satellites. The first five submarines were armed with 16 Polaris A-1s with a 2,200-km (1,189-nm) range; later subs carried Polaris A-2 missiles with a 2,800-km (1,513-nm) range and, later still, Polaris A-3s with a 4,600-km (2,486-nm) range.

As for countermeasures, the USA spent a lot of effort and money on setting up a stationary Sound Surveillance System (SOSUS) which became operational on the East Coast in 1954 and on the West Coast in 1958. Some sources claimed SOSUS had a detection range of 1,000-1,200km (540-648nm). Fielding a similar system in the USSR was impossible due to geographical factors, the Soviet industry's lag in hydrophone systems development and the prohibitive costs involved. Thus, aircraft became the only viable option, and the 100 or so obsolete Beriyev Be-6 *Madge* piston-engined flying boats posed no serious threat.

In 1958-60 the Soviet defence research establishments undertook the Vyaz (Elm) and Mozhzhevel'nik (Juniper) research and development programmes. These resulted in the formulation of general operational requirements (GORs) for ASW aircraft and helicopters, as well as for the submarine detection equipment and weapons they were to carry, to be ordered from the defence industry in the near future.

No fewer than six ASW systems were proposed to the Soviet Navy under the *Vyaz* programme, including four fixed-wing aircraft and two helicopters. The Navy amended the projected performance of the Kamov Ka-25 *Hormone* shipboard helicopter, defined ways of improving airborne submarine detection systems – and issued a GOR for a long-range shore-based ASW aircraft based on the IL-18V which is the subject of this story. The aircraft was to have a 2,200-km (1,189-nm) combat radius and an on-station loiter time of 3-3.5 hours; this should enable it to counter submarines armed with Polaris A-1 SLBMs. The principal armament consisted of AT-1 ASW torpedoes (*aviatsi***on***naya tor***ped***a* – aircraft torpedo) which were being tested at the time; conventional and nuclear depth charges were also under consideration.

The *Mozhzhevel'nik* R&D programme, which began almost simultaneously, was of a more specific nature. It culminated in the development of a specific operational requirement (SOR) for the aircraft's sonobuoy system, as well as SORs for the VGS-2 Oka[1] dunking sonar (developed for the Ka-25) and for the Gagara (Loon) infra-red search and track system.

Defence industry proposals set forth in the course of the Vyaz and Mozhzhevel'nik programmes were studied and altered. For instance, the military turned down the offer to develop an airborne early warning system which would detect and track ballistic missiles launched by enemy submarines and allow the latter to be pinpointed and destroyed. The same fate befell a proposed long-range ASW system built around a multi-role amphibian. (History has a way of repeating itself; consider the fate of the Beriev A-40 *Mermaid*...)

The Soviet Navy compared the capabilities of shore-based and amphibian aircraft in the long-range maritime patrol and ASW role. The verdict was that development of an amphibian in this class involved major technical difficulties; the aircraft would be overweight and less aerodynamically efficient than a landplane without giving any tactical advantage. Eventually after carefully analysing all options the Navy selected an ASW derivative of the IL-18 airliner. Interestingly, the US and British military had taken the same approach; Lockheed had derived the P-3 Orion from the L-188 Electra, while the Hawker Siddeley HS 801 Nimrod (later British Aerospace Nimrod) was an evolution of the de Havilland DH.106 Comet 4.

Thus at this stage the Soviet ASW development programme included four basic components – a long-range shore-based ASW system based on the IL-18, a short-range system built around the Beriyev Be-12 Chaika (Seagull/*Mail*) amphibian, a shore-based amphibious helicopter (the future Mil' Mi-14PL *Haze-A*) and the shipboard Ka-25PL *Hormone-A*.

The choice of the *Coot* as the starting point was a good one. Firstly, the IL-18's capacious fuselage offered ample space for the search and targeting system, weapons, crew and the additional fuel required to give the aircraft the specified maximum endurance of 16 hours. Secondly, the airliner's proven airframe, powerplant and systems ensured adequate reliability during long overwater missions in constantly changing weather conditions. Thirdly, the IL-18 had fairly high performance and good manoeuvrability, which meant an ASW aircraft based on it could quickly reach ocean areas located thousands of kilometres from its base and patrol the area at 100-2,000m (330-6,560ft), making tight turns. The aircraft would be capable of autonomous operation from bases with relatively short strips and minimum ground support facilities which were common in the Soviet Navy's North and Pacific Fleets. Finally, development time and costs were greatly reduced as compared to an all-new aircraft; the time issue was especially important, since the US Navy's first nuclear-powered missile submarine, the USS *George Washington* carrying 16 Polaris A-1 SLBMs, had been commissioned in late 1959.

The long-range ASW derivative of the IL-18 was designated IL-38 and included into the Naval Air Arm's re-equipment programme. Later, as the range of SLBMs increased, the long-range ASW role was reassigned to the Tupolev Tu-142 *Bear-F*, which had flight refuelling capability, and the IL-38 was regarded as a medium-range system.

Officially development of the IL-38 at OKB-240 was kicked off by Communist Party Central Committee/Council of Ministers joint directive No 640-261 issued on 18th June 1960. The document said the first prototype was to commence trials in the second quarter of 1962 (minus operational equipment), followed by the fully-equipped second aircraft in the fourth quarter of the same year.

General Designer Sergey V Il'yushin supervised IL-38 development, with his deputy Yakov A Kootepov as chief project engineer. V M Ghermanov and Radiy Petrovich Papkovskiy were also actively involved in the design effort. Advanced project development and issue of manufacturing drawings proceeded in close cooperation with Soviet Navy representatives. The Navy kept making amendments to the SOR which, as a result, was not finalised until April 1961 when prototype construction had already begun.

At the ADP stage the Il'yushin OKB had to work out the ideology and operational tactics of the ASW system together with the Navy and other organisations involved in the IL-38 programme; these included more than a dozen design bureaux and research institutes. The project was completed in just a few months thanks in no small part to the dedication and industriousness of ADP section employees L M Ryabov, Yuriy I Yoodin, V M Sheynin, N P Stolbovoy, R P Papkovskiy, G G Moorav'yov and O N Yelsookova.

The contours of the IL-38's main airframe subassemblies (fuselage, wings, tail unit and engine nacelles) were identical to those of the IL-18. There was also considerable structural and systems commonality; the flightdeck section, the tail unit and all control surfaces, the powerplant (four 4,250-ehp/3,169-ekW Ivchenko AI-20M turboprops driving AV-68M four-blade reversible-pitch propellers of 4.5m (14ft 9⅛in) diameter and a TG-16 auxiliary power unit), the engine control and fire suppression systems, the de-icing system, the landing gear, many electric and hydraulic components were borrowed straight from the airliner.

The wing torsion box was identical to that of the experimental IL-18I, featuring an integral fuel tank in the centre section.The fuel system comprised 25 tanks holding a total of 33,820 litres (7,440.4 Imp gals) – integral tanks in the outer wings and wing centre section and flexible bag tanks in the fuselage – and featured three-point pressure refuelling; gravity refuelling was also possible through eight filler caps. A fuel management system controlled the tank emptying sequence, automatically maintaining CG position within the prescribed limits.

From the outset the IL-38 was developed as part of an ASW complex featuring the Berkoot (Golden eagle) search and targeting system (STS); the latter was built around a 360° search radar of the same name with a large ventral radome located aft of the nose gear unit. The need to accommodate the system's built-in and droppable components, as well as weapons, led to major structural changes. First of all, the combined weight of the mission avionics and weapons caused a significant forward shift in the aircraft's CG; to compensate for this the wings had to be moved 3m (9ft 10⅛in) forward. This changed the aircraft's proportions completely, giving the IL-38 an uncannily dachshund-like appearance.

The air conditioning system was modified to suit the greatly reduced pressure cabin volume and because of the need to heat the weapons bays and heat/cool certain modules of the Berkoot radar set. Hence two heat exchangers in fairly large elongated fairings were mounted high on the fuselage sides just ahead of the wings. Installation of the mission equipment also led to changes in the electric system.

Two weapons bays closed by twin doors were located ahead and aft of the wing centre section; the aft bay was for weapons proper, while the forward bay held sonobuoys. Because of the aircraft's military role the IL-18's port side entry doors and most of the cabin windows were deleted. The crew entered via a forward-opening ventral hatch immediately aft of the radome opening into a sloping passage; this doubled as an escape chute, the access hatch cover acting as a slipstream deflector. The remaining windows (except the one nearest to the flightdeck on each side) were reduced in size.

Apparently the Berkoot search radar had a secondary weather reconnaissance function, since the IL-18's RPSN-1 Emblema weather radar was deleted and the nose radome replaced by a fairing of identical shape, only the forward half of which was dielectric. Finally, the fuselage terminated in a long tapered boom carrying the sensor of the magnetic anomaly detector (MAD).

The prototype IL-38 was built by cutting up an IL-18 airframe. It is seen here during State acceptance trials. Interestingly, the ventral radome was of all-metal construction on the prototype, as this aircraft was purely an aerodynamics test vehicle. Yefim Gordon archive

The IL-38 prototype with both weapons bays open. Yefim Gordon archive

Front and rear views of the IL-38 prototype from the State acceptance trials report.
Yefim Gordon archive

Internally the fuselage was divided into two sections. The forward section was the pressure cabin, also divided into two parts – the flight-deck (accommodating the captain, first officer, flight engineer, navigator and radio operator) and the rear section where the crew entry hatch was. In the rear cabin, facing aft, sat the search radar operator and the operator of the aircraft receiver/indicator (SPIU – *samo**lyot**noye preeyomno-indi**kah**tornoye oo**stroy**stvo*). The aircraft receiver/indicator (ARI) was a device processing and displaying incoming signals generated by the sonobuoys. There was a well-founded opinion that having the operators face aft would increase crew fatigue, but the limited space of the pressure cabin did not allow their workstations to be redesigned to face forward. Since the sorties would be quite long, the rear cabin featured a galley, a table, a toilet and a rest area with a folding bunk.

The unpressurised rear and lower portion of the fuselage incorporated the two abovementioned weapons bays with a total volume of $30m^3$ ($1,059.4ft^3$). A container for two fuel cells holding a total of 4,200 litres (924 Imp gals) was installed above the forward weapons bay. An avionics bay housing the Berkoot modules, including the digital computer, was located further aft near the CG. The APU was installed aft of the rear weapons bay; its exhaust was located to port about halfway between the wings and tail instead of just ahead of the starboard stabiliser, as on the IL-18. The rear fuselage was accessible in flight from the pressure cabin after the pressure differential between the two had been eliminated; two walkways ran the full length of the rear fuselage.

The avionics suite consisted of three main groups: navigation, communications and targeting equipment. The former group comprised a DISS-1 Doppler speed and drift sensor, an RSBN-2S Svod (Dome) short-range radio navigation (SHORAN) system, an SP-50 ILS, an ARK-11 ADF and an RV-4 radio altimeter. Communications equipment included an SPU-7B intercom, an SGU-15 loudspeaker system, an R-802V command link radio, an R-632 decimetre-waveband command link radio, R-836 Neon and R-847A VHF communications radios, a Peleng (Bearing) HF communications radio and an MS-61 cockpit voice recorder.[2] Finally, the targeting equipment consisted of the Berkoot STS which will be described in more detail later.

The oxygen system enabled the crew to work normally in the event of a decompression at high altitude, to enter the unpressurised rear fuselage in flight and to bail out safely 'if all else failed'. The system comprised 19 stationary 36-litre (7.92-Imp gal) bottles with gaseous oxygen and one 7.6-litre (1.672-Imp gal) portable bottle. When bailing out the crew used KP-23 bottles with enough oxygen for 11 minutes.

Il'yushin engineers had given a lot of thought to crew rescue in the event of a ditching. The IL-38 was tested for ditchability in model form on the Moscow Sea (a large reservoir north of Moscow); impact details and the aircraft's buoyancy were studied and design features ensuring safe evacuation developed accordingly. After ditching the crew was to vacate the aircraft either via a dorsal escape hatch in the middle of the pressure cabin or via an overwing emergency exit on the port side of the unpressurised centre fuselage. A bay with a PSN-6A six-man life raft (*plot spa**sah**tel'nyy nadoov**noy***) was located just aft of the port wing. The bay cover could be opened both from outside and from within by means of cables; the raft then popped out near the wing trailing edge, inflated and was secured by a line to stop it from drifting away. The PSN-6A had a tent to shelter the crew from the elements and carried a three days' supply of food and water for the entire crew, a transceiver, a supply of signal flares, a first-aid kit and a repair kit.

Individual rescue equipment included S-5 parachutes, MLAS-1-OB one-man inflatable rafts and MSK-3M maritime rescue suits (*mor**skoy** spa**sah**tel'nyy kos**tyum***). The latter was an infernal creation of rubberised fabric that could certainly save you from hypothermia in bitterly cold seas but was most inconvenient in ordinary flight, even though it was air-conditioned. Former Tu-126 *Moss* AWACS crews who were issued MSK-3 suits on overwater missions say just thinking of the thing's rubber collar makes your neck itch again – after all these years! The parachute pack contained, apart from the parachute itself, an NAZ-7 survival kit (*nosimyy avareeynyy zapahs*) – a three days' food ration, an emergency radio, signal flares and a few other 'bare necessities'.

Another peculiarity of overwater operations was the corrosive salty environment. Hence the IL-38 featured enhanced corrosion protection of the airframe, engines and systems.

The weapons range comprised AT-1 and AT-2 ASW torpedoes, PLAB-250-120 Lastochka (Swallow), PLAB-50 and KAB-500PL Zagon (Corral) depth charges, mines and a variety of free-fall bombs used against surface ships. Nuclear depth charges could also be carried. About 30 payload combinations were possible and the maximum payload, including RGB-1, RGB-2 and RGB-3 sonobuoys, was 8,000kg (17,640 lb).[3]

Development of the Berkoot STS at NII-131 (LNPO Leninets) in Leningrad began in December 1959 pursuant to Council of Ministers directive No 1335-594 issued on 11th December. Initially the design effort was led by V S Shoomeyko who died in harness and was succeeded by A M Gromov and P A Iovlev. More than ten other research establishments and design bureaux were involved in the IL-38 programme; eg, the sonobuoys were developed by NII-753 in Kiev, a division of the Ministry of Agricultural Machinery (!),[4] while the weapons system's ideology was worked out by the Naval Academy and other institutes.

The STS was designed to detect enemy submarines and provide target data for their destruction. It enabled the IL-38 to automatically go to the required area of the ocean, set up straight and curved barriers made up of sonobuoys in the submarine's anticipated path and then monitor the incoming signals. Theoretically the system enabled the aircraft to return automatically and replace a failed buoy; however, this was impracticable as the barrier quickly lost shape because of the wind and waves, making it impossible to find the failed buoy.

Three models were used: RGB-1 non-directional passive buoys, RGB-2 passive buoys and RGB-3 active/passive buoys. Signals generated by the sonobuoys were picked up by the search radar and processed by the aircraft receiver/indicator (ARI). The latter, besides visualising sonobuoy operation on a display, enabled the operator to listen to noises transmitted by the RGB-1 buoys, determine target bearing relative to the RGB-2 buoys and target bearing/range relative to the RGB-3 buoys. The ARI could perform a brief scan of the dropped sonobuoys in 1 sec or a detailed scan in 60 sec.

Surfaced subs and subs travelling at periscope or snorkel depth could be detected directly by the radar – an inherently archaic approach. Besides the 360° search mode, the radar had 120° and 150° sector viewing modes and handled some navigation tasks as well (it has a ground mapping mode). Finally, the submarines could be detected by the APM-60 MAD (*aviatsionnyy protivolodochnyy magnitometr* – aircraft-mounted anti-submarine magnetometer) which operated independently from the Berkoot system.

The STS also received inputs from numerous air data sensors, the Put'-4B-2K (Way) compass system, the AP-6Ye autopilot, the ARK-B ADF etc. All this equipment was linked into a single suite by the TsVM-264 digital computer (*tsifrovaya vychislitel'naya mashina*) developed under the leadership of V I Lanerdin which was to handle both navigation and tactical tasks. Incidentally, the IL-38 was the first Soviet aircraft to have a mainframe digital computer.

In theory, after the radar operator had entered target data the computer would calculate the chances of a 'kill' with the chosen weapon; when the time was right the weapons bay doors would open and the bombs or torpedoes would be dropped automatically. This high level of automation was no small achievement at the time. In reality, however, some of the system's components proved to be extremely unreliable; unfortunately it took so long to get the system up to scratch that the Berkoot was obsolete by the time it entered service.

The SOR for the IL-38 was approved by the Soviet Air Force C-in-C on 4th May 1961. About the same time the Il'yushin OKB held a review of the advanced project in the presence of numerous invited guests. A Naval Academy representative reported on the IL-38's proposed operational ideology; the aircraft would be used mainly in the Baltic and Norwegian Seas, since the heavy ice in the Arctic Sea rendered it unusable there. The aircraft were supposed to operate singly.

The absence of defensive armament on the IL-38 was a departure from the general Soviet trend, causing many surprised comments. However, Sergey V Il'yushin quickly showed the critics were wrong. Even the simplest twin-cannon tail barbette with ammunition would weigh at least 1,500-1,800kg (3,306-3,968 lb) and require an extra crew member, to say nothing of the ground support personnel. Besides, the aircraft would need a gun laying radar which, apart from adding still more weight, might cause problems with the MAD. Worst of all, all this trouble would not really do anything by way of self-defence. Il'yushin's opinion was that all-aspect air-to-air missiles were the solution but no such weapons were available for heavy aircraft at the time.

Il'yushin was supported by AVMF Commander Air Marshal I I Borzov, Hero of the Soviet Union. Borzov was convinced that ASW aircraft were designed to operate outside the envelope of the enemy's air defences and fly at low level to ensure stealth and thus had no need for defensive armament; to reinforce his point he referred to the IL-38's US counterpart, the Lockheed P-3, which had no such armament. The latter argument was somewhat shaky, since the Orion would be operating within the reach of *friendly* air defences and thus was safe from attack by enemy fighters. However, Borzov was known for his explosive temper and tenacious memory, and no-one commented, feeling that silence would be safer.

The Western press wrote that the P-3 would be armed with McDonnell Douglas Astronautics Company (MDAC) AGM-84 Harpoon air-to-surface missiles with a range of 110-120km (59-65nm) and thus have strike capability. This caused considerable interest among the Soviet military who asked Il'yushin to incorporate similar weapons on the IL-38. Il'yushin promptly replied that the draggy external stores would cause too much stress on the wings, which were heavily loaded as it was.

Despite the fact that the Il'yushin OKB had its hands full with improving the IL-18 and developing the brand-new IL-62 *Classic* long-range

An IL-38 undergoing overhaul at aircraft overhaul plant No 20 in Pushkin. The antenna dish of the Berkoot search radar is laid bare for inspection. Yefim Gordon archive

Another view of the prototype, showing the MAD boom and the glassfibre fin fillet/spine. The aircraft appears to be unpainted and uncoded. Note the dorsal blister fairing (probably associated with test equipment) which was omitted on production aircraft.
Yefim Gordon archive

jetliner, a complete set of technical documents for IL-38 prototype construction was released in a remarkably short time. The prototype was built at the OKB's experimental shop, MMZ No 240, which was conveniently located at Khodynka, right across the field from MMZ No 30 'Znamya Truda'. The work was supervised by the experimental shop's director D Ye Kofman and its chief engineer V A Yoodin.

Prototype construction proceeded quickly, thanks in no small part to the production plant's director P A Voronin; standard IL-18 components were pulled off the 'Znamya Truda' production line and trucked to MMZ No 240 for conversion and assembly. The aircraft was completed without the Berkoot STS, some components of which were then being tested on the IL-18SL avionics testbed (CCCP-75643, see Chapter Two).

On 28th September 1961 – almost a full year ahead of the date set forth in the government directive – the IL-38 took to the air for the first time, powered by AI-20A engines because the intended AI-20Ms were still unavailable. The aircraft was piloted by an Il'yushin OKB crew – captain Vladimir K Kokkinaki (project test pilot), first officer Eduard I Kuznetsov, navigator V F Voskresenskiy and radio operator I S Siliminov. Armament specialist M S Gol'dman was appointed engineer in charge of the flight test programme. This included performance and handling trials, primarily in flight modes typical for an ASW aircraft; weapons bay door operation and the possibility of bailing out through the ventral entry hatch were also checked. The designers of the Berkoot STS were to analyse the actual conditions in which their system would work later on.

Initial flight tests continued until June 1962 at Il'yushin's flight test facility at LII's airfield in Zhukovskiy south of Moscow. Meanwhile, between 15th May and 12th June 1962 the so-called mock-up review commission (a body composed of industry and customer representatives which comes into the picture at the ADP stage) reviewed the IL-38-plus-Berkoot STS advanced development project. The Navy group headed by Air Marshal I I Borzov was keen to see the aircraft enter service as soon as possible and thus provided valuable assistance in every possible way throughout the trials.

Early test flights revealed that the IL-38 rotated more easily on takeoff than the IL-18 and behaved differently in a crosswind because the wings were located further forward. Originally the ventral radome was hemispherical, generating turbulence which caused minor but constant and annoying vibration of the forward fuselage. The problem was quickly rectified by extending the rear portion of the radome aft and the result was a quasi-spherical shape. Generally the IL-38 was easy to fly, inheriting the benign handling characteristics of the airliner from which it was derived. Pilots reported that takeoff and landing presented no problems and the aircraft had adequate stability and controllability throughout its speed, altitude and CG range (the CG limits were 16-30% MAC). The IL-38 showed no handling peculiarities at its never-exceed speed of Mach 0.65 and remained stable with the weapons bay doors open. Rudder and aileron authority was sufficient to counter the yaw and roll in the event an outboard engine failed on takeoff.

The crew evacuation tests also went well. The bottom line was that there were no obstacles to the installation of the Berkoot STS. However, development of the latter was lagging behind schedule and proceeded in unreasonable secrecy, even by Soviet standards of the time. (LNPO Leninets engineers often travelled abroad and met foreign colleagues who offered to sell tried and tested mission equipment for the IL-38, but such a purchase was considered politically unacceptable.)

Upon completion of the manufacturer's flight tests the prototype was fitted out with all components of the STS except the TsVM-264 digital computer. The original AI-20A engines had run out of service life and so were replaced by improved AI-20Ks rated at 3,945ehp (2,942ekW). Stage 2 of the trials programme began on 10th March 1963 – almost a year behind schedule; the objective was to verify the Berkoot STS jointly with LNPO Leninets and to check out the possibility of carrying various models of sonobuoys.

Stage 2 was largely performed at Kirovskoye airbase near Feodosiya on the Crimean peninsula which was managed by Section 3 of NII VVS. This time Aleksandr M Tyuryumin (also Il'yushin OKB) was project test pilot, while M S Gol'dman remained engineer in charge. The tests showed that the search radar, the ARI and other elements of the STS worked well but the RGB-1 sonobuoys were ill-suited to cold-weather operation. At this stage the IL-38 logged 369 hours in 147 flights. Valuable assistance was provided by a crew from the Naval Air Arm's 33rd Combat and Conversion Training Centre named after Ye N Preobrazhenskiy (33rd TsBP i PLS – *Tsentr boye**voy** podgo**tov**ki i pere**oo**chivaniya **lyot**novo sos**tah**va*) in Nikolayev, the Ukraine – pilot Maj A P Sharapov, navigator Capt N P Soorkov, radar operator Capt Igonin etc.

Now the final step had to be taken: the computer had to be installed and all systems integrated and debugged. This stage of the work dragged on until February 1964; Stage A of the NII VVS trials began in March at Kirovskoye AB under the supervision of NII VVS team chief Col Ye B Polyakov. While Soviet Navy test pilots confirmed the company's high opinion of the aircraft proper, the Berkoot system started misbehaving. Avionics failures were encountered virtually in every single flight; the TsVM-264 computer and the sonobuoys proved particularly troublesome.

Despite the fact that some 200 defects discovered at this stage still had to be eliminated, Stage B began on 14th May 1965. This time the NII VVS team was headed by Col O A Voron'ko; Lt Col A K Kiryukhin was engineer in charge of the Berkoot STS and the aircraft was flown by Col S M Sookhin and Lt Col A F Stepanov.

The State acceptance trials finally took place from 6th June to 15th December 1965; the IL-38 logged 287 hours in 87 flights. 80% of the time the aircraft worked in co-operation with surface ships and submarines used as practice targets. The results of the trials were analysed between 1st and 4th December by the State commission chaired by Air Marshal I I Borzov. The commission ruled that the IL-38 should be put into production at MMZ No 30 under the in-house product code *izdeliye* 8 but not officially included into the Naval Air Arm inventory for the time being because mission avionics reliability was unsatisfactory. Other deficiencies discovered during the State acceptance trials included 2,600kg (5,372lb) of excess weight and excessively high noise levels in the cabin caused by the forward shift in wing position.

Another problem facing the engineers was the need for a full-authority automatic propeller feathering system (the AI-20's system enabled autofeathering only at 70% nominal power or more). Sergey V Il'yushin approached a British manufacturer (probably Hamilton Standard) with the request to supply such a system, but the sale was either vetoed or the cost proved prohibitive.

Besides his temper, I I Borzov was also known for his progressive approach to ASW development; he was positive that 'sub hunter' aircraft were a valuable asset and did much to see this class of weapons put into service. As chairman of the State commission reviewing the IL-38 and its Berkoot STS he aided a lot in speeding up the aircraft's development and overcoming bureaucratic snags.

As was customary at the Ilyushin OKB, a full-time team was formed for turning the aircraft over to the 'customer'. The team was headed by M S Gol'dman and included test pilots Aleksandr M Tyuryumin and G N Volokhov, flight engineers V Lebedev and Yuriy Grevtsev. Coincidentally, IL-38 project chief Yakov A Kootepov formed a team of engineers which would quickly solve problems coming up during the trials.

MMZ No 30 began gearing up for IL-38 production in 1965. Production was managed by N I Sood'yin and A I Khmel'kov.

The AT-2 homing ASW torpedo, which was to be the IL-38's main weapon, was doing much better than the STS. Its State acceptance trials were also held at Kirovskoye AB under the supervision of Capt 2nd Grade (the naval equivalent of Lt Col) L G Golubev. During the trials 63 live torpedoes and 19 inert torpedoes were dropped from the IL-38 prototype, as well as from a Tupolev Tu-16T *Badger-A* torpedo bomber and a Mil' Mi-4 *Hound* helicopter. The trials were successfully completed in late 1964 and the AT-2 was recommended for acceptance by the AVMF.

Meanwhile, trials and debugging of the IL-38's STS continued. The first six months of 1966 were devoted to correction of the numerous deficiencies noted during the State acceptance trials and installation of new mission avionics with improved reliability. Concurrently the interim AI-20K engines were replaced by the more powerful, reliable and fuel-efficient AI-20Ms as originally intended. This permitted an increase in takeoff weight from 63,500kg (139,991 lb) to 66,000kg (145,502 lb) and led to a new series of performance tests.

In was not until October 1966 that manufacturer's tests of the improved Berkoot STS and armament began. Once again the test crews were facing numerous problems, as evidenced by the test log:

'...On 27th October a complex sortie was flown with the objective to seek and destroy a submarine with input from RGB-2 sonobuoys. On one of the five buoys dropped the parachute failed to open, on two others the microphone and heading indication system failed to separate after splashdown; objective not completed.

On 29th October a complex sortie was flown with the objective to seek and destroy a submarine with input from RGB-2 sonobuoys. The mission was aborted because of a TsVM-264 [computer] *failure.*

'...On 18th November a complex sortie was flown with the objective to seek and destroy a submarine with input from RGB-2 sonobuoys; objective not completed due to sonobuoy failure.

'...On 3rd December a complex sortie was flown with the objective to seek and destroy a submarine with input from RGB-2 sonobuoys; objective not completed due to failure of the Berkoot STS sensors.

'...On 24th December a complex sortie was flown with the objective to seek and destroy a submarine with input from RGB-2 sonobuoys; mission aborted due to radar transmitter failure.

On 26th December a complex sortie was flown with the objective to seek and destroy a submarine with input from RGB-2 sonobuoys. Aiming accuracy was unsatisfactory – the weapon dropped 5-7km [2.7-3.78nm] *from the target.'*

Through most of 1967 LNPO Leninets kept working on improving the reliability of the prototype's STS, in particular the TsVM-264. On the ground the computer was inspected and checked for hours and hours, and some modules which did not meet the specifications were replaced. Still, subsequent flights showed that the computer's reliability remained as low as ever. Finally, in late October 1967 the STS trials with the prototype were halted.

Later the aircraft served for various tests which were not included into the State acceptance trials programme for some reason. These proceeded under the management of Lt Col S K Apollonov; the aircraft was flown by NII VVS pilots Col S M Sookhin, Lt Col Ye M Nikitin, Lt Col A F Stepanov and Lt Col M V Vlasenko. As a rule, these were highly complicated missions demanding high flying skills, such as high-alpha tests, rejected takeoff (RTO) with simulated engine failure and icing tests. The safety of operations in autopilot mode at 500m (1,640ft) and MAD efficiency at 100m (328ft) were also investigated.

The main part of the STS improvement and debugging work was performed on the first production IL-38 built in 1967 (tactical code unknown, c/n 087010106).

(Note: IL-38 construction numbers follow the same system as those of the IL-18/IL-20, except for a few peculiarities. For example, '71 Red', c/n 082011207; 08 = *izdeliye* 8 (a zero is added at the front to keep the nine-digit format), 2 = year of manufacture 1972, 0 = MMZ No 30, Batch 112, seventh aircraft in the batch. The curious aspect is that, while the IL-38s was also built in batches of five, these were 'add-ons' to existing batches; thus, individual aircraft numbers in each batch ran from 6 to 10 (except for the 112th and final batch which consisted of three aircraft)! Quite possibly IL-38 production was intentionally begun with batch 101 because, if you disregard the first digit, the number conveniently becomes 01 and so forth.

Interestingly, the prototype manufactured in 1967 fitted beautifully into the system, since batch 101 of the IL-18 was also built in 1967 (c/ns 187010101 through c/ns 187010105). However, later aircraft started 'falling behind', ie, the years of manufacture for IL-18/IL-38 batches with the same number are different; cf. IL-18D CCCP-74263 (c/n 188010905) built in 1968 and IL-38 '21 Red' (c/n 081010910) built in 1971.

The IL-38 rarely carries the c/n visibly. If it does, the c/n is stencilled in the same locations as on the IL-18/IL-20/IL-22.)

Between October 1967 and January 1968 the prototype was based at Kirovskoye AB where additional State acceptance trials of the Berkoot system were held under the supervision of Maj P K Zamyshevskiy; Il'yushin OKB navigators V I Melekhin and R B Voronov played an important part at this stage. Unfortunately the fickle autumn weather, storms on the Black Sea and inadequate availability of support ships caused the trials to take rather longer than expected. Yet periods when the weather was good were used to the full, with a lot of flying.

The results were encouraging and the reliability of the STS had clearly improved, even if the Berkoot still left a few things to be desired. This enabled Air Marshal I I Borzov to clear the IL-38 for service. As the Soviet Navy accumulated operational experience with the type, the engineers at LNPO Leninets analysed the most common defects and developed countermeasures. Finally the painstaking work of the system's designers began to yield results. On 17th January 1969 the IL-38 was formally included into the AVMF inventory – after seven and a half years of testing. The official name ran as follows: 'ASW complex – IL-38 aircraft with the Berkoot search and targeting system'.

In November 1971 Sergey V Il'yushin, A I Zhukovskiy, D I Koklin, M A Kootepov and A V Shaposhnikov received the State Prize for the development and refinement of the IL-38. Other Il'yushin OKB employees who were actively involved in the programme included Yakov A, Kootepov, Radiy P Papkovskiy (currently IL-38 chief project engineer), Valeriy Afrikanovich Borog, Ye I Sankov, V I Smirnov and V M Ghermanov. A major contribution was made by AVMF Commander I I Borzov, naval test pilots and engineering staff, the personnel of military research establishments and service units. Many of them received due recognition for this; at NII VVS alone 19 aircrew members received government awards in 1969 for their part in the IL-38 programme.

Full-scale production was originally planned at aircraft factory No 166 in Omsk (currently OAPO 'Polyot') or aircraft factory No 126 in Komsomol'sk-on-Amur (currently KnAAPO).[5] However, Sergey V Il'yushin thought this was completely illogical. He addressed the Soviet government and the Communist Party Central Committee, requesting that IL-38 production be assigned to MMZ No 30 (which, as already mentioned, was building the IL-18 and would have no trouble producing its ASW derivative), and eventually got his way. The first production IL-38 took off in September 1967; production continued until 22nd February 1972 when the 58th and final aircraft was rolled out. (Some sources say 65 IL-38s were built.) Interestingly, the original acquisition plan signed in 1962 envisaged no fewer than 250 aircraft.

The West became aware of the IL-38's existence around 1970 and the aircraft received the

A trio of *Mays* makes a flypast at Moscow-Domodedovo airport during the May 1967 air fest. Yefim Gordon archive

codename *May* (in the 'miscellaneous' category). The first three-view drawings of the aircraft published in the Western press were highly inaccurate, showing the radome located *ahead* of the nose gear.

In the early 1970s a new trend originated: setting up a 'field' of sonobuoys was considered a more effective tactic of hunting submarines than the customary sonobuoy barriers. In practice this meant that buoys were evenly distributed over the entire suspicious area rather than in a line across the sub's anticipated path. Hence the TsVM-264 computer received new software for working in this mode.

In an attempt to give the IL-38 ECM capability a single aircraft was fitted experimentally with SPS-151 and SPS-153 *Seeren'* (Lilac) active jammers. The modified aircraft underwent tests at LII (Zhukovskiy) and at the NII VVS facilities in Akhtoobinsk (near Saratov in southern Russia) and Kirovskoye AB in 1971-72. While the ECM gear eventually was not fitted to production *Mays*, some IL-38s were retrofitted with the Vishnya COMINT system. Outwardly such aircraft could be identified by two small dielectric blisters on each side of the forward fuselage linked by a thin conduit stretching all the way aft to the wing leading edge and two more blisters at the root of the MAD boom to give 360° coverage. Some aircraft featured guidance equipment for KAB-500PL guided depth charges. In 1974-75 the Berkoot STS was upgraded by the addition of the ANP-3V automatic navigation device (*avtomaticheskiy navigatsionnyy pribor*) which enabled the IL-38 to manoeuvre more accurately while tracking a submarine.

An interesting development was the attempt to increase the IL-38's range without any hardware modifications. Having learned that US Navy P-3 crews often shut down one or two engines to save fuel during ocean patrol missions, Il'yushin engineers decided to investigate the possibility of doing the same on the IL-38. To this end a test programme was held at Kirovskoye AB in 1970-72 under the supervision of Col Apollonov. In the course of the tests NII VVS test pilots determined the maximum all-up weights at which the *May* could fly safely on three or two engines and worked out a reliable relight technique for the AI-20 (which, incidentally, was totally different from the one in the flight manual). A crew captained by Col Ye M Nikitin investigated the worst possible scenario (when the aircraft was flying on the two outboard engines and one of them failed) and developed recommendations for service crews what to do in this situation. (It should be noted that even this scenario did not cause imminent danger of a crash.)

Generally the results were encouraging – shutting down one engine increased on-station loiter time by 20-30%. A side effect of this technique was that the pitch change mechanisms had to be filled with a higher-grade oil with better low-temperature performance to make sure it would not congeal when the propellers were feathered. Still, this method did not find practical application with the IL-38.

IL-38M and IL-38MZ Development Aircraft

Another (and obvious) way of increasing range and endurance was to give the IL-38 flight refuelling capability. Work in this direction began at the Il'yushin OKB in the autumn of 1971 under the leadership of the new General Designer Ghenrikh Vasilyevich Novozhilov (previously Sergey V Il'yushin's deputy; Il'yushin had retired in 1970 due to poor health). The actual design effort was led by V M Ghermanov.

The probe and drogue refuelling system already used by the Soviet Air Force's strategic bomber arm, DA (***Dahl'****nyaya avi****ah****tsiya* – Long-Range Aviation), and the AVMF was selected. Two versions of the aircraft were developed simultaneously: the IL-38M (*modifit****see****rovannyy* – modified) and the IL-38MZ (*modifi****tsee****rovannyy/zap****rahv****schchik* – modified/tanker). The former was merely a receiver aircraft equipped with a fixed refuelling probe ahead of the flightdeck glazing on the port side, while the IL-38MZ also had a 'buddy-buddy' tanker capability. To this end additional fuel tankage and a UPAZ-38 hose drum unit (*ooniver****sahl'****nyy podves****noy*** *agre****gaht*** ***zaprahv****ki* – versatile suspended refuelling unit) were installed in the weapons bays. The HDU was a product of OKB-918 led by Guy Il'yich Severin. This bureau later became the Zvezda (Star) company best known for the K-36 ejection seat which is fitted to almost all current Russian combat aircraft.

To ensure accurate rendezvous even in the most adverse weather both aircraft were fitted with the RSBN-20V SHORAN system. The IL-38M's maximum all-up weight (*not* the maximum takeoff weight!) increased to 69,000kg (152,120 lb). Any production *May* could be easily converted to IL-38M/IL-38MZ standard if need arose.

Stage A of the IL-38M's State acceptance trials proceeded at Kirovskoye AB between October 1974 and June 1975, followed by Stage B in November 1976 – May 1977. The aircraft clocked 271 hours in 117 test flights. The NII VVS team holding the trials was headed by Lt Col V F Sil'veystrook. Col G K Yefimov was project test pilot; besides, many test flights on the IL-38M and IL-38MZ were made by Col V I Yefimov, Col Ye M Nikitin, Col V M Grishin, Col V S Avershin, Lt Col A V Poglazov and Lt Col P I Mazikin.

Various locations for the refuelling probe and associated piping were tried; the aircraft sported black phototheodolite calibration markings on the forward and centre fuselage and the inboard engine nacelles. The refuelling equipment was evaluated at various speeds and altitudes in various weather conditions, day and night; incidentally, the IL-38M was the first Soviet aircraft to make a night-time fuel replenishment, using the probe and drogue system. Various refuelling system failures were simulated and a new and simpler refuelling technique developed.

Tests showed that the system increased the IL-38's on-station loiter time by 1.5 to 2 hours; the fuel transfer rate was 1,000 litres/min (220 Imp gals/min). The IL-38M and IL-38MZ were recommended for AVMF service but never got there. Officially the reason was that the Navy didn't care for the idea of having part of the IL-38 fleet (which was small as it was) converted to tankers and made unavailable for ASW duties. However, there were persistent rumours that Andrey N Tupolev had had a hand in the matter to clear the way for the Tu-142. There were reasons for this suspicion, as Tupolev was known to trip up competitors in this fashion, using his influence (the Sukhoi T-4 bomber which was rejected in favour of the less advanced Tu-22M *Backfire* is a case in point).

IL-38 Test and Development Aircraft

One IL-38 was used by LNPO Leninets as a testbed for the Korshoon search radar developed for the Tu-142M *Bear-F Mod*. This had previously been tested on the company's SL-18P avionics testbed (see Chapter Two).

Another IL-38 'fell victim' to *konversiya* (the adaptation of military assets for civilian needs). In the late 1990s the Leninets Holding Company converted it into a geophysical survey aircraft equipped with the IKAR (Icarus) multi-mode survey system developed under F F Zolotookhin; the IKAR acronym stands for *izmeritel'nyy **kom**pleks **a**erogeogra**fi**cheskoy* *raz**vedki*** – aerogeographical prospecting measurement suite.

The IKAR system enables all-weather high-resolution scanning of land and sea, making the demilitarised *May* suitable for such varied tasks as geological mapping, oil and ore prospecting, magnetic and laser scanning of the coastal seabed, searching for submerged objects such as sunken ships, maritime search and rescue, fishery reconnaissance etc. The latter is important, since both IL-18DORRs (CCCP-75462 and CCCP-74268) had been reconverted to IL-18D standard, leaving the long-range ocean fishery reconnaissance role sorely 'understaffed'.

IL-38N (IL-38SD) ASW Aircraft

Given the protracted economic crisis in Russia which effectively rendered development and production of 'clean sheet of paper' ASW aircraft impossible, an increase in the nation's ASW and maritime patrol capabilities could only be obtained by an in-depth modernisation of the IL-38. This implied first and foremost installation of a new mission avionics suite and possibly new engines (the same ideology as used by Lockheed in its P-7 Super Orion project).

In 2000 the Leninets Holding Company completed development of the Morskoy Zmey (Sea Dragon) new-generation STS. Unlike the Berkoot STS, the new system was not centralised but featured a 'federal structure' with several independent data gathering/processing subsystems integrated into a single whole and governed by an Argon mainframe computer. The system's modular structure with duplicated hardware and software maximises the chances of mission success even if some of the components fail.

The Morskoy Zmey (also known under the code letter 'N') comprised a new search radar with a mechanically scanned slotted-array antenna and a 25-kilowatt transmitter. The installation of the new radar did not necessitate any changes to the radome's shape. The radar exhibited high resistance to jamming and had track-while-scan capability. Targets with a radar cross-section of about 1m^2 (10.75ft^2) could be detected at a maximum range of 30-35km (18.6-21.7 miles).

The new radio/hydroacoustic subsystem had 96 channels for receiving and processing data supplied by sonobuoys. The system was worked by two operators and consisted of two modules, each of which could work with up to eight buoys at a time. These comprised RGB-41 passive non-directional buoys, RGB-48 passive directional buoys and RGB-58 active emitting buoys. The buoys weighed from 10 to 16kg (22-35 lb), with a diameter of 120-150mm (4¾-6in) and a length of 1-1.26m (3ft 3⅜in to 4ft 1⅛in). Special RTB-91 magnetometric buoys were also envisaged; according to other data, the system was to include a new MAD with a detection range of 900m (2,950ft).

An ELINT package was installed in a flattened pod mounted on lattice-like struts above the forward fuselage. It had a 360° field of view, scanning signals with a frequency range of 0.5-40GHz. It also featured a gyrostabilised optical/thermal imaging system for detecting and tracking surface targets. The system could be trained manually or automatically.

The mission crew consisted of two tactical information processing system operators and a third operator for the search radar and ELINT system. The joint workstations were equipped with a large colour liquid-crystal display and LCD control panels.

The uncoded prototype of the IL-38/Morskoy Zmey upgrade (c/n unknown) entered flight test at Pushkin in 2000. The new system was advertised at the MAKS-2001 airshow in Zhukovskiy (14-19th August 2001). Due to the system's English name the upgraded *May* is also known as the IL-38SD (for Sea Dragon).

IL-38 Upgrade Projects

Several attempts to enhance the IL-38's capabilities were made over the years. As early as 30th April 1969 MAP issued an order to upgrade the *May*'s mission equipment with a view to enhancing the aircraft's search capabilities and extending its combat radius; the SOR for this was signed on 5th September. The upgrade was dictated by the changing conditions in which the IL-38 was to operate and early operational experience with the type.

A little earlier, in March 1969, the Navy had formulated an SOR for the Tu-142M's Korshoon-1 STS and it was deemed advisable to install this promising new system on the IL-38 as well. The version for the *May* would be slightly different and bore the designation Korshoon-M. The IL-38 would be armed with RGB-75 infrasonic sonobuoys having a detection range of 20-30km (10.8-16.2nm) instead of the RGB-1's 1-2km (0.54-1.08nm). These would be used in conjunction with explosive sound sources (ESS) – bomblets creating an acoustic signal which was reflected from the target and picked up by the buoys, allowing 'quiet' submarines to be detected.

The Korshoon STS featured eight parallel data processing channels, allowing incoming

Seen from a shadowing NATO aircraft, IL-38 '16 Red' drops a sonobuoy retarded by a drogue parachute from the forward weapons bay.
Yefim Gordon archive

signals from the buoys to be analysed much quicker. Finally, it included a tactical information display subsystem (TIDS) which used a preset range of commands stored in the mainframe computer.

The APM-60 MAD would be replaced by the more advanced Bor-1S (Pine forest) model then under development (this later received the service designation APM-73S). New navigation and flight instrumentation systems would be fitted; the standard electric/hot air de-icing system would be replaced by an electro-pulse de-icing system and the extremely inconvenient electric engine starting system by air starters (which, incidentally, would give a 340-kg/749-lb weight saving). No crew members would be 'automated away' during the upgrade; however, since mission time would increase the crew would work in shifts (ie, a relief crew would be carried).

Additionally, the aircraft would have an automatic flight control system (AFCS) replacing the Put'-4B-2K compass system and AP-6Ye autopilot, a Roomb (Cardinal point) attitude and heading reference system (AHRS) replacing the TsGV-10 vertical gyro, hydrological survey equipment, ASO-2B chaff/flare dispensers etc. In reality, however, only the MAD was replaced. The reason was that the existing TsVM-264 computer was incompatible with the Korshoon system and would have to be replaced, which involved a lot of work. It's a shame that the work was never done; there was sufficient funding available and the result would have been worth the effort.

The next attempt at upgrading the IL-38 centered on introducing new sonobuoys – this time RGB-16 broad-band (5Hz-5kHz) non-directional passive buoys which had passed their tests in 1984; these were also to be used in conjunction with ESS. Replacing the Berkoot STS completely turned out to be too expensive and it was decided to add new components instead. This resulted in the Izumrood (Emerald) system which included a 68-channel Volkhov receiver (named after a Russian river), hydroacoustic data processing and display equipment, an interface with the existing Berkoot STS and RGB-16 sonobuoys.

The State acceptance trials programme included 20 flights totalling 58 hours. It showed that detection range was increased several times but the ESS bomblets were extremely impractical. Part of the IL-38 fleet was later upgraded to this standard, with encouraging results.

The *May* in Service

IL-38 deliveries to the newly-formed 24th OPLAP DD (*ot**del'**nyy protivo**lod**ochnyy **avi**a**polk dahl'**nevo **dey**stviya* – Independent Long-Range ASW Regiment) stationed at Severomorsk-1 AB near Murmansk began in March 1968. This North Fleet unit was then commanded by Lt Col V P Potapov, who went on to become Colonel General and Commander of the AVMF. Initial operational capability (IOC) was achieved by August. Amazingly, no sonobuoys were supplied at first because the damn things were still classified (they were not declassified until June 1969)!

At the same time conversion training got under way at the 33rd TsBP i PLS. To cut costs and hasten the training process – which was important, since the new weapons system still had lots of bugs to be ironed out – the engineers of the Centre's research and development section (Col V V Achkasov, Oleg K Denisenko and Capt Magadeyev) devised the Bereg-38 (Shore-38) simulation system which enabled IL-38 crews to practice using the Berkoot STS on a land test range without actually dropping sonobuoys or weapons. (Perhaps it was this system that gave rise to the jocular expression about the 'submarine in the Ukrainian steppes' meaning 'something non-existent'!)

In July-August 1969 the Pacific Fleet's 77th OPLAP DD based at Nikolayevka AB near Vladivostok, also a new unit, was equipped with the IL-38. The Baltic Fleet followed suit three years later when the *May* became operational with the 145th OPLAE (*ot**del'**naya protivo**lod**ochnaya **avia**eskad**ril'**ya* – Independent ASW Squadron) based at Riga-Skulte. Additionally, two aircraft were retained by the 33rd TsBP i PLS, one to three aircraft were perpetually used for test and development work by GK NII VVS; finally, one retired aircraft went to the Loogansk Military Navigator School as a ground instructional airframe.

As usual, the introduction period was accompanied by problems; suffice to say that it took *five days* to prepare enough torpedoes for 18 aircraft in a unit! The reason was the long charging cycle (27 hours) of the torpedoes' silver-zinc batteries and the fact that only eight battery chargers were issued to a unit. (Incidentally, each battery contained 16kg/35 lb of silver.) Later, however, disposable battery packs became available, cutting mission preparation time by a factor of 15.

Another operational difficulty was caused by the strict working temperature limits of the computer and the radar interface. During the preflight procedure each aircraft had to be connected to an air conditioning cart which heated the avionics bay for 1.5 to 2 hours in winter and cooled it in hot areas. Later the problem was alleviated by making changes to the computer which expanded its working temperature range.

'07 Red', a grey-painted production IL-38 operated by the 24th OPLAP DD at Severomorsk-1 AB. Note the open entry hatch just aft of the radome. Yefim Gordon archive

The squadron leader briefs the crew of a 24th OPLAP DD *May* before a mission. Yefim Gordon archive

With the rugged hills of the Soviet Far East and a sister ship providing a backdrop, a 77th OPLAP DD IL-38 taxies out for take-off.

The crew of a Pacific Fleet Air Arm IL-38 gets ready to 'kick the tires, light the fires and go fly'. Note the quasi-civil IL-18 across the taxiway.
Both Yefim Gordon archive

The first real target (that is, foreign submarine) was detected by a North Fleet *May* in 1968 in the Barents Sea. The Pacific Fleet opened up the score in 1974 in the Sea of Japan and the Baltic Fleet two years later in the Indian Ocean. Initially the crews were not very successful in their efforts: only 17 foreign submarines were detected in 1968-73, all of them by North Fleet aircraft. (By comparison, Be-12 crews detected 12 subs within the same time frame. It should be noted that detections were carefully documented by the Navy General HQ and the fleet headquarters at the time.)

Between 1969 and 1981 the Soviet Navy IL-38s made 4,095 sorties, logging 24,540 flight hours. The total number of foreign submarines detected during this period was 172. Operational costs from 1977 to 1983, including the cost of the dropped sonobuoys, was 161 million roubles, or 936,000 roubles per submarine.

The IL-38 was very much a 'pilot's airplane' and popular with flight and ground crews alike for its excellent reliability and ease of maintenance. No fatal crashes occurred with the type in Soviet times, and only one aircraft was written off in a non-fatal accident – and even that was caused by pilot error. On 9th December 1987 the captain of a 77th OPLAP DD *May*, Capt V N Koshkin, decided the aircraft was accelerating too lazily and would not manage to become airborne. At 250km/h (135kts) he aborted the takeoff – too late. The IL-38 overran the runway at 120km/h (65kts), ran into a ditch more than 300m (984ft) beyond the threshold, collapsing the nose gear, and slithered forward relentlessly, sustaining damage to the main gear and propellers. No one was hurt but the aircraft was a write-off.

As noted earlier, the Berkoot STS and associated sonobuoys were the IL-38's primary means of target detection. For crews transitioning to the *May* from the Be-6 and Be-12 the advertised level of mission automation seemed just too good to be true. The crews' misgivings turned out to be justified: the low reliability of the TsVM-264 computer discovered at an early test stage was to plague the IL-38 throughout its service career. True, reliability *did* improve over the years but still the crews preferred to rely on their skills and experience when hunting submarines. The navigator took over the computer's work, the ARI operator became the 'ears' of the aircraft, while the pilots and radar operator (the latter worked the sonobuoy drop buttons) substituted the automatic control features. As a rule, only RGB-1 buoys were used; both the barrier and the 'field' tactics were used when placing the buoys.

The APM-60 MAD was used only as a secondary instrument, since its detection range was little more than 500m (1,640ft). Besides, it could not tell a submarine from a sunken ship; as a result, the MAD often gave false alarms in the shallow Baltic Sea littered with wrecks and other junk. Conversely, the main opponent of the Soviet Navy's ASW force in the Baltic were West German *Bundesmarine* Project 205 and 206 subs. Putting Second World War *Kriegsmarine* experience to good use, the Germans demagnetised the hulls of these relatively small diesel submarines so well that the APM-60 ignored them altogether – even when the sub was cruising in the surfaced position (as often witnessed by IL-38 crews)!

It was a different story in the northern part of the Indian Ocean where the IL-38 also saw service. The Indian Ocean is much deeper and any sunken ships would be resting far below 500m. Besides, here the *May* was up against US Navy nuclear-powered submarines; these were much larger and could not be demagnetised, so the MAD turned out to be quite useful.

The outrageously expensive and unreliable RGB-3 active sonobuoys were used extremely rarely, while the RGB-2 passive buoys were mostly used during weapons practice. Zeroing in on the target and dropping the weapons in manual mode was no easy task, so the crews had to rely on the Berkoot STS and hope the computer wouldn't break down. Using target bearings indicated by two RGB-2s, the computer pinpointed the submarine, calculated its heading and speed and hence the point where the bombs or torpedoes would be dropped.

Torpedo bombing practice in all Soviet Navy fleets took place at least once or twice a year. Since real operational submarines were usually used as practice targets (disposable old hulks weren't too many, after all), the *Mays* were armed with inert torpedoes. These differed from the AT-2 mainly in having a 50-cm (1-ft 7⅝-in) rubber bumper in lieu of a warhead so they wouldn't cause damage in the event of a direct hit. The torpedoes were set to travel at such a depth as to pass over the target; when they did so they rose to the surface automati-

'Mission accomplished' – a typical Soviet publicity shot. Yefim Gordon archive

Bottom: **This IL-38 (79 Red) seen at Pushkin sports an unusual blade aerial on the nose, plus an *Excellent Aircraft* badge – a maintenance award. The star tracker on the flightdeck roof is protected by a cover.** Yefim Gordon archive

cally, indicating that the target had been 'destroyed'. The training followed a real combat scenario as closely as possible, including the dropping of RGB-1 sonobuoys. The only big difference was that the IL-38 crews usually knew the submarine's diving time and location, depth, speed and heading, which allowed them to score a 'kill' even if the STS failed completely.

In August 1975 the Soviet Navy held the *Plyos* (River Pool) exercise in the White Sea. The exercise was of a test and development nature and consisted of several stages, one of which was intended to verify several models of anti-submarine munitions (albeit in somewhat simplified conditions). A retired diesel submarine was anchored at a depth of 40-50m (131-164ft); it was equipped with a noise generator imitating engine and propeller noises and a buoy providing a radar signature.

ASW cruisers were the first to try their luck, launching anti-submarine missiles which missed the target and self-destructed. Next, a pair of IL-38s appeared on the scene, each toting a single AT-2 torpedo. The Navy had serious doubts as to the usefulness of this weapon in the area, even though the depth was in excess of 100m (328ft). This was because the AT-2 had a few peculiarities, including a complex retarding system with two 0.6-m^2 (6.45ft^2) drogue parachutes and a 5.4-m^2 (58-ft^2) main brake parachute. The direction in which the torpedo entered the water and began its target search circulation could not be stabilised; besides, the centre of the torpedo's circular course was located 40-50m from the point of splashdown and completely unpredictably. All this made scoring a 'kill' a pretty chancy affair.

Secondly, the torpedo's guidance system worked in cycles (active/passive homing) with the active mode taking up some 35% of the time. If the level of extraneous noises exceeded a preset limit the system switched to passive mode; if a positive echo was received during circulation the system remained in active mode and guided the torpedo all the way in.

The sceptics turned out to be wrong: the AT-2 did better than predicted. Having got a good lock-on, the torpedo guided all the way in and impacted 1 min 40 sec after splashdown. The resulting explosion tore a 3-m^2 (32.25-ft^2) hole in the submarine's hull and the submarine sank. This was the only case when a live AT-2 ASW torpedo was used against a real submarine.

Much attention in the IL-38 crews' combat training programme was given to setting up minefields, carrying out practice bomb strikes against surface ships and using the oldest anti-submarine weapon – depth charges. The 250-kg (551-lb) PLAB-250-120 Lastochka depth charges packed quite a punch and were to be dropped in sticks, with each charge set at a different depth. The PLAB-250-120 could inflict heavy damage on a submarine at a distance of up to 10m (33ft) – unless, of course, the sub had time to take evasive action.

The *May*'s weapons range also included the KAB-500PL Zagon guided depth charge; like the AT-2 torpedo, this had an acoustic guidance system and featured small rudders on the nose which enabled it to follow a spiral trajectory after splashdown. When the system got a

target lock-on, a solid-fuel rocket motor was ignited and the depth charge headed towards the target like a torpedo. In reality, however, the KAB-500PL and the Orlan (Sea Eagle) rocket-powered torpedo were exotic weapons which most IL-38 crews had the faintest idea about.

In the event of a serious war threat the Soviet Navy placed high hopes on nuclear depth charges. Using these weapons required concerted action by a group of ASW aircraft. For instance, the Baltic Fleet's 145th OPLAE would enact the following plan. When a 'Red Alert' was declared three IL-38s would be configured for the search role with 216 RGB-1 sonobuoys each. Three other aircraft would have the search/strike configuration with typically 144 RGB-1 sonobuoys, ten RGB-2 buoys and one or two AT-2 torpedoes, while two more *Mays* would head for the weapons storage depots at Bykhov AB or Ostrov AB where they would be armed with one or two nuclear bombs each (the number depended on the weapon's model). Then the IL-38s would rendezvous in their assigned patrol sectors. Once the search aircraft had located a submarine the strike *Mays* would head for the target area and drop their weapons.

The nuclear munitions had a complex safety system and the detonators could only be armed by entering secret codes received en route from the General HQ in Moscow. No full-scale 'nuclear scenario' exercises are known to have taken place in the Soviet Navy; training for this scenario was limited to mission preparation (including engine starting) and, for the two bomb-toting aircraft, the flight to the storage depots where dummy nukes would be fitted. Speaking of which, the nuclear bombs were handled amid extremely tight security measures. The bomb would be carted to the aircraft carefully wrapped in tarpaulins and hoisted into position behind canvas screens. Of the IL-38's crew, only the captain, the navigator and the radar operator had the right to even look at the thing!

Generally the *May* was quite effective as compared to other Soviet ASW assets; for instance, the IL-38 and Be-12 were responsible for up to 80% of the 'enemy' submarines detected by the Soviet Navy in 1989. This high efficiency resulted mainly from intensive crew training (the crews logged up to 300 flight hours per year). Training and 'real' sorties were flown around the clock in any kind of weather; quite often mission time was 12 hours or more.

The *Okean* (Ocean; pronounced *oki**ahn***) exercise held in March-April 1970 became a demonstration of the IL-38's capabilities. The exercise was commanded by Soviet Navy C-in-C Fleet Admiral S G Gorshkov and was unprecedented in its scale, taking place in four oceans and fifteen seas. No fewer than 37 aviation regiments took part in the exercise, including 21 AVMF units, eight DA units and eight Air Defence Force (PVO – *Protivovozdooshnaya oborona*) units. The participating North Fleet Air Arm IL-38s had to operate in extremely adverse weather, and some of the watching top brass even voiced doubts as to the advisability of risking the crews, considering that this was just an exercise. In fact, not all crews were skilled enough to fly in such conditions and had to be assisted by Maj Sharapov's crew from the 33rd TsBP i PLS.

The mission was to hunt down a Blue Force[6] submarine cruising in the Norwegian Sea towards its assigned area of service. The initial search was conducted by setting up 'fields' of sonobuoys dropped by two aircraft on divergent headings. The actual conditions, however, made this tactic fruitless: there were fairly heavy seas (sea state 3 or 4) and the buoys switched to continuous emission mode because of the high noise level.

After some hesitation the commanders of the exercise decided to save face, ordering that the submarine be detected by means of the MAD. This was clearly not the brightest idea under the circumstances, yet two hours after the search had begun an IL-38 captained by the unit's CO managed to detect the sub and shadowed it for another hour and a half, making four not-too-stable contacts. However, another crew which was to take over the chase failed to find the target and the search was abandoned – officially because of deteriorating weather (low clouds and heavy icing).

Three days later the Navy General HQ assigned a missile submarine patrolling between the Lofoten Islands and Jan Mayen Island as the Blue Force sub. This time a field of 50 RGB-1 sonobuoys was set up (the AVMF HQ had 'tipped off' the IL-38 crews about the submarine's position in order to avoid squandering the buoys). The target was duly detected and shadowed for more than seven hours, with several *Mays* working in shifts; eventually the surveillance had to be abandoned because of poor weather, and this time it was true.

Generally the IL-38 made a favourable impression during the *Okean* exercise. In comparison, the Be-12 and Ka-25PL looked much less attractive, leading the AVMF top brass to urge accelerated upgrading of the *Mail* and development of a new-generation shipboard ASW helicopter (which emerged as the Ka-27PL *Helix-A*).

Maritime reconnaissance was also an important part of the IL-38's responsibilities; to this end some aircraft were retrofitted with the Vishnya COMINT system, as noted earlier, and the ARI operator training course amended accordingly. ELINT-configured IL-38s routinely shadowed NATO warships; for example, 77th OPLAP DD *Mays* from Nikolayevka AB prowled around Japan and 145th OPLAE aircraft were regular visitors to the region of the British Isles.

Besides operating from their regular bases, the *Mays* were often deployed to 'friendly nations' to extend their reach. In March 1968 the Soviet Union and the Arab Republic of Egypt signed an agreement enabling a detachment of six Soviet Navy Tu-16R *Badger-E/F* ELINT aircraft to operate from Egyptian bases, gathering intelligence for both nations. The *Badgers* were later joined by An-12BK-IS *Cub-C* ECM aircraft, Be-12s and IL-38s. The aircraft were flown by Soviet crews but wore Egyptian Air Force markings for appearance's sake, complete with four-digit Arabic serials as used by the EAF (eg, *Badger-Es* '4376' and '4381', *Badger-Fs* '4380' and '4384', An-12BK-IS '4371' and IL-38 '4299').

Originally the entire Soviet detachment – the 90th ODRAE ON (*otdel'naya dahl'nyaya razvedyvatel'naya aviaeskadrilya osobovo naznacheniya* – Independent Special Mission Long-Range Reconnaissance Squadron) was based at Cairo-West, a major EAF base; later some of the aircraft moved to Mersa Matruh AB. Starting in the autumn of 1970, the quasi-Egyptian IL-38s performed ELINT duties over the Mediterranean. On the pretext of filling the need for more sophisticated equipment the original pair of *Mays* was supplemented in June 1971 by two more aircraft, while the Be-12s returned to the USSR. The operating conditions were nothing out of the ordinary and the missions presented no problems; moreover, the crews were accommodated in comfortable lodges with amenities and facilities they never dreamed of back home.

No submarines were detected initially, but soon the crews grew bolder and started reporting encounters with subs increasingly more often. Makes you wonder how many of those reports were true: checking the claims outside the USSR was impossible.

Flying over the Med, especially at low altitude, did bring a few surprises after all. For instance, the crews reported that salt crystallised copiously on the flightdeck windows during prolonged flights over 'the briny' at 70-100m (229-328ft) and *an alcohol windscreen*

Several Soviet Navy IL-38s wearing Egyptian markings for appearance's sake, including this aircraft serialled 4299, operated from Egyptian bases in the mid-1970s. Yefim Gordon archive

flushing system was urgently required! It's easy to guess the windscreens would not be the only thing to be flushed!

In late 1972 relations between Moscow and Cairo soured, making the detachment's further stay in Egypt impossible, and all 90th ODRAE ON aircraft returned to the USSR. In the course of their operations from Mersa Matruh the *Mays* had 20 reported contacts with submarines. More foreign deployments followed in the early 1980s, involving IL-38 crews from all fleets. IL-38s were regularly deployed to Libyan and Syrian bases; among other things they kept an eye on US Navy battleships (USS *New Jersey* etc) taking part in the 1982 war in Lebanon.

From 9th January 1980 onwards Baltic Fleet/145th OPLAE IL-38s were deployed in South Yemen (People's Democratic Republic of Yemen). Originally they were based in Aden, patrolling the Arabian Sea and the northern part of the Indian Ocean; in 1983 they moved to the major airbase at El Anad. Two Soviet Navy IL-38s were stationed in South Yemen at any one time; the aircraft and crews were rotated on a bi-monthly basis. On the way from Riga the *Mays* made a single stop at Tashkent and thence staged to the Arabian Peninsula via Iranian territory. In 1985 the Pacific Fleet's 77th OPLAP DD relieved the Baltic airmen. (It has to be said here that the living conditions in South Yemen were horrendous as compared to the ones in Egypt.)

Sorties were flown every five or six days and always in pairs. The IL-38s were mainly tasked with locating and monitoring US Navy carrier groups. As a rule, the crews would be briefed from Moscow on the whereabouts of the ships to be shadowed which were usually heading for the Persian Gulf. The *Mays* tried to approach the target stealthily, flying at 100-150m (328-490ft) and maintaining radio silence. If the carrier could not be spotted within a 50 to 60-km (27 to 32-nm) radius, the aircraft would climb to 6,000-7,000m (19,685-22,965ft) and try to detect it by radar. The Berkoot radar could make out a large ship at 260-280km (140-151nm) range but could not tell an aircraft carrier from a supertanker – of which there are many in the area.

After getting the target on their radar scopes the IL-38s descended again and passed over the carrier group, photographing it. Then, if enough fuel remained, they would move out 60-80km (32-43nm) ahead of the group and try to detect the nuclear-powered submarine which invariably accompanied the group. The MAD was used for initial search, which meant the aircraft had to go down to 50m (164ft) or even 30m (98ft); the crew of the wingman could clearly see the lead aircraft leaving a wake on the sea like a speedboat. In this mode all flying was strictly manual; engaging the autopilot was strictly forbidden and the pilots were well aware of the consequences of an autopilot failure at this altitude. The *Mays* flew a shuttle pattern across the submarine's anticipated course; if the MAD gave a positive reading, four RGB-1 sonobuoys were often dropped in a cruciform pattern to determine the submarine's heading. Once the presence of a submarine had been confirmed, the crew reported to Moscow or to the Baltic Fleet HQ, which usually ordered them to return to base. The reason was that the cost of an RGB-1 was comparable to that of a colour TV and continuing the search could cost a pretty penny.

It should be noted that Soviet submarines were also active in the area and the orders to 'go home' after detecting a sub were not based on cost grounds alone. Quite simply, the submarine could turn out to be a 'friendly' one and the *Mays* could reveal its location to the Americans by circling over it long enough. Speaking of which, the IL-38's target identification capabilities were, putting it mildly, modest. Unlike the P-3's mission equipment, the Berkoot STS had no IFF function. Theoretically the ARI operator could identify the type of the sub by the noise it emitted (and thus tell the 'good guys' from the 'bad guys'), but this required an operator with excellent musical skills and such individuals were few in the Navy; most operators could no more than tell a diesel submarine from a nuclear-powered one. Still, the obligatory report to the higher command clarified the situation, since the HQ knew well enough where the Soviet submarines would be.

If ordered to continue the search, the crew would set up sonobuoy barriers. In doing so they never trusted the Berkoot; even if the STS worked properly at the moment, there was no guarantee that it would not fail in the next minute. If it did, the mission would go down the drain because the IL-38 had no precision navigation system; in a manner of speaking, only the aircraft knew where it had dropped the buoys and you couldn't ask where! Hence the crew relied on the old-fashioned map, logarithmic rule and stopwatch.

Since the submarine's course was by then more or less clear, two semi-circular barriers of RGB-1s would be set up consecutively, giving a much better idea of where it was heading as it crossed the barriers. For instance, while the first four buoys indicated the submarine's heading with an error margin of ±45°, the first barrier reduced the error to ±22° and the second barrier to ±10-11°. Working in this fashion required good crew co-operation and placed high demands on the navigator's skill; it took about five years for a navigator school graduate to reach the required skill level.

From 29th January 1981 onwards, 145th OPLAE IL-38s were also periodically deployed to Asmara, Ethiopia (or, to be precise, the Eritrea province which later gained independence), where the living conditions were even worse. From there they likewise flew reconnaissance sorties and patrolled the Arabian Sea. Aircraft of the 77th OPLAP DD also visited Asmara. One such temporary deployment in 1984 ended in disaster when Eritrean separatists overran the base and destroyed everything in sight, including two *Mays*.

The Soviet Union's burning desire to maintain a military presence in the Mediterranean (apart from the helicopter-carrying warships which the West did not seem to be taking seriously anymore) led to the signing in 1982 of an agreement with Libya enabling two IL-38s to operate from Maitiga AB. During this deployment the IL-38s flew only a few sorties, detecting one submarine with the help of the mission equipment and another visually (ie, catching it in the surfaced position). The Soviet Navy commanders were deluded into thinking the Libyan government would take the *Mays*' presence as a gift of fate and supply the fuel at a nominal price or even free of charge. However, they were in for a disappointment; Libya charged a neat US$ 180,000 for the fuel, board and lodging for the crews. Still, IL-38s were periodically deployed in Libya in later years.

The *May* was also used in the maritime SAR role. To this end a large teardrop-shaped KAS rescue capsule (*kas**se**ta avar**ee**no-spa**sah**tel'naya*), misidentified as a second radome by Western observers, could be carried under the forward weapons bay and paradropped to people in distress. For example, 77th OPLAP DD aircraft were called upon to assist a Soviet ship carrying lumber from the port of Nakhodka to Japan. They also participated in the huge search operation following the tragic incident on 1st September 1983 when a Korean Air Lines Boeing 747-230B registered HL7442 (c/n 20559, f/n 186) intruded into Soviet airspace and was shot down into the Sea of Okhotsk off Moneron Island by a Sukhoi Su-15TM *Flagon-F* interceptor, killing all on board.

North Fleet IL-38s participated in the failed attempt to save the crew of the nuclear-powered missile submarine SNS[7] *Komsomolets* (K-278) which suffered a catastrophic fire in the Norwegian Sea on 7th April 1989. Three and a half hours after the submarine's SOS had been received the *Mays* appeared on the scene. Having spotted the *Komsomolets* in the surfaced position and established radio contact with her, the IL-38s dropped rescue capsules and circled over the area, acting as communications relay aircraft and guiding surface ships towards the stricken sub. The capsules fell some 20m (66ft) from the submarine but the exhausted and frozen sailors were unable to use them. (The media published at least two other accounts of the incident based on the stories of the survivors, claiming that the capsules had fallen too far away or that the inflatable life rafts were within reach but had capsized, which made them unusable.) Speaking of which, the IL-38 had no optical sight and it was no wonder that the capsules often fell way off the mark.

In a more humorous vein, the IL-38 has sometimes been called a 'springboard' for C-in-Cs, generals and other big brass. There was some truth in this statement: V P Potapov who had commanded the very first *May* unit in the North Fleet became AVMF Commander in 1988. He flew the IL-38 a lot, even in the worst imaginable weather conditions (for example, he took part in the aforementioned *Okean* exercise), and contributed a lot to the development of operational tactics for the type. Baltic Fleet Air Arm Commander Lt Gen V P Proskoorkin had also commanded the same IL-38 unit in his time.

The intensity of IL-38 operations mounted steadily, reaching its peak by the early 1990s. The demise of the Soviet Union, however, changed things abruptly. Some of the *May*'s bases and training centres were located in Latvia and the Ukraine – ie, outside Russia which happened to be the only republic of the Soviet Disunion interested in the type.

The 145th OPLAE was the first to fall victim to this situation. In 1992 the unit vacated Riga-Skulte AB and was disbanded the following year; most of its aircraft were transferred to the 24th OPLAP DD at Severomorsk-1 AB and the 77th OPLAP DD at Nikolayevka AB. A few went to the newly-formed Russian Naval Aviation Training Centre at Ostrov AB some 40km (25 miles) from Pskov, ie, the 240th GvOSAP (*Gvar***dey***skiy ot***del'***nyy* ***smesh****annyy* ***a****via****polk*** – Guards Independent Composite Air Regiment). There were changes in the Pacific Fleet, too. After the Be-12s based at Yelizovo (a combined civil airport and military base near Petropavlovsk-Kamchatckiy) had been written off as time-expired, several 77th OPLAP DD IL-38s were permanently stationed there.

The Ukraine had no further use for its *Mays* and the aircraft based in Nikolayev and at Kirovskoye AB have mostly been scrapped. The example in Loogansk has had a happier fate. Although the local navigator school was closed in 1997, the IL-38 and the other ground instructional airframes were donated to a museum founded by the local aircraft overhaul plant; however, the museum is located in restricted territory and hence is not exactly crawling with visitors.

The prolonged economic crisis in Russia has had a devastating effect on the combat readiness of the Russian Naval Air Arm and its ASW component; at one time the entire arm was all but grounded for economic reasons (fuel shortages etc). The all-too-few flying hours and the resulting drop in proficiency were the most probable cause of the crash which occurred on 3rd February 1994. A North Fleet IL-38 captained by Maj V Goloschchapov was on short finals in poor weather after a routine training sortie. At 11:40 PM Moscow time the aircraft struck the ground with a wingtip and crashed, killing all seven on board, including senior inspector pilot Lt Col Ye V Roobtsov who was the co-pilot on this flight.

On the other hand, the much-improved relations with the NATO (that is, until Operation *Allied Force* in March-June 1999; it took a lot of time to rebuild them afterwards) and the Partnership for Peace programme have made possible things which would have been unthinkable just ten years ago – and, had they happened then, would have been regarded as high treason. Since the mid-1990s IL-38s have begun visiting airbases of the former 'potential adversary'. For instance, in 1995 a *May* coded '71 Red' (c/n 082011207) made an official visit to Elmendorf AFB, Alaska. The standard overall light grey colour scheme was livened up for the occasion, with a Russian double-headed eagle superimposed on a large Russian flag on the lower half of the fin and *Aviatsiya VMF Rossii/Russian Navy* titles on the forward fuselage. In the summer of 1996 another IL-38 coded '22 Red' (c/n 08*011006) participated in that year's Royal International Air Tattoo at RAF Fairford, one of the most prestigious military airshows. This aircraft sported different tail art in the form of a seagull carrying a fish in its beak on a blue background.

When Yevgeniy M Primakov was appointed Prime Minister of Russia, the Russian armed forces stepped up their activities somewhat. This also applied to the IL-38s which resumed their patrol sorties. Considerable importance was attached to the latter during the war in Yugoslavia which caused a cooling off in the relations between Russia and the NATO, with some Russian observers predicting a new Cold War. Predictably, the *Mays*' renewed activities promptly gave results; for instance, on 15th April and 14th May 1999 a 77th OPLAP DD aircraft captained by Lt Col Zemlyanov and based at Yelizovo detected a submarine near Kamchatka – presumably a US Navy submarine out on a reconnaissance mission.

On 28th July 1996 a grand naval parade was held on the Neva River in St Petersburg on occasion of the Russian Navy's 300th anniversary. The parade included an impressive flypast staged by the Air Force (the 76th Air Army commanded by Lt Gen Basov) and the Naval Air Arm. The first to appear, at noon, was a trio of Mi-8 *Hip* helicopters carrying the Russian state flag, Russian Air Force banner and Russian Navy banner. Next came a Tu-142 escorted by a pair of MiG-29 *Fulcrum* fighters, an IL-38 from Ostrov AB escorted by two Su-27 *Flanker-Bs*, then a pair of Be-12s (also from Ostrov) and an An-124 Ruslan (*Condor*) big lifter accompanied by two Su-27s from the Roosskiye Vityazi (Russian Knights) display team from Kubinka AB. These were followed by a pair of Tu-22M3 *Backfire-Cs* in echelon starboard formation piloted by regiment CO Col Anan'yev and deputy squadron leader Maj Ivanov. The display was concluded by four Su-27s from the *Novorossiyskiy* fighter regiment led by squadron leader Lt Col Spiridonov.

On 31st July 1998 the North Fleet held a major exercise led by Defence Minister Marshal Igor' Sergeyev. The exercise also involved several *Mays* which dropped depth charges on the place where a Blue Force submarine was supposed to be.

The IL-38 may be long in the tooth, having first flown 38 years ago, but it is no dead lion. The *May*'s service life (40,000 hours) is immense by Russian military aviation standards and the aircraft has an excellent reliability record; thus the IL-38 is guaranteed to soldier on for a few more years. Besides, as already mentioned, the Leninets Holding Co is developing upgrades for the type, including the Morskoy Zmey new-generation STS which will meet the latest demands; airframe changes and new engines are also under consideration. Still, it all depends on whether sufficient funding can be found.

This *May* coded '22 Red' is unusual in having the dielectric nose fairing and ventral radome painted white. Note the blue-tipped spinners.
Yefim Gordon

Chapter Seven

May We Have a Look Inside? The IL-38 in Detail

Type
Four-engined anti-submarine warfare aircraft. The airframe is of all-metal construction and is broadly similar in design to that of the IL-18. The crew of seven comprises the captain, co-pilot, navigator, flight engineer, radio operator, search radar operator and aircraft receiver/indicator operator.

Fuselage
Basically similar to that of the IL-18; the skin thickness varies from 1.2 to 2mm (0.047 to 0.078in). The duralumin reinforcement plates on the fuselage sides in the propellers' plane of rotation are attached between frames 10-12.

The *centre fuselage* (frames 3-56) is divided into two sections by a flat pressure bulkhead at fuselage frame 10. The front portion, together with the *forward fuselage*, makes up the pressurised crew section comprising the flightdeck and the aft-facing mission equipment operators' workstations; it also accommodates the radar set of the Berkoot search radar. Instead of the IL-18's glassfibre radome, frame A of the unpressurised nose fairing grafted onto the forward fuselage carries a two-piece nosecone, only the front portion of which is dielectric.

The flightdeck roof features an emergency exit measuring 0.55 x 0.8m (1ft 9⅝in x 2ft 7½in) offset to port between frames 4-6 to be used in the event of a wheels-up landing. An entry/escape hatch measuring 0.7 x 1m (2ft 3½in x 3ft 3⅜in) with a hydraulically-actuated forward-opening door (doubling as a slipstream deflector for bailing out) and a sloping chute, both of which incorporate recessed steps, is located on the forward fuselage underside between frames 8-10; in flight the upper end of the chute is closed by twin doors forming part of the crew section floor. Immediately ahead of the entry/escape hatch (frames 3-7) is a faired mounting ring for the revolving antenna of the search radar and associated radome. Two full-size windows of 400mm (1ft 3¾in) diameter are located on each side of the forward fuselage aft of the flightdeck between frames 5-6 and 7-8 to provide natural lighting for the mission equipment operators; the rear pair is placed higher up.

The rest of the fuselage is unpressurised, accommodating the weapons and systems, and is accessible from the crew section via two doors in the rear pressure bulkhead leading to walkways running along the sides. Access is possible in flight after the pressure has been equalised. Two weapons bays closed by double doors are located in the centre fuselage fore and aft of the wing torsion box carry-through structure between frames 10-19 and 27-40; the doors are hydraulically actuated and may be manually opened 180° on the ground for maintenance. The space between frames 12-17 above the No 1 weapons bay is occupied by two fuel tanks. Five smaller windows of 300mm (11¾in) diameter are located on each side of the centre fuselage. The port side features an oval emergency exit (incorporating the No 3 window) located over the port wing between frames 24-25 and a life raft bay closed by a downward-opening cover aft of the wing trailing edge (and the No 5 window) between frames 32-34. A dorsal spine fairing made of glassfibre runs along the fuselage from frame 31 all the way to the fin fillet.

The *rear fuselage* is almost identical to that of the IL-18, except that it terminates in a tapered boom 5.59m (18ft 4in) long attached at frame 72 and featuring 17 frames; it carries the MAD sensor enclosed by a glassfibre fairing. The former rear baggage compartment door measuring 1.28 x 0.9m (4ft 2⅜in x 2ft 11½in) to starboard between frames 58-61 is retained, acting as a service door during maintenance and as a second emergency exit.

Wings
As for the IL-18, except that the wing torsion box is moved 3m (9ft 10⅛in) forward, the spars being attached to centre fuselage mainframes 20, 23 and 26.

Tail Unit As for the IL-18.

Landing Gear
As for the IL-18, except that the wheelbase is shortened from 12.755m (41ft 10⅛in) to 9.755m (32ft 0in).

Powerplant
As for the IL-18D – ie, four Ivchenko AI-20M (AI-20 Srs VI) turboprops with a take-off rating of 4,250ehp and a cruise rating of 2,700ehp driving AV-68I propellers – except that the APU is installed on the port side of the centre fuselage between frames 41-46 (the exhaust is located between frames 46-47 and surrounded by a heat-resistant steel plate).

Control system
As for the IL-18, except for some changes in the control runs.

Fuel System

25 fuel tanks with a total capacity of 33,820 litres (7,440 Imp gals). Integral tanks are housed in the centre fuselage above the forward weapons bay, in the outer wing torsion boxes and the wing centre section; the inner wings accommodate bag-type tanks (fuel cells). The fuel system automatically maintains CG position as fuel is burned off.

The IL-38 has single-point pressure refuelling. Fuel grades used are Russian T-1, TS-1 or T-2 jet fuel.

An inert gas pressurisation system is provided to pressurise the fuel tanks and reduce the hazard of explosion if hit by enemy fire. Nitrogen for the system is supplied by four OSU-5 bottles installed on the starboard side near the rear entry door (frames 55-57).

Electrics

The electric system serves for engine starting and operates the avionics, mission equipment, part of the de-icing system, fuel system components etc.

Main DC power (27V) is supplied by eight engine-driven STG-12TMO-1000 starter-generators; backup DC power is provided by four 12SAM-28 batteries. 115V/400Hz single-phase AC is supplied by four SGO-12 generators (three main units and one back-up); the AC circuits feature AZP-1SD automatic overload protection devices (*avto**maht** zas**chchit**y ot pere**grooz**ki*). On the ground electric power is provided by the APU which drives a GS-24A starter-generator; a ground power receptacle is provided.

Hydraulics

The hydraulic system operates the landing gear, nosewheel steering mechanism, wheel brakes, weapons bay doors, propeller feathering actuators and windscreen wipers. It features two NP-25-5 pumps driven by the inboard engines, two hydraulic reservoirs for the general system and two separate hydraulic reservoirs for the wheel brakes. The system uses AMG-10 oil-type hydraulic fluid; nominal pressure is 210kg/cm² (3,000psi).

Nitrogen System As for the IL-18.

De-icing System

As for the IL-18. Additionally, the windscreen features an alcohol spraying system to wash away the salt building up during overwater operations at 30-50m (100-165ft).

Photographs on the opposite page:

The nose of a standard IL-38 awaiting overhaul at the Russian Navy's ARZ No 20. Note the 'Excellent aircraft' badge. Yefim Gordon

Close-up of the radome of the Berkoot search radar. Note the open entry hatch immediately aft and the mudguard on the nosewheels protecting the radome from damage by loose stones and the like. Yefim Gordon archive

Oxygen System

The oxygen system ensures crew survival in the event of decompression at high altitude and permits entry into the unpressurised centre/rear fuselage in flight. Gaseous oxygen is stored in 19 KB-1 oxygen bottles (*kislo**rod**nyy bal**lon***) installed between frames 26-30 and 53-55, each holding 36 litres (7.92 Imp gals), and a 7.6-litre (1.67-Imp gal) portable bottle. KP-23 breathing apparatus with an 11 minutes' supply of oxygen are provided for the crew to ensure survival in the event of bailing out.

Fire Suppression System

Three groups of fire extinguisher bottles charged with carbon dioxide for each engine. The first shot is triggered automatically by flame sensors in the engine nacelles; the second and third shots are fired manually. A separate fire extinguisher is provided for the APU.

Air Conditioning & Pressurisation System

The crew section is pressurised by engine bleed air. Pressurisation air is cooled by heat exchangers located in the wing roots. A mobile air conditioning unit may be connected to the aircraft on the ground.

Armament

The IL-38 can carry AT-1 and AT-2 anti-submarine torpedoes, PLAB-250-120 Lastochka, PLAB-50 and KAB-500PL Zagon depth charges, mines and a variety of free-fall bombs. Nuclear depth charges can also be carried. RGB-1, RGB-2 and RGB-3 sonobuoys are carried for detecting submarines and homing in on them. About 30 payload combinations are possible; the maximum payload, including sonobuoys, is 8,000kg (17,640 lb).

Avionics and Equipment

The IL-38 is fully equipped for all-weather day/night operation, including automatic flight assisted by an autopilot.

Navigation and piloting equipment: The navigation suite includes an RSBN-2S Svod (Dome) short-range radio navigation system with flush antennas built into the fin, a DISS-1 Doppler speed/drift sensor system with flush circular antennas under the wing centre section, an SP-50 Materik instrument landing system, an RV-4 radio altimeter, an ARK-11 automatic direction finder with a ventral strake aerial under the wing centre section offset to starboard etc. An astro-sextant star tracker is mounted on the flight-deck roof between frames 4-5.

Communications equipment: R-847A and R-836 communications radios, a Peleng (Bearing) HF comms radio, and R-802V and R-632 command radios, served by dorsal and ventral blade aerials on the forward and centre fuselage. An SPU-7B intercom and an SGU-15 loudspeaker system are provided for communication between crew members.

IFF system: SRO-2 or SRO-2M Khrom IFF transponder. The IFF aerials are located ahead of the flightdeck glazing, under the aft fuselage.

Electronic support measures (ESM) equipment: Sirena-3 radar homing and warning system (RHAWS) with aerials on the forward/aft fuselage sides and wingtips. Alternatively, some upgraded Indian Navy IL-38s have two pairs of streamlined ESM/ECM fairings on the sides of the extreme nose, at the root of the MAD boom.

Mission equipment: Berkoot search and targeting system (STS) built around a 360° search radar of the same name with a secondary weather reconnaissance function; the radar is installed in a large quasi-spherical glassfibre radome aft of the nose gear unit. APM-60 magnetic anomaly detector mounted on a boom on the rear fuselage. Some IL-38s were equipped with the Vishnya communications intelligence system with four pairs of ESM or active jamming antennas mounted ahead and aft of the wings, with external wiring conduits for the two forward pairs.

Data recording equipment: Standard Soviet MSRP-12-96 primary FDR, K-3-63 backup FDR and MS-61B CVR.

Exterior lighting: Port (red) and starboard (green) navigation lights at the wingtips; white tail navigation light under the root of the MAD boom. Retractable landing/taxi lights on the sides of the nose and under the wingtips. Red SMI-2 anti-collision strobe lights under the rear fuselage (frame 49) and at the top of the fin. Three EKSP-39 signal flare launchers on the starboard side of the aft fuselage between frames 43-46.

Survival Equipment

A PSN-6A six-man life raft in a bay just aft of the port wing. The crew is provided with S-5 parachutes, MLAS-1-OB one-man inflatable life rafts and MSK-3M maritime rescue suits. Each parachute pack features an NAZ-7 survival kit.

IL-38 Specifications

Wing span, m (ft)	37.4m	(122' 8½")
Length overall, m (ft)	40.1m	(131' 6¾")
Height on ground, m (ft)	10.1m	(33' 1⅝")
Wing area, m² (ft²)	140.0	(1,505.37)
MTOW, kg (lb)	66,000	(145,500)
Max landing weight, kg (lb)	52,000	(11,643)
Maximum payload, kg (lb)	8,400	(18,520)
Fuel capacity, litres (Imp gals)	33,820	(7,440.4)
Fuel load, kg (lb)	26,550	(58,530)
Top speed, km/h (kts)		
at 6,000m (19,685ft)	650	(351)
Min loiter speed, km/h (kts)	350	(189)
Range, km (nm)	9,500	(5,135)
Combat radius, km (nm)		
with 3 hours' on-station loiter	2,200	(1,189)
Service ceiling, m (ft)	10,000	(32, 810)
Endurance		
(incl flight from/to base)	12 hours	
Take-off run, m (ft)	1,700	(5,580)
Landing run, m (ft)	1,070	(3,510)
Ordnance load, kg (lb)	8,400	(18,520)
Crew	7	

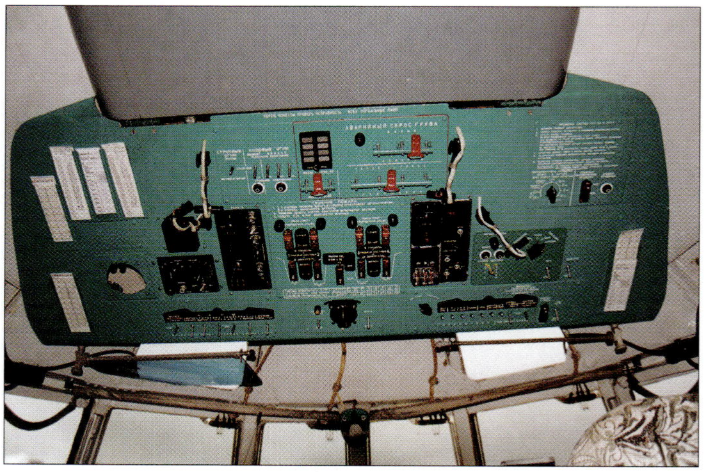

Left, top to bottom: **The flightdeck of IL-38 '25 Red' (c/n 081010907) at Pushkin in mid-2001. The aircraft has conventional electromechanical flight instruments and the centre panel is occupied by engine control instruments. Note then nosewheel steering handwheel on the captain's control column and the dished seat pan for the parachute.** Dmitriy Komissarov

Above top to bottom: **The captain's side console of IL-38 '25 Red', with the rubber-bladed cooling fan typical of Soviet heavy aircraft. The cover inscribed 'Keep closed in flight' conceals the engine starting panel.** Dmitriy Komissarov

The co-pilot's side console of IL-38 '25 Red' with de-icing system and pressurisation system controls and the like. Dmitriy Komissarov

The overhead circuit breaker panel of IL-38 '25 Red'. Among other things, it has guarded switches for emergency warload jettisoning. Dmitriy Komissarov

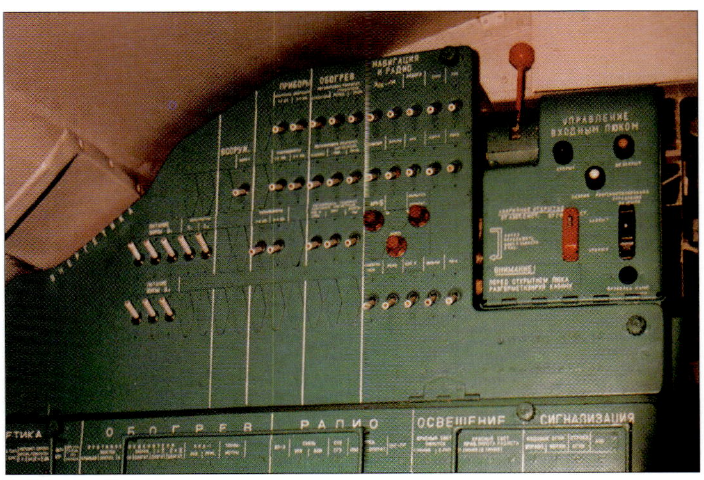

Above, top to bottom:

The navigator's port side workstation of IL-38 '25 Red'. Dmitriy Komissarov

The radio operator's workstation of IL-38 '25 Red' on the starboard side. Dmitriy Komissarov

The rear bulkhead of the IL-38's flightdeck is occupied by circuit breaker panels. Dmitriy Komissarov

Right, top and centre: **The aft-facing workstations of the search radar operator; and the operator of the aircraft receiver/indicator on IL-38 '25 Red', showing the four circular cathode-ray tube displays.** Dmitriy Komissarov

Right, bottom: **The antenna of the Berkoot radar with the radome removed. The stencil on the radar reads *Nastroyeno – ne trogat'* (Tuned – Do not touch).** Yefim Gordon archive

Post-Soviet *Coots* at Home (CIS Operators)

After the break-up of the Soviet Union and the dissolution of the sole national carrier into hundreds of new airlines, ex-Aeroflot and ex-Soviet Air Force IL-18s found themselves in several of the new CIS republics, finding new owners. For each republic, operators are listed in alphabetical order, with each airline's two-letter International Air Transport Association (IATA) designator and three-letter International Civil Aviation Organisation (ICAO) designator where applicable. Aircraft no longer operated by the respective carrier are shown in italic script in the fleet lists (except in cases when the airline itself no longer exists – or when *all* of its IL-18s have been sold or retired).

BELARUS' (BELORUSSIA)

The **BELORUSSIAN AIR FORCE** inherited a single IL-22M-11 airborne command post, CCCP-75916 (c/n 2964017102), which was assigned to the Belorussian Defence District. As in Soviet times, the aircraft was operated by Squadron 1 of the 50th OSAP at Machoolischchi AB near Minsk in full Aeroflot colours; for some reason it never received the Belorussian EW- prefix. In June 1998 CCCP-75916 was sold to the Latvian carrier Concors Airlines and demilitarised, receiving the registration YL-LAO.

KAZAKHSTAN

Whereas the former Kazakh Civil Aviation Directorate had two united air detachments operating IL-18s, none of these aircraft survived long enough to see the break-up of the Soviet Union. All *Coots* operated in the CIS republic of Kazakhstan (or Kazakstan, to use the indigenous name) were of foreign provenance.

In the summer of 2000 the Kazakh flag carrier **AIR KAZAKSTAN** [9Y/KZK] operated IL-18E 3C-KKK (c/n 186009202) on lease from Air Cess. The aircraft was returned to the lessor in October 2000.

AVIATRACK [–/KAK], an airline based in Almaty, occasionally leased IL-18D EL-AKQ (c/n 187010204) from Air Cess, starting in the autumn of 1996. The leases continued after the aircraft was transferred to the Swazi register as 3D-SBQ in 1998 and then to the Central African register as 3C-KKL; the last known lease ended in February 1999.

In 1998 **GVG AIR COMPANY** [–/GDR], another airline from Almaty, purchased IL-18Gr (IL-18V) YR-IMF (c/n 184007105) from Alfa Line. At first the aircraft retained its former Romanian registration, but on 4th June 1998 it was entered into the Kazakh register as UN 75111; the registration was somewhat unusual, as the 751xx registration block was not used in Soviet times.

Initially UN 75111 sported basic TAROM colours (a white fuselage and a dark blue tail) with a new tail logo and red/blue AIR *GVG* COMPANY titles. In June 2001 it was leased to the Russian carrier Tret'yakovo Airlines and repainted in basic Aeroflot colours with Tret'yakovo titles/logo. Its fate following the demise of the lessee is as yet unknown.

The **KAZAKH AIR FORCE** operated a single IL-22M-11 'Zebra' airborne command post (UN-75915,[1] c/n 2964017101) left over from the Central Asian Defence District. Since the republic had no real need for an ABCP, the aircraft was converted to a luxurious VIP configuration, gaining the full livery of the then national flag carrier, **KAZAKHSTAN AIRLINES (KAZAKHSTAN AUE ZHOLY)** or **KAZAIR** [K4/KZA], for the sake of convenience. Unfortunately, just a few days after the extensive (and expensive!) refit had been completed the aircraft was damaged beyond repair in a ground collision at Pavlodar in January 1995.

In September or October 2002 an airline called **YUZHNAYA AIR-COMPANY LTD** [–/UGN], also based in Almaty, obtained an IL-18D which received the unusual registration UN 75001 (ex-Romavia YR-IMM, c/n 187009904); again, the 750xx registration block was not used in Soviet times.

An unknown airline(s) operated IL-18E UN 75002 (ex-Air Cess 3C-KKR, c/n 185008603), UN 75003, UN 75004 and UN 75005 (ex-Air Cess 3C-KKL, c/n 187010204) in December 2002/January 2003. The former aircraft is all-white without titles, whereas UN 75003 and UN 75004 retain basic Air Cess colours; this means they are ex-3C-KKJ/3C-KKK (c/ns 184006903 and 186009202) but it is not known which one is which.

KYRGHYZSTAN

Again, IL-18s had been in service with the Kirghiz CAD in Soviet days but all of them had been retired by the time the Kirghiz SSR became the Republic of Kyrghyzstan, and all *Coots* now registered in the republic are imported ones.

AEROVISTA AIRLINES [–/AAP] based at Sharjah, UAE, operated IL-18V ER-ICM (c/n 182004804) jointly with Star Airlines (see next chapter/Somalian section) in late 2001.

PHOENIX AVIATION [–/PHG] operated ten IL-18s, mostly in passenger or combi configuration. It should be noted that the aircraft are based at Sharjah, United Arab Emirates, and the airline itself was until recently listed in reference books as a UAE carrier, being a division of the Phoenix Free Zone Enterprise. However, the 2002 edition of the *JP Airline-Fleets International* yearbook lists Phoenix Aviation in the Kyrghyz section. Some of the airline's IL-18s have non-standard three-digit registrations related to the c/n.

Registration	*Version*	*C/n*	*Notes*
EX-105	IL-18D	188011105	Ex-Nadym Tyumen' Airlines RA-74267, bought 1999; 102-seater
EX-201	IL-18D	188011201	Ex-Nadym Tyumen' Airlines RA-74268, bought 1999; 102-seater
EX-405	IL-18V	184007405	Ex-BIO Air Company/Phoenix T9-ABB, reregistered ?-01; combi configuration
EX-601	IL-18E	185008601?	Ex-Santa Cruz Imperial EL-ALD?; 110-seater
EX-75427*	IL-18V	183005905	Ex-Inter Tropical Airlines LZ-BFU, bought 10-00; leased to Daallo Airlines; 105-seater
EX-75442	IL-18D	187009702	Ex-Ramair RA-75442, converted IL-18D 'Tsiklon'; bought early 1999; combi configuration
EX-75449	IL-18D	187010004	Ex-Ramair RA-75466, converted IL-24N; bought early 1999; combi configuration

EX-75466	IL-18D	187010403	Ex-Ramair RA-75466, converted IL-24N; bought early 1999. Converted to IL-18GrM at Sharjah 1/3-3-01, received cargo door of crashed IL-18V/F (SCD) RA-75554
EX-75825	IL-18V	182004904	Ex-Santa Cruz Imperial EL-ALW, bought ?-99; ex-RA-75825. Reregistered ?-01 as, see next line
EX-904			Combi configuration. Damaged beyond repair Neghazi, Angola, 15-9-02?
EX-75905†	IL-18E?	186008905?	Ex-Bulgarian Airlines LZ-ZAH? (c/n unconfirmed), bought 2000; combi configuration

* LZ-BFU started life back in February 1963 as CCCP-75870, so logically it should have become EX-75870, not EX-75427.

† Most of the CIS republics have chosen to retain the Soviet 1959-standard five-digit numeric registration format. Unfortunately, however, violations of the rule that aircraft in different CIS republics cannot have the same registration digits (with different nationality prefixes) at the same time occur with annoying regularity. Thus the registration EX-75905 was allocated despite the fact that an IL-22 registered CCCP-75905 (now RA-75905, see Russian Air Force) existed since 1985!

Additionally, Phoenix Aviation leases the aircraft to other airlines as required. Thus, EX-75442 was on short-term leases first with Sudan Airways in 2000 and then with the Nigerian airline Fresh Air in 2001.

An IL-18V (ex-Santa Cruz Imperial EL-ADY No 2, c/n 182004804) was placed on the Kyrghyzstani civil register in 1999 with the non-standard registration EX-7504. Despite this, it was operated by the Somalian carrier Star Airlines. At the end of the year the aircraft was transferred to the Moldovan register as ER-ICM.

An IL-18V registered EX-011 and operated by an unknown airline was noted at Sharjah in February 2003. Judging by the colour scheme, it is very probably the former ER-ICM (c/n 182004804).

MOLDAVIA (MOLDOVA)

AIR MOLDOVA [6U/MLD], the Moldovan flag carrier, leased IL-22M-11 ER-75929 from Vichi Airlines (see below) in 1993. The aircraft wore basic Aeroflot colours with large red AIR MOLDOVA titles and the national flag on the tail.

RENAN [–/RAN], an airline based in Kishinyov (Chişinău), operated three ex-TAROM IL-18s. Depending on the seasonal demand they flew in 109-seat or cargo configuration.

Registration	*Version*	*C/n*	*Notes*
ER-ICG	IL-18V	184007301	Ex-YR-MG, bought 3-98
ER-ICJ	*IL-18E*	186009102	Ex-YR-MJ, bought ?-97. Was operated illegally for West African Air Services 7/8-2000 as EL-ALY. Sold to Tavria-MAC late 2000 as UR-TMD (see next line!); bought back by 4-03
ER-ICL	*IL-18D*	*187009903*	*Ex-YR-JML, bought 3-98. Sold to unknown operator by 12-00 (reported in error as to UR-TMD!)*
ER-ICM	IL-18V	182004804	Bought 1-2000; ex-Phoenix Aviation EX-7504, ex-Santa Cruz Imperial EL-ADY No 2, ex-Air Pass 3D-ALQ, ex-TAROM YR-IMD. Sold to unknown operator by 3-03 as EX-011?

The **MOLDOVAN AIR FORCE** (FARM – *Forţele Aeriene de Republica Moldova*), operated a single IL-22M-11 (ER-75929, ex-CCCP-75929, ex-IL-18D CCCP-74251, c/n 187010505) inherited from the Carpathian Defence District after the break-up of the USSR. Since Moldova had no use for an ABCP, the aircraft was promptly transferred to **VICHI AIRLINES** [–/VIH], the commercial division of FARM, and converted into passenger/cargo configuration, shedding the aerials on the fuselage and the ventral fairing characteristic of the *Coot-B* (the fin top 'cigar' and TA-6A APU remained, of course). In 1995-97 Vichi leased the aircraft to the Romanian carrier Acvila Air.

RUSSIA

The **AIR TRANSPORT SCHOOL** (*Shkola vozdooshnovo trahnsporta*) [–/AIS] operated IL-18D RA-74296 (c/n 188010603) from Zhukovskiy in July 1994. The aircraft wore basic Aeroflot colours without titles, lacking the company logo composed of the Cyrillic letters ShVT. On 28th December 1994 it was transferred to Tret'yakovo Airlines.

ATLANT-SOYUZ AIRLINES [3G/AYZ] established on 8th June 1993 operate scheduled and charter passenger and cargo services from Moscow (Sheremet'yevo-2 and Domodedovo) and Chkalovskaya AB, mostly with aircraft leased from other carriers and the Russian, Belorussian and Ukrainian air forces as required. In 1995 these included two IL-18D 'Salons' leased from the 223rd Flight Detachment State Airline resident at Chkalovskaya AB – RA-74295 (c/n 187010602) and RA-75478 (c/n 189011302). The *Coots* operated in full Aeroflot colours.

The now-defunct charter carrier **ALAK** [J4/LSV] based at Moscow-Vnukovo purchased a single IL-18E registered RA-75850[2] (ex-75711, ex-LOT Polish Airlines SP-LSK, c/n 185008503) in 1994; two years later it was sold to the Ukrainian airline Kryla as UR-75850.[3] The name stands for *aktsionernaya lizingovaya aviakompahniya* (joint-stock leasing airline).

The **BYKOVO AIR SERVICES COMPANY** (BASCO – ie, ARZ No 402) reportedly owned IL-18B RA-75703 (c/n 189001505), a former avionics testbed bought from NPO Vzlyot after 1992. However, the aircraft, which retained basic Aeroflot colours without titles, does not appear to have moved since 1994.

The all-Il'yushin fleet of the **DOMODEDOVO CIVIL AVIATION PRODUCTION ASSOCIATION** [E3/DMO] included quite a few IL-18s belonging to the 212th Flight, but only three of them survived long enough to see the break-up of the USSR and the ensuing deregulation. In 1998 the Domodedovo CAPA was rebranded **DOMODEDOVO AIRLINES/DOMODEDOVSKIYE AVIALINIÏ**, but none of the IL-18s received the new titles.

Registration	*Version*	*C/n*	*Notes*
RA-74267	IL-18D	188011105	Ex-IL-18D-GAL geophysical survey aircraft; blue cheatline with red pinstripe. Sold to Nadym Tyumen Airlines ?-99
RA-74268	IL-18D	188011201	Ex-IL-18DORR; blue cheatline with red pinstripe. Sold to Nadym Tyumen Airlines ?-99
RA-75462	IL-18D	187010304	Ex-IL-18DORR; basic Aeroflot Polar c/s. Registration cancelled 1-3-96; derelict Moscow-Domodedovo

ELF AIR [E6/EFR],[4] the commercial flying division of avionics designer NPO Vzlyot, operates a mixed bag of passenger and cargo aircraft from Zhukovskiy, including four IL-18Vs.

Registration	*Version*	*C/n*	*Notes*
RA-75431	*IL-18V*	*180002003*	*Ex-geophysical survey aircraft; transferred from LII 23-8-94. Basic Aeroflot c/s, no titles. Sold to Compagnie Africaine d'Aviation, DRC, as 9Q-CHB 9-00*
RA-75811	IL-18Gr	182004504	Converted IL-18E, bought from Ilavia late 1997 or 1-98. Basic Aeroflot c/s, Elf Air titles
RA-75851	IL-18Gr	182005501	Converted IL-18V, ex-avionics testbed. Basic Aeroflot c/s, no titles. Leased to IRS Aero
RA-75894	IL-18V	182004801	Avionics testbed, full Aeroflot c/s

In 2003 Elf Air was reportedly renamed *Aviakom**pah**niya **im**eni Grizo**doo**bovoy* (Grizodoobova Aviation Co) in honour of the female test pilot Valentina S Grizodoobova.

The overwhelming mix of test and development aircraft used in the multifarious activities of the **FLIGHT RESEARCH INSTITUTE NAMED AFTER MIKHAIL M GROMOV (LII)** in Zhukovskiy included several IL-18s. Some aircraft used in Soviet times are included for the sake of completeness.

Registration	*Version*	*C/n*	*Notes*
CCCP-Л5821	IL-18LL	188000201	Converted IL-18A, de-icing systems testbed. Reregistered CCCP-75637, data link systems testbed
CCCP-27220	IL-18SIP	188000401	Converted IL-18A CCCP-75647. Retired by 8-88, derelict Zhukovskiy; scrapped by 1995
RA-75423	IL-18V	182005601	Environmental survey aircraft. Reconverted to standard and sold to IRS Aero ?-97
CCCP-75743	IL-18V	181002901	Ex-Polar version. Derelict Zhukovskiy by 8-92; scrapped
CCCP-75884	IL-18V	183006202	Derelict Zhukovskiy by 8-92
RA-78732	IL-18V	181004103	Ex-CCCP-78732/avionics testbed, ex-CCCP-75794. WFU Zhukovskiy by 8-01

GOSNII GA (the State Civil Aviation Research Institute) [–/ISP] at Moscow/Sheremet'yevo-1 used seven IL-18s for research purposes.

Registration	*Version*	*C/n*	*Notes*
CCCP-74267	IL-18D-GAL	188011105	Geophysical survey aircraft; Polar c/s. Reconverted to standard and sold to Domodedovo CAPA as RA-74267
CCCP-74268	IL-18DORR	188011201	Polar c/s. Reconverted to standard IL-18D and sold to Domodedovo CAPA as RA-74268
RA-75442	IL-18D 'Tsiklon'	187009702	Weather research aircraft. Reconverted to standard and leased to Ramair 20-7-97. Sold to Phoenix Aviation as EX-75442
RA-75449	IL-24N	187010004	Reconverted to IL-18D and leased to Ramair 1-1-97
CCCP-75462	IL-18DORR	187010304	Polar c/s. Reconverted to standard IL-18D and sold to Domodedovo CAPA as RA-75462
RA-75466	IL-24N	187010403	Reconverted to IL-18D and leased to Air Transport Office 9-93. Sold to Phoenix Aviation as EX-75466
RA-75598	IL-18E 'Meteor'	186008802	Weather research aircraft; converted to geophysical survey aircraft. Reconverted to standard and leased to IRS Aero 9-4-97

ILAVIA [–/ILV], the Il'yushin Design Bureau's 'house airline' established in 1994, operated two IL-18Vs – RA-75811 (c/n 182004504) and RA-75834 (c/n 182005104). Both aircraft gained the airline's smart red/white colour scheme in 1996 and 1997 respectively. By early 1998, however, RA-75811 had been sold to Elf Air and converted to an IL-18Gr, while RA-75834 was stripped of Ilavia titles/logos by August 1999 and sold to Titan Aero in October.

Established in 1997, **IRS AERO** [LD/LDF, later 5R/LDF] operated a mix of own and leased aircraft, including four IL-18s. Despite the fact that the airline's registered office was in Moscow, persistent rumours circulated that IRS Aero was in fact owned by the Chechens (with obvious negative implications).

Registration	*Version*	*C/n*	*Notes*
RA-75423	IL-18V	182005601	Full c/s. Bought from the Gromov Flight Research Institute ?-00, ex-environmental survey aircraft; combi configuration
RA-75598	IL-18E	186008802	Full c/s. Leased from GosNII GA 9-4-97, ex-IL-18E 'Meteor' weather research aircraft. Type incorrectly painted on as IL-18D
RA-75840	IL-18V	182005301	Full c/s. Bought from the Russian Navy, ex-IL-18RT; D/D 2-10-98. Crashed 15km SE of Kalyazin 19-11-01
RA-75851	IL-18Gr	182005501	Basic Aeroflot c/s, no titles. Modified IL-18V, leased from Elf Air ?-00

The crash of RA-75840 and the ensuing investigation, during which IRS Aero's operational standards were placed under close scrutiny, was a sort of fate's warning – which the airline did not heed. On 15th September 2002 IL-18V RA-75423 overran when landing at Moscow-Bykovo; the investigation that followed unearthed several grave breaches (mostly forged paperwork), which eventually led the Russian CAA to withdraw IRS Aero's operating licence later that year. The fate of the aircraft remains unknown.

The now-defunct charter airline **KORSAR** [6K/KRS] based at Moscow/Vnukovo-3 reportedly operated all three IL-18Ds remaining in service with the Russian Government flight (GTK Rossiya) – RA-75453, RA-75454 and RA-75464 (see next page). *Kor**sar*** means 'corsair' in Russian, but the airline's name had nothing to do with pirates; it was derived from the names of its founders, Korovin and Sarzhveladze.

NPP MIR [–/NPP], the commercial flying division of the Leninets Holding Co (NPP = *na**ooch**no-proiz**vod**stvennoye predpri**yah**tiye* – scientific & production enterprise), operated four former SL-18 avionics testbeds from Pushkin. By the time of their official transfer to the new airline three of the four aircraft (which had already been performing cargo flights for some time while outwardly retaining their respective SL configurations) had been completely – or almost completely – stripped of the non-standard external fairings.

Registration	*Version*	*C/n*	*Notes*
RA-75411	IL-18E	186009205	Ex-SL-18P. Transferred 24-12-95; Korshoon radar mounting initially retained but later removed. Full c/s
RA-75713	IL-18D	186009403	Ex-SL-18D. Transferred 16-6-95; full c/s with large Cyrillic 'Olimpiada-2004' titles added. Reportedly became 75713 (no prefix) in 1999
RA-75786	SL-18V	181003905	Converted IL-18V. Never received NPP Mir titles; WFU Pushkin by 8-01 as 75786 (no prefix), still with Tu-160 radar nose and APM-60 MAD in tail
RA-75804	IL-18V	182004305	Ex-SL-18I. Transferred 24-4-95; full c/s, combined SLAR/search radar radome and dorsal fairing removed but 'droop snoot' nose radome still in place

The airline **RAMAER** or **RAM AIR** [–/RMY] based at Chelyabinsk in Russia and Sharjah (UAE) operated four IL-18s.

Registration	*Version*	*C/n*	*Notes*
RA-75442	IL-18D	187009702	Ex-IL-18D 'Tsiklon' weather research aircraft. Lsf GosNII GA 20-7-97 to 31-12-98; basic Aeroflot c/s with red Cyrillic 'Ramaer' titles; later full c/s with blue 'Ram Air' titles. Later sold by GosNII GA to Phoenix Aviation as EX-75442 in Ramair c/s
RA-75449	IL-18D	187010004	Ex-IL-24N. Leased from GosNII GA 1-1-97, basic Aeroflot c/s with red Cyrillic 'Ramaer' titles; was subleased to Daallo Airlines; returned by 6-98
RA-75466	IL-18D	187010403	Ex-IL-24N. Bought from GosNII GA 7-97; Domodedovo CAPA-style cheatline, all-blue tail, red Cyrillic 'Ramaer' titles; named 'Nadym'. Sold to unknown operator 11-98
RA-75554	IL-18GrM	185008404	IL-18V/F (SCD). Bought from Viola-Aviatrans ?-96, basic German European Airlines c/s with Cyrillic 'Ramaer' titles and 'Cargo' subtitles. DBR Johannesburg-Int'l 17-12-97; scrapped 2001, cargo door to IL-18D EX-75466

The 235th Independent Air Detachment (the Soviet Federal Government flight) operated a steady succession of VIP-configured *Coots* over the years. Only three such aircraft remained on strength when the 235th IAD became **GTK ROSSIYA** (*Gosoo**dar**stvennaya **trahns**portnaya kom**pah**niya 'Ros**si**ya'* – Russia State Transport Company) [R4/SDM] on 3rd December 1993, the VIP transport role having long since been taken over by jets. Known examples used by the Federal Government flight are listed below.

Registration	*Version*	*C/n*	*Notes*
RA-75453	IL-18D 'Salon'	187010103	Converted to communications relay aircraft
RA-75454	IL-18D 'Salon'	187010104	Converted to communications relay aircraft
RA-75464	IL-18D 'Salon'	187010401	Converted to communications relay aircraft. Sold to Bykovo-Avia 1999, converted to freighter?
CCCP-75563	*IL-18V 'Salon'*	*1850078O2?*	*Crashed Moscow-Domodedovo 1-7-67*
CCCP-75705	*IL-18S*	*189001702*	*Crashed near Kiev 17-8-60*
CCCP-75716	*IL-18V 'Salon'*	*180001902*	*Transferred to Polar Directorate and converted to weather research aircraft*
CCCP-75717	*IL-18V 'Salon'*	*180001904*	*First IL-18V with AI-20A (AI-20 Srs 3) engines*
CCCP-75823	*IL-18V 'Salon'*	*182004902*	*Transferred to Far Eastern CAD and converted to standard 'V before 1970*

As already mentioned, the last three IL-18Ds on strength (CCCP-75453, CCCP-75454 and CCCP-75464) were fitted out as communications relay aircraft supporting HF communication between the Kremlin and suitably equipped VIP aircraft. Originally they retained full Aeroflot colours; later the Aeroflot titles and logo were painted out, small italic 'Rossiya' titles in blue were applied on the nose where the pinstripe under the cheatline was broken, and the vertical tail was painted grey to make the white stripe of the Russian flag stand out.

Later, RA-75464 received a white fuselage with a blue/red cheatline below the windows, red 'Rossiya' titles in ancient Cyrillic script above them and the Russian 'double eagle' coat of arms with elaborate gold trim on the fin. This stylish livery (dubbed *gherbo**va**ya ras**krah**ska*, 'coat-of-arms colour scheme', by spotters) was also worn by a large portion of GTK Rossiya's fleet. Unfortunately it was later replaced by an altogether uninspiring overall grey livery with a curved white/blue/red cheatline in Russian flag colours. This colour scheme has earned the disdainful nickname ***se**ryy **oo**zhas* ('grey horror'), though 'Grey Ghost' would sound better perhaps, and RA-75454 has this 'Grey Ghost' scheme.

The three IL-18s continued in service until the turn of the century; RA-75453 was last noted at Moscow-Vnukovo in August 2001.

The Soviet Air Force (VVS) was a large-scale user both of the basic passenger/VIP version of the IL-18 and of specialised military derivatives. Most of these aircraft were taken over by the **RUSSIAN AIR FORCE**, serving in several of the Defence Districts (DDs) into which the nation's territory is divided. Until 1994, when Russian troops ultimately pulled out of post-unification Germany, several aircraft were on temporary deployment to the Group of Soviet Forces in Germany (GSVG)/Western Group of Forces (ZGV).

Quasi-civil aircraft are listed in Part 1 of the table, and some aircraft operated in Soviet days are also included for the sake of completeness.

Registration	*Version*	*C/n*	*Notes*
CCCP-33569	*IL-18S*	*188000801*	*Ex-Soviet AF '001 Red', registration is former ATC callsign; non-standard c/s with 'lightning bolt' cheatline. Reregistered CCCP-75479*
RA-54460	?	?	Reported in flight safety report in 1995; possibly callsign only and the aircraft had a standard registration in the 75xxx block!
RA-74295	*IL-18D 'Salon'*	*187010602*	*Ex-CCCP-74295, ex-CCCP-75472. Moscow DD/8th ADON. To 223rd Flight Detachment 25-8-94, see below*
74626	IL-18D	?	Red/white c/s; registration is ATC callsign
RA-75478	IL-18D (ABCP)	189011302	HF comms gear. Moscow DD/8th ADON. To 223rd Flight Detachment 25-11-93, see below
CCCP-75472	*IL-18D 'Salon'*	*187010602*	*Reregistered CCCP-74295 by 3-73, see above*
CCCP-75479	*IL-18S*	*188000801*	*Ex-CCCP-33569, ex-'001 Red' (see part 2); same non-standard c/s*
CCCP-75480	IL-20RT	173011405	May still be operational as such!
CCCP-75481	*IL-20RT*	*173011503*	*To Russian Navy, see below*
RA-75482	IL-20RT	173011505	
RA-75483	IL-20RT	173011601	
RA-75496	IL-18D (ABCP)	189011303	C/n painted on in error as 185011303. HF comms gear. Moscow DD/8th ADON. To 223rd Flight Detachment 25-11-93, see below
CCCP-75497	*IL-18D 'Salon'*	*187010204*	*Sold to UralAviali by 1994 as RA-75497*
RA-75498	IL-18D 'Salon'	187009804	Moscow DD/8th ADON. To 223rd Flight Detachment 25-8-94, see below
RA-75499	IL-18D 'Salon'	188011004	Moscow DD/8th ADON. To 223rd Flight Detachment 25-8-94, see below
RA-75516	IL-18V 'Salon'	183006604	HF comms gear. Moscow DD/8th ADON
CCCP-75530	*IL-18V 'Salon'*	*184006903*	*Sold to Air Zory 11-93 as LZ-AZC*
RA-75591	IL-18V 'Salon'	185008003	Out-of-sequence registration (may have had CCCP-75571 allocated). Based at Ulan-Ude/Vostochnyy
CCCP-75601	*IL-18S*	*189000905?*	*Out-of-sequence registration*
RA-75602	IL-18V (ABCP)	182004203	Out-of-sequence registration (may have had CCCP-75795 or CCCP-75798 allocated). HF comms gear. Moscow DD/8th ADON. To 223rd Flight Detachment 25-11-93, see below
CCCP-75604	*?*	*?*	*Out-of-sequence registration*
RA-75606	IL-18V (ABCP)	182004405	Out-of-sequence registration. HF comms gear. Moscow DD/8th ADON. To 223rd Flight Detachment 25-11-93, see below
CCCP-75614	*?*	*?*	*Out-of-sequence registration*
CCCP-75676 #1	*IL-18S*	*189001001*	*Crashed 2-9-59; see next line*
RA-75676	IL-18E ('Salon'?)	185008605	Ex-CCCP-75676 #2, out-of-sequence registration; was reported in error as ex-CCCP-75552 (c/n 184007404) or ex-YR-IMZ #1 (c/n 185008602). To 223rd Flight Detachment 25-11-93, see below
CCCP-75749	*IL-18S*	*189001601?*	*Out-of-sequence registration; non-standard c/s with 'lightning bolt' cheatline*
CCCP-75774	*IL-18V 'Salon'*	*181003604*	
CCCP-75825	*IL-18V 'Salon'*	*182004904*	*Sold to SPAir as RA-75825*
CCCP-75894	*IL-18V 'Salon'*	*182004801*	*Out-of-sequence registration. Transferred to Ministry of Electronics/NPO Vzlyot ?-82, converted to avionics testbed*
RA-75895	IL-22 (IL-18D-36)	0393607850	
RA-75897	IL-22 (IL-18D-36)	0393607430	Moscow DD/8th ADON. WFU Chkalovskaya AB by 8-99 as CCCP-; RA- registered in 2003
CCCP-75898	IL-22 (IL-18D-36)	0393607950	
RA-75899	IL-22 (IL-18D-36)	?	
CCCP-75900	IL-22 (IL-18D-36)	*039360...40?*	C/n off poor-quality photo
RA-75901	IL-22	0393609935	Moscow DD/8th ADON, Chkalovskaya AB
75902	IL-22 (IL-18D-36)	0393610226	Transbaikalian DD, based in Irkutsk. Converted to new version (designation unknown).
RA-75903 #1*	IL-18D (ABCP)	172011402	Modified IL-20 airframe; ex-CCCP-75903 #2
RA-75903 #2*	IL-22 (IL-18D-36)	0393610235	Moscow DD/16th Air Army, Kubinka AB. Operated as CCCP-75903 #1 until 1997. Converted to VIP transport by 8-03
CCCP-75904	IL-22 (IL-18D-36)	?	Moscow DD/8th ADON, Chkalovskaya AB
RA-75905	IL-22 (IL-18D-36)	?	Moscow DD/8th ADON, Chkalovskaya AB
RA-75906	IL-22 (IL-18D-36)	0393610501	Moscow DD/8th ADON, Chkalovskaya AB. Also reported as converted from IL-18D CCCP-75469 (c/n 187010501) but this is also reported as scrapped, and last five digits of c/n fit IL-18D-36 sequence nicely

IL-18D EX-75466 with Phoenix Aviation titles at Sharjah in the days before it was retrofitted with the cargo door salvaged from the wreck of IL-18GrM RA-75554. While the red/blue/white colour scheme was originally that of Ramaer (Russian flag colours, after all!), it has now come to be recognised as the Phoenix livery. Peter Davison

The Domodedovo Civil Aviation Production Association (now Domodedovo Airlines) was one of the last Russian carriers to operate the IL-18 as a passenger aircraft. Here, IL-18D RA-74268, one of the airline's last two remaining *Coots*, rests between flights at Moscow-Domodedovo on 11th November 1998. It was sold to Tyumen' Airlines shortly afterwards. Dmitriy Komissarov

IL-18V RA-75840, one of the four *Coots* operated by IRS Aero, begins its take-off run from Zhukovskiy's runway 12 on 13th August 2001, one day before the opening of the MAKS-2001 airshow. Just over three months later this aircraft crashed, killing all on board. Dmitriy Komissarov

IL-18D RA-75442 in full Ram Air colours tucks up its undercarriage as it makes a typically smoky take-off. The projection on the flightdeck roof and the bulged nose gear doors are the only indications this used to be the IL-18D 'Tsiklon' weather research aircraft. Yuriy Kirsanov

The 235th Independent Flight Detachment (Soviet government flight) operated several IL-18Ds fitted out as communications relay aircraft. This is one of them at Moscow-Vnukovo in post-Soviet days as RA-75454 (c/n 187010104). Note the additional HF communications aerials on the forward and rear fuselage. Yuriy Kirsanov

Seen here around 2001, combi-configured IL-18V RA-75834 displays the simple but nevertheless attractive livery of Moscow-based Titan Aero. Yuriy Kirsanov

IL-18Gr (IL-18D) RA-74296 in full Tret'yakovo Air Transport Company colours parked at Moscow/ Vnukovo-1 on 18th June 2002. This particular aircraft has been christened *Moskva*, echoing the IL-18's original popular name. Dmitriy Komissarov

CCCP-75907?	IL-22 (IL-18D-36)?	?	Existence not confirmed but likely
CCCP-75908	IL-22M-11	0394011091	Converted to new version (designation unknown)
RA-75909	IL-22M-11	0394011092	Converted to new version (designation unknown) by 3-1999. Moscow Defence District/8th ADON, Chkalovskaya AB
CCCP-75910	IL-22M-11	0394011094	
RA-75911	IL-22M-11	0394011096	
RA-75912	IL-22M-11	0394011097	Moscow DD/8th ADON, Chkalovskaya AB
RA-75913	IL-22M-11	0394011098	GSVG/16th Air Army, Oranienburg AB, later Sperenberg AB; to Moscow DD/8th ADON, Chkalovskaya AB, ?-90
RA-75914	IL-22M-11	0394011099?	
RA-75917	IL-22M-11	187010105	Ex-CCCP-75917, ex-IL-18D CCCP-75451. Moscow DD/8th ADON, Chkalovskaya AB; later to AEW&C unit, Ivanovo-Severnyy AB
CCCP-75919	IL-22M-11	2964009805	Ex-IL-18D CCCP-75780 #2 (c/n 187009805)
RA-75920	IL-22M-11	2964017551	Moscow DD/8th ADON, Chkalovskaya AB
CCCP-75921	IL-22M-11	2964017553?	
RA-75922	IL-22M-11	2964017552	Moscow DD/8th ADON, Chkalovskaya AB
RA-75923 #1	*IL-22M-11*	*187010305*	*Ex-CCCP-75923, ex-IL-18D CCCP-75463. Moscow DD/8th ADON, Chkalovskaya AB. WFU Chkalovskaya AB by 8-99; scrapped 1-03*
RA-75923 #2	IL-20M (mod)	173011501	Ex-'90 Red', registered by 8-99; GLITs, Chkalovskaya AB
RA-75924	IL-22M-11	2964017554	Moscow DD/8th ADON, Chkalovskaya AB
RA-75925	IL-22M-11	2964017557	Moscow DD/8th ADON, Chkalovskaya AB
CCCP-75926	IL-22M-11	188011203	Ex-IL-18D CCCP-74270. C/n painted on as 2964011203 in error for a while but then changed back to 188011203. WFU Kubinka AB by 8-03
RA-75927	IL-22M-11	2964017558	Moscow DD/8th ADON, Chkalovskaya AB
CCCP-75928	IL-22M-11	187010003	Ex-IL-18D CCCP-75448. WFU Pushkin/ARZ No 20 by 7-94

* Even though the existing registration system is designed to preclude simultaneous use of the same registration on several operational aircraft, 'things happen' sometimes and there were two CCCP-75903s in service at the same time. Even more curiously, 'IL-18D' (IL-20) c/n 172011402 registered on 2nd February 1988 was the second aircraft to be registered CCCP-75903 because IL-22 production had ended by then, which means IL-18D-36 c/n 0393610235 bearing the same registration already existed. However, the Russian re-registration order was reversed – IL-20 CCCP-75903 No 2 became RA-75903 No 1 in 1994, and it was another four years before IL-22 CCCP-75903 No 1 became RA-75903 No 2!

The second part of the table deals with aircraft wearing overt military markings. Because of the Soviet/CIS tactical code system, only positively identified aircraft are listed (in construction number order).

C/n	*Version*	*Tactical code*	*Notes*
188000801	*IL-18S*	*'001 Red'*	*May have had CCCP-75667 allocated; callsign CCCP-33569. Was the aircraft of Defence Minister Marshal Gheorgiy M Grechko. To CCCP-33569, later CCCP-75479*
173011404	IL-20M	'88 White'	
173011501	*IL-20M*	*'90 Red'*	*GK NII VVS (later GLITs), Chkalovskaya AB. To RA-75923 #2 before 8-99*
173011502	IL-20M	'20 Red'	GSVG/390th OAO, Sperenberg AB (1975-94); to Leningrad DD/Levashovo AB, later to Smolensk
173011504	IL-20M	'21 Red'	GSVG/390th OAO, Sperenberg AB (1975-94); to Leningrad DD/Levashovo AB, later to Smolensk
173011602	IL-20M	no code	
174011603	IL-20M	no code	
174011604	IL-20M	no code	
175011701	IL-20M	no code	Also reported as '07 Red'?
176011707?	IL-20M	'34 Red'	C/n unconfirmed (possible misquote for 176011702)

The Russian Air Force also has commercial divisions. The 223rd OSAP (*otdel'nyy **smesh**annyy **avia**polk* – independent composite air regiment), one of the units making up the 8th ADON at Chkalovskaya AB, was transformed into an 'airline' called **223RD FLIGHT UNIT STATE AIRLINE** [–/CHD]. Among other things it operated eight IL-18s, all of them former VIP or staff transport aircraft, in full Aeroflot colours.

Registration	*Version*	*C/n*	*Notes*
RA-74295	*IL-18D*	*187010602*	*Transferred 25-8-94. Crashed near Batumi 25-10-2000*
RA-75478	IL-18D	189011302	Transferred 25-11-93
RA-75496	IL-18D	189011303	Transferred 25-11-93
RA-75498	*IL-18D*	*187009804*	*Transferred 25-8-94; registration cancelled 12-3-01 (sold or retired)*
RA-75499	*IL-18D*	*188011004*	*Transferred 25-8-94. Believed sold to Phoenix Aviation by 1-01 as EX-75499*
RA-75602	*IL-18V*	*182004203*	*Transferred 25-11-93; registration cancelled 12-3-01 (sold or retired)*
RA-75606	*IL-18V*	*182004405*	*Transferred 25-11-93; registration cancelled 12-3-01 (sold or retired)*
RA-75676	IL-18E	185008605	Transferred 25-11-93

The **RUSSIAN NAVAL AIR ARM** (AVMF – *Aviahtsiya voyenno-morskovo flota*) also inherited a number of – well no, mostly not *Coots* but IL-38 ASW aircraft.

Registration	*Version*	*C/n*	*Notes*
CCCP-75481	IL-20RT	173011503	Russian Navy Combat & Conversion Training Centre, 240th GvOSAP, Ostrov AB; used as trainer
CCCP-75500	IL-18V 'Salon'?	183006303?	Last reported 1995 in incident report
CCCP-75528	*IL-18RT*	*183006901*	*Reconverted to IL-18V; Pacific Fleet*
CCCP-75666	*IL-18B (ABCP)*	*188000705*	*HF comms gear; Soviet Navy Commander's aircraft*
CCCP-75840	*IL-18RT*	*182005301*	*Transferred from Aeroflot (Moscow Territorial CAD/Vnukovo UFD/65th Flight) 10-2-65 as IL-18V, conversion date unknown. Reconverted and sold to IRS Aero 2-10-98 as RA-75840*

Again, because of the Soviet/CIS tactical code system the aircraft in overt military markings are listed below in c/n order.

C/n	*Version*	*Tactical code*	*Notes*
08...10808	IL-38	'21 Red'	
081010907	IL-38	'25 Red'	
081010910	IL-38	'21 Red'	
08...11006	IL-38	'22 Red'	Russian Navy Combat & Conversion Training Centre, 240th GvOSAP, Ostrov AB
082011106	IL-38	'79 Red'	
082011207	IL-38	'71 Red'	'VMF Rossii/Russian Navy' titles and Russian flag tail colours
not known	IL-38	'06 Red'	Russian Navy Combat & Conversion Training Centre, 240th GvOSAP, Ostrov AB
not known	IL-38	'07 Red'	
not known	IL-38	'08 Red'	
not known	IL-38	'09 Red'	
not known	IL-38	'10 Red'	
not known	IL-38	'11 Red'	
not known	IL-38	'19 Red'	
not known	IL-38	'20 Red'	Russian Navy Combat & Conversion Training Centre, 240th GvOSAP, Ostrov AB
not known	IL-38	'23 Red'	

SPAIR AIR TRANSPORT CORPORATION [–/PAR] (*Aviatrahnsportnaya korporahtsiya Spaer*)[5] based at Yekaterinburg-Kol'tsovo and mainly concerned with cargo carriage acquired two IL-18s in 1994. The letters SP in the carrier's name are derived from the name of its director, Valeriy Spoornov.

The first aircraft was IL-18D RA-75497 (c/n 187010204) purchased from the defunct UralAviali Company. In 1995 it was sold to Air Cess, receiving the Liberian registration EL-AKQ (see next chapter). The other *Coot* was IL-18V RA-75825 (ex-IL-18V 'Salon' CCCP-75825, c/n 182004904) bought from the Soviet Air Force; in 1997 it was sold to Santa Cruz Imperial as EL-ALW.

The cargo airline **TITAN AERO** [–/RTT] established in 1999 operated IL-18V RA-75834 (c/n 182005104), an ex-Ilavia machine, from Zhukovskiy. In 2002, however, the aircraft was sold to Tret'yakovo Air Transport Company.

In 1999 the Nadym branch of **TYUMEN' AIRLINES** / *Tyumenskiye avialinii* [7M/TYM] bought the last two IL-18Ds remaining operational with Domodedovo Airlines – RA-74267 and RA-74268. Since they were operated jointly with Phoenix Aviation, both aircraft received the flashy blue/red/white Phoenix livery, except for the titles which were those of Tyumen' Airlines, with additional 'Nadym' titles. Eventually RA-74267 and RA-74268 were sold to Phoenix Aviation and reregistered, becoming EX-105 and EX-201 respectively.

The **TRET'YAKOVO AIR TRANSPORT COMPANY** [–/TKO] which was established in 1997 operated three IL-18s based at Moscow-Domodedovo and Lookhovitsy-Tret'yakovo, the factory airfield of MAPO. The aircraft wore basic Aeroflot colours with large Cyrillic 'TRET'YAKOVO' titles.

Registration	*Version*	*C/n*	*Notes*
RA-74296	IL-18D	188010603	Originally operated without titles/logo; named 'Moskva' by 6-02
UN 75111	IL-18V	184007105	Ex-YR-IMF; leased from GVG Air Company ?-01
RA-75834	IL-18V	182005104	Bought from Titan Aero ?-02

Unfortunately, one of the airline's aircraft (IL-62M RA-86452) crashed on landing at Bishkek-Manas airport on 23rd October 2002. The investigation that followed revealed such an appalling picture that the Tret'yakovo Air Transport Company's operating licence was revoked by the authorities in early 2003. The fate of the airline's IL-18s remains unknown at the time of writing.

Chelyabinsk-based **URALAVIALI COMPANY** [–/URL] purchased IL-18D CCCP-75497 from the Russian Air Force on 19th April 1993. The aircraft was reregistered RA-75497 and sported a rather unusual 'negative' colour scheme with a blue top, a grey belly and a white cheatline! In late 1994, however, the airline went bankrupt and its sole aircraft was sold to SPAIR.

An airline named **VIOLA-AVIATRANS** bought two IL-18Vs from German European Airlines in 1995. The first aircraft, D-AOAO (ex-DDR-STF, ex-DM-STF No 2, ex-CCCP-75553, c/n 184007405), regained its original identity on 20th December 1995 as RA-75553, while IL-18GrM D-AOAP (ex-DDR-STI, c/n 185008404) became RA-75554 – a 'ghost ship' because CCCP-75554 (c/n 184007501) was retired and preserved at Moscow/Sheremet'yevo-1 by then!

Incredibly, both aircraft, which were delivered on 22nd and 23rd December 1995 respectively, retained the complete livery of their previous owner (including the German European Airlines titles) until they were sold. During 1996 RA-75553 became LZ-AZO with the Bulgarian carrier Air Zory, whereas RA-75554 was sold to the Russian charter carrier Ramair.

THE UKRAINE

AIR UKRAINE/AVIALINIÏ UKRAÏNY [6U/UKR], the Ukrainian flag carrier, leased IL-18V UR-75475 from Kryla (see below) in 1993-98.

In 1992 the L'vov-based airline **KRYLA** [–/KRL],[6] a subsidiary of the Elektron Concern, bought IL-18E 75711 (ex-LOT Polish Airlines SP-LSK, c/n 185008503) from an unidentified owner. Reregistered UR-75850, the aircraft initially retained basic LOT colours. Curiously, both of its new identities were 'revisited', there having previously been IL-18Vs CCCP-75711 (c/n 189001802) and CCCP-75850 (c/n 182005405)! The following year Kryla doubled its fleet by purchasing another re-export *Coot*, ex-BerLine IL-18V D-AOAQ (c/n 184007401) which became UR-75475 – again a reused registration (see East German section in Chapter 9).

From September 1994 to January 1996 UR-75850 was on lease with the Russian carrier ALAK as RA-75850, regaining the Ukrainian prefix upon return from lease. The other aircraft was operated by Air Ukraine and thus did not receive Kryla titles and tail colours until 1999. In 2001, however, the airline suspended operations; both IL-18s were sold to the Angolan airline ALADA, UR-75475 and UR-75850 becoming D2-FAM and D2-FDY respectively.

In 1999 **LVIV AIRLINES** (***L'vivs'ki avialinii***) [5V/UKW], the former L'vov branch of Air Ukraine ('Lviv' is *their* pronunciation) took delivery of a surplus IL-20M registered UR-BXD (former tactical code unknown, c/n 172011401).[7] The aircraft has been completely demilitarised and wears 'IL-18D' nose titles, but the two windows in the forward cabin on each side instead of three and the unusually small door of the No 2 baggage compartment reveal this is no ordinary *Coot*. Interestingly, UR-BXD has Lviv Airlines titles to port and Ukraine West titles to starboard.

SEVASTOPOL'-AVIA leased IL-18E UR-TMD from Tavria-MAC in early 2001 (see next entry).

TAVRIA-MAC JOINT-STOCK CO [BE/TVM], an airline based in Simferopol' (the 'capital' of the Crimea Region), leased IL-18E ER-ICJ (c/n 186009102) from Renan in November 1999, adding its titles to the owner's livery. The following year the airliner was purchased outright, changing its identity to UR-TMD. (This aircraft is sometimes reported in error as ex-ER-ICL, c/n 187009903.) In the spring of 2003, however, the aircraft was resold to Renan, regaining is previous identity.

The **UKRAINIAN AIR FORCE** (UAF, or VPS – *Voyenno-povitryany seely*) operated a pair of *Coot-B* airborne command posts – IL-22 *sans suffixe* 75896 (later UR-75896, c/n 0393607150) and IL-22M-11 75918 (c/n unknown). Both were based at Vinnitsa and wore UAF roundels and shield-and-trident tail insignia in addition to the civil registrations. Apparently there was no real need for two IL-22s (or, more probably, financial problems forced the UAF to sell part of its fleet). In early 2003 UR-75896 was demilitarised and sold to the Angolan airline ALADA as D2-FFR.

The UAF also had at least two IL-20Ms. One was coded '09 Red'; the c/n has been reported as 17...011409 but this is doubtful. The other aircraft (tactical code unknown, c/n 172011401) was completely demilitarised by the end of 1999 and sold to Lviv Airlines as UR-BXD.

Besides Russia, the Ukraine was the only other CIS republic to operate the IL-38; the **UKRAINIAN NAVY** appears to have inherited at least three of the type. However, funding shortfalls and other problems have grounded the fleet. Two *Mays* – '09 Red' (c/n unknown) and '10 Red' (c/n 087010106) – remained at the 33rd Naval Aviation Combat & Conversion Training Centre named after Ye N Preobrazhenskiy at Nikolayev-Kul'bakino; the latter aircraft never flew after the Ukraine gained independence and was ultimately scrapped in the late 1990s.[8]

A Ukrainian Navy IL-38 coded '78 Red' has been reported with the c/n 70321. However, since this is obviously wrong, the number is probably an ATC callsign (UR-70321).

IL-18 Operators Abroad

Over the years the IL-18 has seen service in more than 30 nations in Europe, Asia, the Middle East, Africa, Central and South America (excluding the new CIS republics) – an impressive figure by any standards. Of these, 20 countries took delivery of new or, more rarely, used *Coots* directly from the Soviet Union. Of course, some nations have obscure operators flying well-used aircraft – many of which change their country of registry frequently (a case of the notorious 'flag of convenience' registrations).

Again, IATA/ICAO codes are stated as appropriate and aircraft no longer current are marked in italics, except where the airline is defunct or the type has been withdrawn altogether.

AFGHANISTAN

In the summer of 1967 the **AFGHAN AIR FORCE** (*Afghan Hanai Qurah*) acquired an IL-18D 'Salon' (c/n 187010105). Serialled T-001 and named 'Kandahar', it was used by King Mohammad Zakir Shah for state visits.

In February 1979 the aircraft was returned to the Soviet Union; whatever the reason, this doubtless saved it from destruction in the Afghan War which broke out at the end of the year. Upon arrival T-001 was registered CCCP-75451 No 2 (this registration had been worn by IL-18D c/n 187010101 which became OK-WAJ instead). After refit as an ordinary passenger aircraft the *Coot* was delivered to the Krasnoyarsk CAD/1st Krasnoyarsk UAD/214th Flight where it served on until 1983. (As a point of interest, in 1983 the aircraft was 'drafted' again, being converted into an IL-22M-11 and reregistered CCCP-75917.)

ALGERIA

In 1963 the government flight of the **ALGERIAN AIR FORCE** (*al Quwwat al-Jawwiya al Jaza'eriya/Force Aérienne Algérienne*) purchased an ex-Aeroflot IL-18V (CCCP-75764, c/n 181003402), taking delivery of the aircraft on 6th January 1964. Reregistered 7T-VRA, this was the sole IL-18 delivered to Algeria, despite persistent reports of at least four being operated. The *Coot* carried *République Algérienne* titles and an airline-style red/white colour scheme; the cheatline curved down to the radome, giving the aircraft a 'smiling' appearance in a head-on view. Some sources reported that 7T-VRA crashed near Tamanrasset in southern Algeria on 12th June 1971 but this was not the case.

ANGOLA

ALADA – SOCIEDADE DE TRANSPORTES AÉREOS [–/RAD], an airline based in Luanda, acquired four *Coots* in 2001-03.

Registration	*Version*	*C/n*	*Notes*
D2-FAM	IL-18V	184007401	Ex-Kryla UR-75475, bought 12-01. Leased briefly to unknown operator as 3D-SEP
D2-FDL	IL-18E?	186009102?	Bought before 2-03; ex-Tavria-MAC UR-TMD
D2-FDY	*IL-18E*	185008503	Ex-Kryla UR-75850, bought 12-01
D2-FFR	IL-22 (IL-18D-36)	0393607150	Ex-Ukrainian Air Force UR-75896, demilitarised; bought before 2-03

BOSNIA & HERZEGOVINA

In 2000 the Bosnia-Herzegovenian cargo airline **BIO AIR COMPANY** [–/BIO], a division of the Bosnian Investment Organisation, operated IL-18V T9-ABB (ex-EL-ADY No 1, c/n 184007405) jointly with Phoenix Aviation (see previous chapter/Kyrghyzstan). The aircraft retained the titles of previous owner European Airlines for almost a year before finally receiving Phoenix titles. In 2001 it was transferred to the Kyrghyzstani register as EX-405.

BULGARIA

Established in 1949 as a Bulgarian/Soviet joint venture, **TABSO – BULGARIAN AIR TRANSPORT** took delivery of eleven IL-18s in 1962-66. The *Coot* became the first turbine transport in Bulgarian service, replacing piston-engined IL-14Ps on the most important routes (primarily the international ones).

Unlike some other Bulgarian civil aircraft, in the case of the IL-18 the registrations did not run in sequence but were allocated randomly. The airliners wore Bulgarian Air Transport titles, the name TABSO (an acronym for ***Trahns****portno-aviatsi****on****noye* *bol****gar****sko-so****vet****skoye* ***o****bschchestvo* – Bulgarian-Soviet Aviation Transport Co) being incorporated into a stylised 'bird' logo on the tail.

Registration	*Version*	*C/n*	*Notes*
LZ-BED	IL-18E	186009002	D/D 24-4-66. Crashed on approach to Zürich 18-1-71
LZ-BEG	IL-18E	186009101	D/D 19-5-65. Crashed on approach to Bourgas 3-9-68
LZ-BEK	IL-18V	182004603	D/D 25-4-62. Transferred to Balkan
LZ-BEL	IL-18V	182004601	D/D 18-3-62.* Transferred to Balkan
LZ-BEM	IL-18V	182005602	D/D 6-3-63. Crashed on approach to Moscow-Sheremet'yevo 3-3-73
LZ-BEN	IL-18V	184007101	D/D 30-3-64. Crashed near Bratislava 24-11-66
LZ-BEP	IL-18V	185008105	D/D 27-3-65. Transferred to Balkan
LZ-BER	IL-18V	184007203	D/D 5-64. Transferred to Balkan
LZ-BES	IL-18V	185008104	D/D 12-2-65. Crashed near Sofia 21-12-71
LZ-BET	IL-18E	186008904	D/D 24-5-66. Transferred to Balkan
LZ-BEV	IL-18V	185008201	D/D 11-6-65. Transferred to Balkan

* According to some sources, TABSO's first IL-18 was delivered on 30th March 1962.

The introduction of the IL-18 opened new perspectives for TABSO. On 28th May 1963 the airline started flying to Northern Africa, opening up the Sofia – Algiers service; on 2nd November 1963 a TABSO IL-18 crossed the equator for the first time on a scheduled flight from Sofia to Nairobi. On 11th April 1964 a Bulgarian *Coot* inaugurated the Sofia - Lima service (via Africa with several refuelling stops).

On 8th September 1964 a TABSO IL-18 made the first scheduled flight on the 5,000-km (3,100-mile) route from Sofia to Khartoum via Istanbul, Ankara, Damascus and Cairo. Eighteen days later a sister aircraft covered the 11,000km (6,830 miles) from Sofia to Tokyo, carrying the Bulgarian Olympic team to the XVIII Olympic Games. 5th April 1965 saw the inauguration of TABSO's 17th international service (to London-Heathrow).

In 1968 TABSO sprouted a division named **BULAIR** and tasked with charter flights. In addition to cargo charters flown by Antonov An-12 transports, it performed passenger charters to Bulgarian holiday resorts on the Black Sea, using seven *Coots* – LZ-BED, -BEG, -BEK, -BEL, -BEM, -BEP and -BET. Unfortunately, the venture was shortlived and in 1971 Bulair was merged into Balkan Bulgarian Airlines (see next entry).

On 1st April 1968 TABSO was renamed **BALKAN BULGARIAN AIRLINES** [LZ/LAZ]; however, it was some time before a new livery was introduced. The blue TABSO cheatline with 'feathers' at the front and the blue tail logo gave way to a simple red cheatline and Bulgarian flag tail colours, with the same 'bird' logo – now in white and carrying the legend BALKAN – superimposed on the green stripe of the flag. Taking over the six surviving *Coots* of its predecessor, the new Bulgarian flag carrier eventually increased the number to 13 – among other things, by snapping up IL-18s disposed of by LOT Polish Airlines.

During the late 1980s Balkan Bulgarian Airlines introduced a new livery matching the latest fashion, with curved red/white/green stripes on a white background, which gave the vintage airliner a completely new look. LZ-BEU was one of those to be repainted. However, the IL-18 could not match the Tu-134 and Tu-154 (introduced in 1968 and 1972 respectively) in terms of passenger comfort or with the An-12 in terms of cargo handling efficiency. By the end of the 1990s Balkan had retired the type.

Registration	*Version*	*C/n*	*Notes*
LZ-BEA	IL-18D	188010802	Ex-Bulgarian Air Force? Converted to IL-18Gr, date unknown; WFU Varna by 9-97
LZ-BEH	IL-18E	186008905	Ex-SP-LSI bought by end 1988; basic LOT c/s, Balkan titles. Converted to IL-18Gr, date unknown, received 'CARGO' titles instead of Balkan titles. WFU Sofia-Vrazhdebna by 6-99; sold to Bulgarian Airlines by 2-00 as LZ-ZAH
LZ-BEI	IL-18V	181002805	Ex-SP-LSC, D/D 18-4-88; WFU Varna by 9-97
LZ-BEK	IL-18V	182004603	Scrapped Varna around 1985
LZ-BEL	IL-18V	182004601	Returned to the Soviet Union and was later destroyed by fire
LZ-BEO	IL-18D	186009602	Ex-CCCP-75438, D/D 27-5-70. Preserved as a café 75km from Sofia
LZ-BEP	IL-18V	185008105	DBR Sana'a 16-6-84
LZ-BER	IL-18V	184007203	
LZ-BET	IL-18E	186008904	WFU Sofia-Vrazhdebna by 5-92
LZ-BEU	IL-18V	183005905	Ex-Yemen Airways 4W-ABO, D/D 24-9-84. WFU Varna by 6-98; sold to Bulgarian Flying Cargo ?-99 as LZ-BFU
LZ-BEV	IL-18V	185008201	Preserved in playground Sofia
LZ-BEW	IL-18E	185008601	Ex-SP-LSF, bought before 1991. WFU Sofia-Vrazhdebna by 1-98; sold to Southern Cross ?-98 as 3D-ALD
LZ-BEZ	IL-18E	185008603	Ex-SP-LSG, bought before 1991; basic LOT colour scheme, Balkan titles. Sold to Santa Cruz Imperial 26-2-98 as EL-ARK

The Sofia-based airline **AIR ZORY** [–/MZA] operated three IL-18s.

Registration	*Version*	*C/n*	*Notes*
LZ-AZC	IL-18V	184006903	Ex-Soviet Air Force CCCP-75530, bought 11-93. Sold to Air Cess 7-97, initially as LZ-AZC but reregistered 3D-SBC 3-98
LZ-AZR	IL-18D	188010904	Ex-BerLine D-AOAU, bought before 1-94. Sold to Aero Caribbean 12-95 as CU-T131
LZ-AZZ	IL-18GrM	188010805	IL-18D/F (SCD), ex-BerLine D-AOAS, bought before 1-94. Sold to Aero Caribbean 12-95 as CU-T132

Sofia-based **BULGARIAN FLYING CARGO** or **BF CARGO** [FN/BFB] bought IL-18V LZ-BEU from Balkan in late 1999. In keeping with company standards the aircraft received a new registration, LZ-BFU (Bulgarian Flying Cargo also had six An-12s registered in the BF block). Despite the airline's name, the IL-18 was operated not only as a pure freighter but also in 105-seat and combi configurations as per necessity. By September 2000, however, LZ-BFU had been sold to an airline called Inter Tropic Airlines (see end of chapter).

An unknown airline acquired another ex-Balkan *Coot*, IL-18E LZ-BEH, in late 1999 or January 2000. Reregistered LZ-ZAH, the airliner wore basic LOT colours and Bulgarian Airlines titles; the latter *did not mean anything at all*, being in fact a remnant of the original BALKAN BULGARIAN AIRLINES titles!

By 5th March 2000 the spurious markings had vanished; five days later LZ-ZAH was already proudly wearing Phoenix titles. The Bulgarian registration was removed on 30th September and a few days later the aircraft became EX-75905.

CHILE

From 20th February to June 1992 an airline called **EAST WEST CHILE** leased IL-18D CU-T1270 (c/n 187010301) from Cubana.

CHINA

China was the first foreign customer for the IL-18. In 1959-67 the **CIVIL AVIATION ADMINISTRATION OF CHINA** (CAAC) took delivery of 13 *Coots*. At first they wore simple three-digit registration numbers; in 1972 the newly adopted B- nationality prefix was added for clarity, though it took several years to apply it to the entire fleet. One more aircraft was transferred from the military, while three other IL-18s were operated on short-term dry leases from TAROM, bringing the total number operated to 17.

Registration	*Version*	*C/n*	*Notes*
B-200	IL-18B	189001504	
B-202	IL-18B	189001401	D/D 21-11-59; crashed Canton 24-12-82
B-204	IL-18B	189001602	Ex-XT-804? DBR Shenyang ?-2-77
B-206	IL-18B	189001601	Ex-206, ex-XT-806? WFU Tientsin 4-84
B-208	IL-18E 'Salon'	185008701	Ex-208 (see B-224 #1!); Mao Tse Tung's aircraft; scrapped Xian by 10-88
B-210	IL-18V	184007602	Ex-210 (B- prefix added 1974). GIA in Tianjin Technical School
B-212	IL-18V	184007702	
B-214	IL-18E	185008505	
B-216	IL-18E	185008704	Preserved in a park in Chengdu
B-218	IL-18V	184007601	Ex-218 (B- prefix added 1974). WFU 29-6-90; transported to Langzhou ?-98 and preserved
B-220	IL-18D	187009605	D/D 5-5-67
B-222	IL-18D	187009901	Ex-222. Crashed near Chongqing 18-1-88
B-224 #1	IL-18V	184007805	Ex-PLAAF '50854 Red', transferred between 10-86 and 10-88
B-224 #2 (?)	IL-18D	187009703	Preserved PLAAF Museum, Changping, by 3-92; painted as '208' by 4-00 to represent Mao Tse Tung's aircraft
B-230 #1	IL-18V	181003602	Ex/to YR-IMA, leased from TAROM 9-85 to 30-3-87; see China United Airlines
B-232	IL-18V	183006205	Ex/to YR-IME, leased from TAROM 9-85 to 3-86
B-234	IL-18V	184007105	Ex/to YR-IMF, leased from TAROM 28-3-85 to 21-3-86
B-240	IL-18V	184007604	To PLAAF as '50850 Red', transferred between 9-85 and 10-86

The **PEOPLE'S LIBERATION ARMY AIR FORCE** (PLAAF, or *Chung-kuo Shen Min Taie-Fang-Tsun Pu-tai*) operated seven IL-18s.

Serial	*Version*	*C/n*	*Notes*
'50850 Red'	IL-18V 'Salon'	184007604	Ex-CAAC B-240, transferred between 9-85 and 10-86
'50851 Red'	IL-18V 'Salon'	184007605?	To China United Airlines by 5-89 as B-230 #2, c/n 184007605 confirmed for this
'50852 Red'	IL-18E 'Salon'	185008604?	C/n stated in Chinese register as 184007701 but this is CCCP-75569
'50853 Red'	IL-18D 'Salon'	187009704	To China United Airlines. WFU Nanyuan AB 6-94; to ground instructional airframe
'50854 Red'	IL-18V 'Salon'	184007805	To CAAC between 10-86 and 10-88 as B-224 #1 (see CAAC table)
'50855 Red'	IL-18E 'Salon'	185008702	To China United Airlines by 1988 as B-228
'50856 Red'	IL-18D 'Salon'	187009902	To China United Airlines by 1993 as B-226

The meaning of PLAAF serials is obscure, but the first two digits may be a defence district code, the fourth digit a unit code within the district, while the third and fifth digits make up the individual number of the aircraft in the unit.

Starting in 1988, **CHINA UNITED AIRLINES** [–/CUA], the commercial division of the PLAAF, operated four IL-18s – B-226, B-228, B-230 and '50853 Red' (see above). The aircraft were based at Nan Yuan airbase near Beijing. Upon retirement B-226 was preserved in the PLAAF Museum at Datang Shan AB in Changping near Beijing, while B-228 went to the China Agricultural Museum in northern Beijing.

CONGO-BRAZZAVILLE

The **CONGO AIR FORCE** (*Force Aérienne Congolaise*) operated a single IL-18 serialled TN-105 as a VIP transport; the version and c/n are unknown.

CONGO-KINSHASA – SEE ZAÏRE

CUBA

The fleet of the Cuban flag carrier **EMPRESA CONSOLIDADA CUBANA DE AVIACIÓN**, commonly known simply as **CUBANA** [CU/CUB], included eight IL-18s, most of which were second-hand examples transferred from Aeroflot. The aircraft were mostly in 89-seat configuration.

Registration	*Version**	*C/n*	*Notes*
CU-T830	IL-18V	182004905	D/D 1964; crashed near Cienfuegos, Cuba, 10-7-66. Ex-Soviet registration not known; reported as ex-CCCP-75826 but this was seen after CU-T830 crashed!
CU-T831	IL-18V	182005202	CCCP-75836, bought 12-9-63; named 'Capitán Fernando Alvaréz'
CU-T832	IL-18V	182005501	Ex-CCCP-75851, bought ?-64; resold to the Soviet Union 2-85 as CCCP-75851
CU-T899	IL-18D	188011102	8-11-68; crashed near San José de los Lajas near Havana 19-1-85
CU-T900	IL-18D	188011104	D/D 3-12-68. Converted and reregistered 1991 as, see next line
CU-C900†	IL-18Gr		Transferred to Aero Caribbean by 12-91
CU-T1268	IL-18D	188010704	Ex-CCCP-74256, Krasnoyarsk CAD/1st Krasnoyarsk UAD/400th Flight. D/D 1-4-86; transferred to Aero Caribbean by 5-92
CU-T1269	IL-18V	184007801	Ex-CCCP-75562, Turkmen CAD/Ashkhabad UAD/369th Flight. D/D 12-6-87; transferred to Aero Caribbean ?-92
CU-T1270	IL-18D	187010301	Ex-CCCP-75459, D/D 26-5-87; transferred to Aero Caribbean ?-92
YR-IMG	IL-18V	184007301	Leased from TAROM ?-2-90 to 7-8-90

* Contrary to some sources, Cubana never operated the IL-18B.

† The Cuban system of civil aircraft registrations resembles the pre-1959 Soviet system in that it features role identifier letters. Thus, passenger aircraft were registered in the CU-Txxxx series (T for *transporte*); so were freighters initially, but after a while they were reregistered in the CU-Cxxx block (C for *carga*). Similarly, agricultural aircraft are registered in the CU-Exxx series and helicopters in the CU-Hxxx series.

A curious feature of the Cuban *Coots* was that larger nosewheels from the Bristol Britannia Srs 318 – a type also operated by Cubana – were fitted to reduce vibration levels during the take-off run, which required bulges to be made in the nose gear doors. This measure was probably necessitated by poor-quality runways which even the IL-18 found hard to handle.

The type was utilised primarily on long-haul international routes, including flights from Havana-José Marti to Moscow, Prague, Berlin, Managua, Lima, Montréal, Madrid, Mexico City etc. However, when Cubana took delivery of IL-62M and Tu-154B-2/Tu-154M jet airliners, the IL-18 was relegated to routes of secondary importance and cargo transportation duties.

Occasionally the IL-18s also performed non-scheduled flights in the interests of the Cuban government. One such flight in 1964 ended in a very embarrassing situation. On 14th October an IL-18V which staged through Montréal-Dorval, carrying Cuban Foreign Minister Raúl Roa who was returning home after attending a Non-Aligned Movement conference in Cairo, was impounded because Cubana had an outstanding debt of US$ 953,858 to Worldwide Airways, Inc. The authorities managed to play down the scandal and the aircraft was allowed to continue the journey after a bail of US$ 1 million had been paid.

During 1991 Cubana transferred its remaining IL-18s to **AERO-CARIBBEAN** [7L/CRN], a charter subsidiary formed in 1982. The purchase of two more examples from Air Zory increased the airlines's IL-18 fleet to six.

Registration	*Version*	*C/n*	*Notes*
CU-T131	IL-18D	188010904	Ex-Air Zory LZ-AZR, bought 12-95; basic BerLine c/s. Reregistered ?-2000 as, see next line
CU-T1532			
CU-T132	IL-18GrM	188010805	IL-18D/F (SCD), ex-Air Zory LZ-AZZ, bought 12-95. Basic BerLine c/s, CARGO titles. Reregistered CU-C132 soon after delivery; Aero Caribbean titles finally applied by 11-96. Reregistered again ?-2000 as, see next line
CU-C1515			WFU Havana-José Marti by 9-01
CU-C900	IL-18Gr	188011104	Converted IL-18D. WFU Havana-José Marti by 4-98
CU-T1268	IL-18D	188010704	Reregistered again ?-2000 as, see next line
CU-C1517*			WFU Havana-José Marti by 1-03
CU-T1269	*IL-18V*	*184007801*	*'Sun, sea and palm trees' c/s. WFU Havana-José Marti by 11-96; scrapped 11-99*
CU-T1270	IL-18D	187010301	*Crashed into Isabel de Torres hill, Dominican Republic, 15-11-92*

* CU-T1268 was also reported as converted to IL-18Gr configuration and reregistered CU-C1268!

An Aero Caribbean IL-18 registered CU-T1552 was also reported in March 2003 but this may have been a mis-sighting for CU-T1532.

An unidentified Cuban airline bought 'IL-18D' (demilitarised IL-22M) YL-LAO from Concors (see Latvian section) in November 2003; the aircraft was reregistered CU-T1539.

A line-up of three People's Liberation Army Air Force IL-18s at Nanyuan AB, with IL-18E 'Salon' '50852 Red' foremost. Yefim Gordon archive

'50851 Red', another PLAAF example – an IL-18V 'Salon'. It was later registered B-230. Yefim Gordon archive

IL-18D CU-T900 was one of only two *Coots* delivered to Cubana as brand-new aircraft in 1968. It was subsequently reregistered CU-C900 to reflect its new freighter role. Yefim Gordon archive

Seen from under the belly of sister aircraft registered OK-PAE, a ČSA Československé Aerolinie IL-18V in 1959-standard colours rests between flights. Yefim Gordon archive

CZECHOSLOVAKIA

The nation's flag carrier **ČSA ČESKOSLOVENSKÉ AEROLINIE** (Czechoslovak Airlines) [OK/CSA] operated 12 IL-18s of its own and five leased examples between 1960 and 1984. The airliner was operated in 80-seat tourist-class or 78-seat mixed-class layout.

Registration	*Version*	*C/n*	*Notes*
OK-NAA	IL-18B	189001604	D/D 8-1-60, named 'Piešťany'. Damaged beyond repair in a ground collision Prague-Ruzyne 2-1-77; donated to Military Museum (VM VHÚ), Prague-Kbely, 8-3-79
OK-NAB	IL-18B	189001605	D/D 8-1-60, named 'Košice'. Crashed on approach to Bratislava 28-7-76
OK-OAC	IL-18V	180002101	D/D 31-5-60, named 'Sliačkupe'; retired 30-5-80
OK-OAD	IL-18V	180002102	D/D 31-5-60. Crashed near Nürnberg 28-3-61
OK-PAE	IL-18V	181002902	D/D 22-4-61, named 'Karlový Vary'. Retired 18-5-80
OK-PAF	IL-18V	181002904	D/D 22-4-61. Crashed near Casablanca 12-7-62
OK-VAF	IL-18E	186009004	Ex-Czech Federal Government IL-18E 'Salon' OK-BYZ, transferred 6-12-77. Retired, delivered to Nürnberg airport 23-1-90 for use as a fire trainer
OK-PAG	IL-18V	181004201	D/D 4-1-62, named 'Výsoké Tatry'. Retired 21-2-80; to Slusovice as a restaurant
OK-PAH	IL-18V	181004202	D/D 4-1-62, named 'Marianské Lazně'. Retired 31-10-79 and broken up 8-81
OK-WAI	IL-18D	187009705	D/D 19-5-67. Crashed near Gander 5-9-67
OK-PAI	IL-18V	181003105	Ex-Czech Federal Government IL-18V 'Salon' OK-BYP, transferred 21-1-77. Retired; sold to *Auto & Technik Museum* (Sinsheim, Germany) 1-90
OK-WAJ	IL-18D	187010101	D/D 21-6-67. Retired 25-10-84, preserved Bakov, Czech Republic
DM-STC	IL-18V	180002202	Leased from Interflug 2-1-77 to 11-2-77 (as a substitute for the wrecked OK-NAA?)
DM-STD	IL-18V	180002302	Leased from Interflug 27-2-77 to 6-3-77
DM-STF #2	IL-18V	184007405	Leased from Interflug 12-2-77 to 17-2-77
DM-STG	IL-18V	182004402	Leased from Interflug 4-4-77 to 16-4-77
DM-STH	IL-18V	184007305	Leased from Interflug 14-11-76 to 27-12-76

Incidentally, the Czechs and Slovaks have a unique civil aircraft registration system. The first letter following the nationality prefix denotes the year of registration, except B (see next entry) and Q (which is never placed first); eg, O = 1960, Z = 1970 etc. The second letter denotes the type (eg, A = IL-18, later reused for Airbus A310-308) and the third letter is individual. In the case of both ex-Government *Coots* the last letter was reused for some reason. However, in both cases the first letter of the new registrations *did not match the year*: OK-BYP and OK-BYZ should have become OK-HAI and OK-HAF respectively.

Starting off with high-density domestic routes, ČSA was soon using the type on routes to Western Europe and other destinations. Thanks to its long range the *Coot* could carry passengers from Prague to Bamako and Conakry (via Rabat and Dakar in both cases), Baghdad (via Athens and Damascus), Helsinki (via Stockholm) and Ankara. When the airline took delivery of two long-range IL-18Ds in 1967, these were used on the Pragu-Havana service, allowing the two Bristol Britannia Srs 318s leased from Cubana (OK-MBA and OK-MBB) to be returned.

Originally the IL-18s sported a red cheatline beneath the windows, with 'feathers' at the front, and a red pinstripe running at the window centre level; unlike the old TABSO livery, the 'feathers' were below cheatline level rather than above it, which gave rise to the nickname *fousy* ('mustache' in Czech). The vertical tail was white with a small Czech flag. In 1969 ČSA introduced a new livery with a simple red cheatline running across the windows and a red tail proudly bearing a large flag and the letters OK. This, as the reader remembers, was both Czechoslovakia's nationality prefix and the airline's flight code; however, the coincidence with the universally known slang word made a positive selling point – 'that's an OK airline!' In the 1970s most of the IL-18s were named after Czechoslovakian cities and towns.

Interestingly, while the Tupolev Tu-134A introduced by ČSA in December 1971 was originally viewed as an 'IL-18 replacement', the two types served on alongside each other until the IL-18 was finally withdrawn in October 1984. IL-18D OK-WAJ performed the type's last revenue flight in ČSA service on 24th October; it was retired the following day with 23,578 hours on the clock and 20,829 cycles. The aircraft was eventually to be preserved.

The **CZECH FEDERAL GOVERNMENT FLIGHT** (LOMV – *Letecký oddíl ministerstva vnitra*; also called LS FMV – *Letecká společnost federalního ministerstva vnitra*, Airline of the Federal Ministry of the Interior) operated two VIP-configured IL-18s. The first of these was IL-18V OK-BYP (c/n 181003105) delivered in early July 1962; the other one was IL-18E 'Salon' OK-BYZ (c/n 186009004) delivered on 9th May 1966. It has to be said here that all of the unit's fixed-wing aircraft were registered in the BYx block; the B did not denote the year of registration but stood for [*Veřejná*] *bezpečnost* – 'public security' (ie, police).

Both IL-18s were sold to ČSA in 1977 and refitted with tourist-class interiors when the government flight took delivery of new IL-62 'Salon' VIP aircraft. OK-BYP became OK-PAI on 21st January, while OK-BYZ was delivered on 6th December as OK-VAF. The registration OK-BYP later passed to a Tu-154M 'Salon' (c/n 90A858) and OK-BYZ to an IL-62M 'Salon' (c/n 2647737).

As already noted in Chapter Two, the **VZLÚ** research and development organisation (*Výzkumní a zkušební létecký ústav*) purchased IL-18V DDR-STC (c/n 180002202) from Interflug in late October 1987. Reregistered OK-018 on 30th November, the airliner was converted into the IL-18LL testbed for the Walter (Motorlet) M-602 turboprop. Upon completion of the trials the IL-18LL was withdrawn from use and finally scrapped in January 1996.

DJIBOUTI

In mid-2000 **AIR DJIBOUTI – RED SEA AIRLINES** [–/DJU] operated IL-18E 3C-KKK (c/n 186009202) wet-leased from Air Cess.

DAALLO AIRLINES [D3/DAO] operated a total of six IL-18s leased from other carriers as required.

Registration	*Version*	*C/n*	*Notes*
EX-75427	IL-18V	183005905	Leased from Phoenix Aviation 10-00, full colour scheme; reported in error as RA-75427
RA-75449	IL-18D	187010004	Sub-leased from Ramaer 5-98 to 6-98
RA-75713	SL-18D	186009403	Converted IL-18D; leased from LNPO Leninets 10-93 to 10-94
75804	SL-18I	182004305	Converted IL-18V; leased from LNPO Leninets 7-93. Aeroflot cheatline, green tail, Daallo titles/logo
SP-FNW	IL-18E	186009202	Leased from Polonia Airways 3-96 to 8-96
4R-EXD	IL-18GrM	187009802	IL-18D/F (SCD). Leased from Expo Aviation 3-03

DOMINICAN REPUBLIC

TAINO AIRLINES, S.A. [–/TIN] based at Santo Domingo-Las Americas airport leases IL-18s from Aero Caribbean as required. For instance, it operated IL-18D CU-T1268 (c/n 188010704) in March 2000; the aircraft was in Aero Caribbean colours with additional Taino Airlines titles.

EGYPT

UNITED ARAB AIRLINES [MS/MSR], the Egyptian flag carrier, took delivery of four late-production IL-18Ds in 1968-69. The name was a leftover from the United Arab Republic (the union of Egypt and Syria which existed from February 1958 to September 1961); for some reason it persisted until 1972 when the airline was rebranded **EGYPTAIR**. Unfortunately next to nothing is known about Egyptian IL-18 operations.

Registration	Version	C/n	Notes
SU-AOV	IL-18D	188011003	D/D 8-68. Crashed Nicosia 29-1-73
SU-AOX	IL-18D	188011005	D/D 18-10-68. Returned to the USSR as CCCP-75430 (Ukrainian CAD/L'vov UAD)
SU-AOY	IL-18D	188011101	D/D 19-12-68. Returned to the USSR before 8-75 as CCCP-75429 (Krasnoyarsk CAD/1st Krasnoyarsk UAD/ 214th Flight, to Urals CAD/1st Sverdlovsk UAD/ 120th Flight; believed to Moscow Territorial CAD/ Domodedovo UAD/212th Flight)
SU-APC	IL-18D	188011301?	D/D 4-3-69; crashed Aswan 20-3-69
YR-IMK	IL-18E	186009104	Leased from TAROM. Crashed into Red Sea 9-12-74

Sometime in the late 1980s or early 1990s the charter airline **AIR LUXOR** leased IL-18V LZ-BEU (c/n 183005905) from Balkan Bulgarian Airlines.

While Egypt was strictly a civil operator of the IL-18, it is worth noting that Soviet Navy IL-38s wearing **EGYPTIAN AIR FORCE** (*al Quwwat al-Jawwiya il-Misriya*) markings for appearance's sake operated from Egyptian bases in the mid-1970s when the country was still on good terms with the USSR. Only one aircraft serialled 4299 is known.

EQUATORIAL GUINEA

CESSAVIA [–/CSS], later rebranded **AIR CESS** (not to be confused with its earlier Liberian namesake – see page 113), operated four IL-18s transferred from Air Pass (see Swaziland section) in November 1998. It should be noted that all three airlines were controlled by the same man – an expatriate Russian named Victor Bout[1] who has gained notoriety as a weapons trafficker; the aircraft were simply transferred to a different country of registry and the airline incorporated under a new name when things became too hot for comfort! The aircraft were based at Sharjah, UAE, and periodically leased to various operators on the African continent (where, it should be noted, armed conflicts continue unabated and there is a consistent demand for weapons).

Registration	Version	C/n	Notes
3C-KKJ	IL-18V	184006903	Ex-Air Pass 3D-SBC; full c/s. Sold to unknown operator early 2003 as UN 75003 or UN 75004
3C-KKK	IL-18E	186009202	Ex-Air Pass 3D-SBW; full c/s. Sold to unknown operator early 2003 as UN 75003 or UN 75004
3C-KKL	IL-18D	187010204	Ex-Air Pass 3D-SBQ, named 'Zlatoust'. Sold to unknown operator 9-02 or 10-02 as UN 75005
3C-KKR	IL-18E	185008603	Ex-Santa Cruz Imperial EL-ARK. Sold to unknown operator early 2003 as UN 75002

GERMANY (EAST GERMANY & POST-REUNIFICATION GERMANY)

Civil operations of the type in the former East Germany (German Democratic Republic) may seem a bit confusing because *two different airlines used the same aircraft for a while*. Founded in early 1955, **DEUTSCHE LUFTHANSA** (DLH) ordered its first three IL-18Vs in October 1959 to replace the slow IL-14P on high-priority routes.

(Yes – Deutsche Lufthansa! Before the reader decides that either he or the authors have had 'one too many', we hasten to explain that in the time frame between 1955 and 1963 there were *two Deutsche Lufthansas* – one in East Germany and the other (the one which still exists) in West Germany. Both considered themselves to be the rightful successors of the original (ie, pre-war) Lufthansa – and, to complicate things further, both used the same 'flying crane' logo! Total confusion (with potentially disastrous results, such as an unsuspecting passenger boarding a flight of *the wrong Lufthansa* and ending up in Karl-Marx-Stadt instead of Hannover, much to the relish of the infamous Stasi!) was avoided solely because the East German airline operated only domestic flights.)

The first aircraft was delivered on 28th March 1960, the first revenue flight from Berlin-Schönefeld to Moscow-Sheremet'yevo taking place as early as 3rd April. The aircraft was captained by instructor pilot Gerhard Friess, with trainee pilot Herbert Kohl as first officer, Kurt Roerich as flight engineer, Herbert Wittnebel as navigator and Heinz Martin as radio operator. Actually this flight was more of a 'yes, we can' show; the first two years of IL-18 operation were largely devoted to flight and ground crew conversion training and familiarisation. It was not until 1962 that the type really began scheduled flights in East Germany.

To fully realise the importance of the IL-18's introduction, consider that the airline did not order the Tupolev Tu-104 because plans were in hand to manufacture the Baade 152/II four-turbojet medium-haul airliner in Dresden, for which Deutsche Lufthansa was to be the launch customer. However, in late 1960 the East German government pulled the plug on the national aircraft industry, cancelling Baade 152 production; this left the IL-18 as East Germany's only modern airliner. Three aircraft (DM-STA through -STC) have been reported in DLH markings, although, judging by the registration/delivery dates, DM-STF No 1 and DM-STG should have had them as well (see table below).

For over four years Deutsche Lufthansa existed in parallel with **INTERFLUG** [IF/IFL], the second East German airline established on 18th September 1958. As the name implies (a contraction of Internationale Fluggesellschaft), this was originally a charter airline serving exclusively international routes which were off limits to DLH because of its name. (As long as Deutsche Lufthansa flew domestic flights, nobody seemed to mind, but any attempt to venture outside the country would immediately result in a legal spat with its West German namesake.) Interflug leased the IL-18s as required, retaining the blue cheatline but changing the titles and logo; upon return from lease the DLH titles and logo were reapplied.

Then the government decided there was one airline too many. Thus on 1st September 1963 DLH merged with Interflug – officially for economic reasons but actually rather for the purpose of expanding the foreign route network and ending the legal dispute with *the other* Deutsche Lufthansa. The revamped Interflug now had a whole range of divisions responsible for different kinds of aerial work, mirroring the 1950s structure of the Soviet Union's civil aviation. The IL-18s belonged to the largest and most important department – Interflug Verkehrsflug (IF/VF) tasked with passenger and cargo transportation, except for DM-STP which was used by the Inspectorate of Air Traffic Control & Communication (IF/TP-FSNW – Interflug Technische Prüfung – Flugsicherung und Nachrichtenwesen) as a calibrator since 1978.

The red/white Interflug livery was introduced concurrently with this unification. Curiously, there was no standard shade of red; the paint shops used whatever paint was available, with the result that the colour of the cheatline and tail varied from brick-red to pinkish! The lower fuselage and wings, which were originally left unpainted, later became grey; again individual aircraft competed in whose grey paint was the greyest. (The one-of-a-kind calibrator was again an exception to the rule with its Grey Mouse colour scheme.)

The ten IL-18s delivered directly to the two airlines were later augmented by six ex-VIP machines transferred from the East German Air Force (see next page). With the introduction of the IL-62 *sans suffixe* in 1972 and then the IL-62M in 1980 the *Coots* were chiefly relegated to charter flights and so-called 'solidarity flights'; the longest and most gruelling of these was the 11,000-km (6,830-mile) route from Berlin to Hanoi via Moscow, Tashkent and Dacca. Interflug IL-18s were also used for non-scheduled flights to the UK, Iceland and Cuba associated with crew rotation on East German fishing trawlers.

In addition to passenger transportation the *Coots*' activities included VIP flights and numerous relief missions under the auspices of the United Nations Organisation. For instance, in 1964 IL-18V DM-STC made a tour of Indonesia, Cambodia, Burma, Ceylon and India, carrying an East German government delegation. During 1986-88 the IL-18s were used alongside Antonov An-26s chartered from the East German Air Force for famine relief flights in Ethiopia. Incidentally, Interflug compared favourably to the other East European nations operating the IL-18 in terms of flight safety; only one aircraft was lost in a crash – and, ironically, the accident occurred on a relief flight.

In 1981 East Germany changed its nationality prefix from DM- to DDR-, resulting in a unique six-letter format, and the entire civil aircraft fleet was reregistered in the course of the year. Curiously, the DDR- prefix had already been in use from 1952 to 1956. The reason why they reverted to the old prefix may have been that someone in the government perceived the letters DM as a hint at the currency of the ideologically hostile West Germany (Deutschmark)!

After German reunification on 3rd October 1990 the seven IL-18s remaining operational by then changed their identity again, remaining with Interflug for the time being. Usually the D-AOxx registrations[2] were allocated in the same order as the old DDR-STx registrations had run; the calibrator was *again* an exception, receiving an out-of-sequence registration. However, when East and West Germany entered a financial and currency union on 1st July 1990 in preparation for reunification, Interflug began rapidly losing ground to western air carriers. On 7th February 1991 the airline was declared bankrupt and the entire fleet was sold off in due course.

Registration	*Version*	*C/n*	*Notes*
DM-STA*	IL-18V	180001905	Regd 28-4-60. To DDR-STA 22-7-81; WFU 1988, preserved Leipzig-Schkeuditz
DM-STB*	IL-18V	180002001	Regd 2-4-60. To DDR-STB 8-9-81; WFU 1987, preserved *Luftfahrthistorische Sammlung Finow* ?-92. To Berlin-Schönefeld 8-93, preserved *Aeropark Diepensee* 2-95/6-98; to *Automobilmuseum Leipzig-Plagwitz* 5-00
DM-STC*	IL-18V	180002202	Regd 16-7-60. To DDR-STC 6-10-81. Sold to VZLÚ ?-11-87 as OK-018
DM-STD	IL-18V	180002302	Ex-East German AF/IL-18V 'Salon', transferred 3-9-64. To DDR-STD 1-7-81; WFU 1988, used as a restaurant in Harbke near Helmstedt
DM-STE	IL-18V	182005101	Ex-East German AF/IL-18V 'Salon', transferred 3-6-64. To DDR-STE 1-10-81; WFU 1989, preserved Borkheide
DM-STF #1*	IL-18V	181004105	Ex-CCCP-75796, regd 13-9-62. Damaged by fire Moscow-Bykovo ?-1-67 and replaced by, see next line; repaired, to Aeroflot (Krasnoyarsk CAD/1st Krasnoyarsk UAD/214th Flight) as CCCP-75475
DM-STF #2	IL-18V	184007405	Ex-CCCP-75553, regd 7-3-67 as compensation for loss of DM-STF #1. To DDR-STF 21-8-81; reregistered 3-10-90 as, see next line
D-AOAO			Sold to Air Cargo after demise of Interflug
DM-STG*	IL-18V	182004402	Regd 24-3-62. To DDR-STG 19-8-81; WFU 1981, preserved Erfurt
DM-STH	IL-18V	184007305	Ex-East German AF/IL-18V 'Salon', transferred ?-2-70. To DDR-STH 19-8-81; last flight 5-5-90 to Augsburg and preserved at Augsburg airport collection; sold to *Flugausstellung Leo & Peter Junior*, Hermeskeil, in 1994
DM-STI	IL-18V	185008404	Ex-East German AF/IL-18V 'Salon', transferred 24-1-74. To DDR-STI 15-11-81; rereg 3-10-90 as, see next line
D-AOAP			Sold to Air Cargo after demise of Interflug
DM-STK	IL-18E	186009202	Regd 12-7-66. To DDR-STK 12-8-81; reregistered 3-10-90 as, see next line
D-AOAR			Sold to Polnippon 2-11-90 as SP-FNB
DM-STL	IL-18D	186009402	Regd 9-9-66. Crashed Luanda 26-3-79 on a cargo flight
DM-STM	IL-18D	188010805	Ex-East German AF/IL-18D 'Salon', transferred ?-3-74. To DDR-STM 1-1-81; reregistered 3-10-90 as, see next line
D-AOAS			Registration applied 4-10-90! Sold to Air Cargo after demise of Interflug
DM-STN	IL-18D	188010903	Regd 15-4-68. To DDR-STN 25-8-81; reregistered 3-10-90 as, see next line
D-AOAT			Sold to Polnippon 7-12-90 as SP-FNC
DM-STO	IL-18D	188010904	Regd 15-4-68. To DDR-STO 2-11-81; reregistered 3-10-90 as, see next line
D-AOAU			Sold to Air Cargo after demise of Interflug
DM-STP	IL-18V	184007401	Ex-East German AF/IL-18V 'Salon', transferred 1-4-70; converted to navaids calibration aircraft ?-78. To DDR-STP 8-9-81; reregistered 3-10-90 as, see next line
D-AOAQ			Sold to Air Cargo and reconverted to standard after demise of Interflug

* Originally delivered to Deutsche Lufthansa

The **EAST GERMAN AIR FORCE** (LSK/LV – *Luftstreitkräfte und Luftverteidigung der Deutschen Demokratischen Republik* – Air Force and Air Defence Force of the German Democratic Republic) took delivery of six VIP-configured IL-18s. Originally they were operated by **STS-29** (*Selbständige Transportfliegerstaffel* – independent airlift squadron) at Marxwalde AB, Brandenburg District/Vorpommern.[3] The unit performed government VIP flights and hence was commonly known as *die Regierungsstaffel* (the government squadron). On 1st January 1973 STS-29 was reorganised, becoming **TFG-44 'Arthur Pieck'** (*Transportfliegergeschwader* – airlift wing) named after the first General Director of the East German Deutsche Lufthansa (and subsequently of Interflug).

Serial/Reg	*Version*	*C/n*	*Notes*
'493 Black'	IL-18V 'Salon'	180002302	Ex-CCCP-75726; regd 28-10-62, callsign DM-VAX. Registered as, see next line
DM-STD			Regd ?-64; sold to Interflug 3-9-64
'497 Black'	IL-18V 'Salon'	184007401	Regd 6-7-64, callsign DM-VAZ. Registered as, see next line
DM-STP			Regd ?-2-70 (officially reregistered 2-7-70!); sold to Interflug 1-4-70
DM-STE*	IL-18V 'Salon'	182005101	Regd 27-9-62; callsign DM-VAY added 10-10-62. Deutsche Lufthansa colours until 1-9-63. Sold to Interflug 3-6-64
DM-STH	IL-18V 'Salon'	184007305	Regd 12-9-64; sold to Interflug ?-12-70
DM-STI	IL-18V 'Salon'	185008404	Regd 26-11-65; sold to Interflug 24-1-74
DM-STM	IL-18D 'Salon'	188010805	Regd 22-6-68; sold to Interflug 3-74

* IL-18V 'Salon' c/n 182005101 is sometimes reported as '499 Black' with the callsign DM-VAY but this is unconfirmed.

The aircraft were mostly quasi-civil, wearing Deutsche Lufthansa or, after 1st September 1963, Interflug colours; this facilitated flying across the territory of certain countries which denied entry to military aircraft from the Eastern Bloc. Overt military insignia were only carried by the example which was the current official transport of the East German Minister of Defence. Such aircraft wore a colour scheme patterned on Aeroflot's pre-1973 standard for the type, although '497 Black' had red trim instead of blue. They also had phony registrations in the DM-VAx block (for [*Nationale*] *Volksarmee* – National People's Army) allocated as ATC callsigns which were not worn visibly.

After serving with the unit for three or four years the *Coots* were sold to Interflug and replaced by new ones. By May 1974 the IL-18 had been completely replaced by Tu-134K and Tu-134AK VIP jets in LSK/LV service.

Whoever took this picture was extremely lucky – unlike the aircraft he photographed. United Arab Airlines IL-18D SU-APC crashed barely two weeks after it was delivered! Yefim Gordon archive

In 1958-1963 Interflug existed as a charter carrier, leasing IL-18s from the East German Deutsche Lufthansa as required. This is IL-18V DM-STG, the last example delivered to Deutsche Lufthansa, during one of these leases; note the combination of the DLH cheatline (with a characteristic 'horn' aft of the flightdeck) and DLH white tail with Interflug titles and logo.
Yefim Gordon archive

No more Deutsche Lufthansa. Here, an Interflug IL-18V is seen after 1963, before the introduction of grey fuselage undersurfaces and wings which are obviously unpainted in this view.
Yefim Gordon archive

IL-18V HA-MOC in the old colours of MALÉV Hungarian Airlines taxies in at Moscow-Sheremet'yevo after a flight in the mid-1960s; the Nos 1 and 4 engines are already shut down to save fuel. On 28th August 1971 it crashed on approach to Copenhagen-Kastrup.
Yefim Gordon archive

Five ex-Interflug IL-18s were acquired by Berlin/Schönefeld-based **AIR CARGO**. Shortly afterwards (in late 1991) the carrier changed its name to **BERLINE** (a contraction of 'Berlin' and 'airline'). To meet a special order by BerLine the Il'yushin OKB converted two of the *Coots* to IL-18GrM configuration with a large side cargo door. However, on 31st March 1994 the airline ceased operations and filed for bankruptcy.

Registration	*Version*	*C/n*	*Notes*
D-AOAO	IL-18V	184007405	Ex-DDR-STF. Sold to German European Airlines
D-AOAP	IL-18V	185008404	Ex-DDR-STI. Converted to IL-18GrM, aka IL-18V/F (SCD), 2-93/6-93. Sold to German European Airlines
D-AOAQ	IL-18V	184007401	Ex-DDR-STP. Sold to Kryla Airlines 1-93 as UR-75475
D-AOAS	IL-18D	188010805	Ex-DDR-STM. Converted to IL-18GrM, aka IL-18D/F (SCD), 9-92/1-93. Sold to Air Zory 1-94 as LZ-AZZ
D-AOAU	IL-18D	188010904	Ex-DDR-STO. Sold to Air Zory 1-94 as LZ-AZR

In March 1994 **GERMAN EUROPEAN AIRLINES** bought the last two aircraft of the defunct BerLine – IL-18V D-AOAO and IL-18GrM D-AOAP. Apart from the titles and logo, the only change to the colour scheme was an ochre/dark blue cheatline instead of an all-blue one. However, on 20th December 1995 both aircraft were sold to Russian airline Viola Avia Trans as RA-75553 and RA-75554 respectively; delivery took place on 22nd and 23rd December respectively.

GHANA

In September 1960 **GHANA AIRWAYS** [GH/GHA] ordered its first four IL-18Vs, later increasing the order to eight aircraft which were delivered in 1960-62. According to Western press reports, initially the airliners were flown and serviced by Soviet crews while Ghanaian personnel were undergoing training in the USSR – a statement open to doubt, considering that usually the flight crews of customer nations flew the aircraft home themselves after taking the complete course of training at the COMECON Civil Aviation Training Centre in Ul'yanovsk. According to the same source, the airliners were purchased for £670,000 apiece on an eight-year financial lease basis and each aircraft came complete with four spare engines.

The IL-18s were used chiefly on the Accra-Beirut and Accra-Khartoum services; attempts to use them on 400-km (250-mile) domestic routes as well were soon given up as uneconomic. In February 1963 Ghana Airways launched flights from Accra to Moscow via Bamako, Tunis, Zürich and Prague. Soon, however, the *Coots* found themselves in the middle of a scandal. Perhaps Ghana Airways had been a little over-optimistic about traffic volumes when ordering the aircraft, but, as Western trade magazines wrote, IL-18 operations had turned into an economical nightmare. *Aeroplane & Commercial Aviation News* reported on 31st October 1963: *'Since taking delivery of its first Il'yushin IL-18 in 1961, Ghana Airways has achieved an annual utilization of only about 225 hours with each of the eight aircraft it eventually received. As a result, the airline incurred a total loss of about £876,000 in one year with its IL-18 fleet alone. The operating loss was £167,000 and to this was added another £700,000* (sic) *in respect of depreciation and interest charges.'* A spokesman for the Public Accounts Committee was quoted as saying average load factors *'had been only 18% – 14 passengers in a 78-seat aircraft'*.

An article in the same vein had appeared in *Newsweek* back in February 1962. The Soviet Union was not slow in reacting: on 13th April 1962 the ***Nov**oye **vr**emya* (New Time) magazine let go with a broadside, denouncing the *Newsweek* feature as *'a defamatory fib from beginning to end'*. The author of the article (titled *They're Jealous*) maintained that *'the Western aircraft industry monopolies which are losing lucrative business are starting to use the dirtiest tricks of competition, attempting to discredit Soviet civil aviation and its aircraft. It is really amazing what a lot of nonsense concerning Soviet aircraft [...] is currently being disseminated in the Western press. [...] In the Aviaexport All-Union Company [...] we were informed there are numerous extremely positive comments on the performance, maintainability and operating economics of the IL-18.'* Saints'R'Us. Well, the reader should remember that in the Cold War days the East and the West never missed a chance to take a jab at each other and, putting it mildly, the statements on both sides were not always fair. The truth probably lies somewhere in between... Anyway, in 1963 all eight aircraft were returned to the USSR against a partial refund.

Registration	*C/n*	*D/D*	*Notes*
9G-AAI*	180002402	D/D 3-12-60	Returned 14-6-63; to CCCP-75531?
9G-AAJ*	180002405	D/D 3-12-60	Returned 14-6-63, to CCCP-75535 (regd 17-6-65)
9G-AAK*	180002501	D/D 14-2-61	Returned 14-6-63; to CCCP-75533?
9G-AAL*	180002502	D/D 14-2-61	Returned 14-6-63; to CCCP-75534?
9G-AAM	181003305	D/D 21-4-61	Returned 17-6-65, to CCCP-75424
9G-AAN	181003403	D/D 28-9-62	Returned before 9-65, to CCCP-75425 (Urals CAD/1st Sverdlovsk UFD/120th Flight)
9G-AAX	181002803	D/D 8-9-61†	WFU 31-12-63; to USSR as CCCP-75426 (West Siberian CAD/Tolmachovo UAD/384th Flight), regd 20-6-66
9G-AAY	181002804	D/D 8-9-61†	WFU 14-6-63; returned to USSR?

* 9G-AAI through 9G-AAL have been reported in error as ex-CCCP-75731 through -75734.
† Reported as first noted Dakar in June 1961!

GUINEA

AIR GUINÉE [GI/GIB] started IL-18 operations in 1960 with several aircraft leased from Aeroflot pending delivery of its own *Coots*. The first three of these originally had out-of-sequence registrations denoting major cities of the country (3X-KKN for Kankan, 3X-LBE for Labé and 3X-NZE for Nzerekore) but were later reregistered. All in all, Air Guinée operated 12 IL-18s.

The *Coots* were mostly used on scheduled international flights from Conakry to Dakar, Lagos (via Abidjan and Accra) and Robertfield. Occasionally they also carried pilgrims to Jeddah via Kano, Nigeria, and took President Sésé Sekou Touré on his foreign visits. (Inside the country the President used an Avia 14 'Salon' (a VIP-configured Czech-built IL-14P) aptly registered 3X-PRG for Président de la République de Guinée.)

Registration	*Version*	*C/n*	*Notes*
3X-GAT	IL-18D	188010604	Ex-Chosonminhang '834', D/D 8-8-68; named 'Conakry'. WFU Conakry ?-86
3X-GAX	IL-18D	187009803	Ex-Yugoslav Government Flight IL-18D 'Salon' YU-AIB, bought 3-78. Crashed Conakry 3-9-78
3X-GOC	IL-18D	?	Leased from Aeroflot?, dates unknown
3X-GOD	IL-18D	186009601	Leased from Aeroflot 1968-69, ex/to CCCP-75437
3X-GOE	IL-18D	?	Leased from Aeroflot?, dates unknown
3X-GOF	IL-18D 'Salon'	187010103	Lsf Aeroflot (235th Independent Air Detachment) 3-68 to 8-68, ex/to CCCP-75453
3X-KKN	IL-18V	181003703	D/D ?-61; reregistered as, see next line
3X-GAB			Crashed Casablanca 9-7-67
3X-LBE	IL-18V	181003705	D/D ?-61; reregistered as, see next line
3X-GAA			Returned and WFU Moscow-Sheremet'yevo by 10-77; refurbished and delivered to Aeroflot (Moscow Territorial CAD/Domodedovo UAD/212th Flight) as CCCP-74299 by 4-79
3X-NZE*	IL-18V	181003704	D/D ?-61; reregistered as, see next line
3X-GAC			Returned, to Aeroflot as CCCP-75428
75800	IL-18V	182004205?	Leased from Aeroflot 1963-64, ex/to CCCP-75800
75803	IL-18V	182004304	Leased from Aeroflot 1963, ex/to CCCP-75803
75841	IL-18V	182005302	Leased from Aeroflot 1965, ex/to CCCP-75841

* 3X-NZE has also been reported with the c/n 180002004, which makes it ex-CCCP-75719, but this is doubtful.

The transport squadron of the **GUINEAN AIR FORCE** (*Force Aérienne de Guinée*) reportedly used the Air Guinée IL-18s from time to time; 75803 and 75841 were reported as Air Force operated in 1982. According to *Flight International* of 28th November/4th December 2000, two IL-18s were still in service of that date!

GUINEA-BISSAU

Two IL-18Vs registered RA-75431 (c/n 180002003) and RA-75851 (c/n 182005501) were leased from the Flight Research Institute in Zhukovskiy in the autumn of 1992; the lessor was some government organisation, judging by the fact that both aircraft wore Republica da Guinée Bissau titles. Both were returned on 16th December 1992.

HUNGARY

MALÉV HUNGARIAN AIRLINES [MA/MAH] (ie, Magyar Légiközlekedesi Vallalat – Hungarian Air Transport Co) had eight IL-18s.

Registration	*Version*	*C/n*	*Notes*
HA-MOA	IL-18V	180001903	D/D 2-4-60. Converted to IL-18Gr. Last flight and WFU 19-2-87; preserved Férihegy Airport Museum 6-92
HA-MOC	IL-18V	181002903	D/D 1-4-60. Crashed on approach to Copenhagen-Kastrup 28-8-71
HA-MOD	IL-18V	180002002	D/D 1-4-60 as HA-MOB but rereg because of the English word 'moo'! Crashed Paris-le Bourget 23-11-62
HA-MOE	IL-18V	182005505	D/D 27-12-62. Converted to IL-18Gr. Last flight and WFU 22-4-87; preserved *Magyar Repülöstörteneti Muzeum* (Hungarian Air Force Museum), Szolnok
HA-MOF	IL-18V	183006301	D/D 12-6-63. DBR Bucharest-Otopeni 23-11-77
HA-MOG	IL-18V	184007103	D/D 28-3-64. Converted to IL-18Gr. WFU 6-12-88 and handed over to Malév Training Centre at Budapest-Férihegy, preserved
HA-MOH	IL-18V	184007104	D/D 28-3-64. Crashed on approach to Budapest-Férihegy 15-1-75
HA-MOI	IL-18D	187010002	D/D 21-6-67. Converted to IL-18Gr. Last flight to Pápa and WFU 29-8-89; preserved as a restaurant in Györ

The original livery with a 'zig-zag' cheatline and a 'handwritten' M on the tail was replaced in the 1970s by a new one introduced by the Tu-134 *sans suffixe* in 1969. The *Coots* were originally delivered in 84-seat configuration, but in the 1980s the four surviving IL-18s were converted to freighters, their original role taken over by the Tu-154 which Malév introduced in 1973. The last commercial IL-18 flight in Hungary was performed by HA-MOI on 25th January 1989.

INDIA

India was the sole export customer for the IL-38 ASW aircraft. In 1977 the **INDIAN NAVY** purchased five ex-AVMF aircraft. It has to be said the Soviet Navy was rather reluctant to part with the aircraft, as the IL-38s were scarce and production had ended by then. Selling the *Mays* was not as easy as it seems, since all manuals had to be translated into English, all labels and maintenance stencils replaced etc.

Serialled IN301 through IN305 (no c/ns are known), the IL-38s were operated by Indian Navy Air Squadron (INAS) 315 based at Dabolim AB near Goa, the capital of India's smallest state bearing the same name, and accordingly wore DAB tail codes. Conversion training took place at Riga-Skulte, starting in September 1976; the first six trainees completed their course on 17th May 1977 and the entire group did so on 30th August. The first three aircraft were delivered on 2nd September 1977. The IL-38s were supplied with AT-1E export ASW torpedoes.

In 1999 the Indian *Mays* received an upgraded mission avionics/ESM suite with large antenna fairings on the sides of the extreme nose and at the root of the tailboom. Unfortunately on 1st October 2002 the Indian Navy lost nearly half of its IL-38 fleet when IN302 and IN304 collided and crashed during an air fete in Goa.

IRAN

In 1993-95 the charter carrier **KISH AIR** [–/IRK], a subsidiary of the Kish Development Organisation (a free zone enterprise on Kish Island), leased IL-18D YR-IMZ No 2 (c/n 187009802) from Romavia.

LATVIA

In June 1998 the Vilnius-based charter airline **CONCORS** [–/COS] acquired IL-22M CCCP-75916 (c/n 2964017102) from the Belorussian Air Force. After being demilitarised the aircraft was registered YL-LAO, receiving a smart red/white livery with IL-18D nose titles and a three-class interior with a VIP cabin at the rear. In 2003, however, YL-LAO was put up for sale and eventually sold to an unidentified Cuban operator in November as CU-T1539.

LIBERIA

AIR CESS (LIBERIA), INC [–/ACS] purchased IL-18D EL-AKQ (ex-SPAir RA-75497, c/n 187010204) in December 1995; the aircraft was named 'Zlatoust' after a Russian town once famous for its knives and swords. A second *Coot*, IL-18V EL-AHO (ex-Southern Cross 3D-AHO (?), ex-TAROM YR-IME, c/n 183006205) in basic TAROM colours without titles, joined the fleet before May 1998.

There is good reason to believe that the registrations were simply a flag of convenience – among other thins because, though Liberian-registered, the aircraft were actually based at Sharjah, UAE. EL-AHO was cancelled from the Liberian register by July 2001 (its fate is unknown),[4] while EL-AKQ was transferred to the airline's sister company Air Pass (see Swaziland section) in late 1997, becoming 3D-SBQ.

SANTA CRUZ IMPERIAL [–/SCI], another Sharjah-based cargo airline, operated five or six IL-18s.

Registration	*Version*	*C/n*	*Notes*
EL-ADY #1	IL-18V	184007405	Ex-Air Zory LZ-AZO, bought 4-97; still in full European Airlines colour scheme! Registration removed by 3-98; sold to BIO Air 4-98 as T9-ABB. See next line
EL-ADY #2	IL-18V	182004804	Ex-Air Pass 3D-ALQ, bought before 12-98. SCI colour scheme, no titles. Was reported as EL-AHQ in JP Airline-Fleets '98, probably in error! Sold to Star Airlines ?-99 as EX-7504
EL-AHO?	IL-18V	183006205	Ex-Air Cess? Reported in JP Airline-Fleets '98, probably in error!
EL-ALD*	IL-18E	185008601	Ex-Southern Cross 3D-ALD (?), ex-Balkan LZ-BEW; bought 5-98, basic LOT Polish Airlines colour scheme, no titles. Sold to Phoenix Aviation ?-00, believed reregistered EX-601
EL-ALW	IL-18V	182004904	Ex-RA-75825; named 'Lastochka' (Swallow). Sold to Phoenix Aviation ?-99 as EX-75825
EL-ARK	IL-18E	185008603	Ex-Balkan LZ-BEZ, basic LOT Polish Airlines colour scheme (later SCI c/s), no titles. Transferred to Air Cess as 3C-KKR 3-99

* The registration EL-ALD was previously worn by an Antonov An-12BP (ex-RA-11105, c/n 01347801), likewise operated by Santa Cruz Imperial.

In July-August 2000 **WEST AFRICAN AIR SERVICES** operated an IL-18E wearing the illegally applied Liberian registration EL-ALY (ex-Renan ER-ICJ, c/n 186009102); the aircraft was never in the official Liberian register and was still listed with the latter airline at the time. Soon afterwards ER-ICJ was sold to Tavria MAC as UR-TMD.

MALI

When Mali proclaimed independence in 1960, all equipment deliveries from the Western world were promptly cut off and the government of the newly-established republic turned to the Soviet Union for assistance – which was readily extended in the hope of securing another political beachhead in Africa. On 18th March 1961 the Soviet Union and the Republic of Mali signed a trade agreement which, among other things, provided for help in the setting up of a national air carrier, **AIR MALI** [L9/MLI]. Two days later the airline ordered a sizeable number of aircraft, including three IL-18Vs. On behalf of Mali the order was signed by Minister of Home Affairs and Information Madeira Keita.

According to the Soviet press, Air Mali began IL-18 operations in 1962; considering the registration dates of TZ- registered examples (see table), the aircraft in question were obviously wet-leased from Aeroflot and their identities remain unknown as of this writing. Apart from scheduled international flights (the destinations included Paris, Marseilles and Rabat), the IL-18s were used by President Modibo Keita on his trips outside the country. On 19th March 1964 the airline's director Assan Gindo said in an interview to the Soviet TASS News Agency, 'We can say with assurance that it was only due to the delivery of IL-14 and IL-18 airliners and of other necessary equipment by the Soviet Union that the national Air Mali company went into operation. [...] Soviet airliners [...] have shown excellent performance in the difficult conditions of Africa. We get many letters from our passengers who speak highly of the Soviet pilots.'

A further IL-18 was delivered in 1975 to replace a crashed aircraft. Eventually, however, all surviving Air Mali IL-18s were returned to the USSR.

Registration	*Version*	*C/n*	*Notes*
TZ-ABD	IL-18V	181003303	Ex/to CCCP-75762, regd 8-1-65; returned 2-6-70
TZ-ABE	IL-18V	181003304	Ex-CCCP-75763, regd 8-1-65. Crashed Lingomin, Upper Volta, 11-8-74
TZ-ABY	IL-18V	182004505	Returned ?-72, fate unknown
TZ-ADF	IL-18D	187010301	Ex/to CCCP-75459, lsf Aeroflot 2-4-74. WFU/stored Bamako 3-84; returned 31-3-87

MAURITANIA

On 19th April 1969 **AIR MAURITANIE** [MR/MRT] took delivery of IL-18D 5T-CJL (c/n 189011304) – the last-ever passenger-configured IL-18. However, only two years later the aircraft was sold to Háng Không Viêt Nam, becoming VN-B198.

NIGERIA

In July 2001 **FRESH AIR** [–/FRR], an airline home-based at Sharjah (UAE), leased IL-18D EX-75442 (c/n 187009702) from Phoenix Aviation. The aircraft was reportedly still (or maybe *again*) on lease in February 2003.

NORTH KOREA

North Korea is one of the world's most secretive nations, and details of its aviation are notoriously hard to get. It is known that the national airline **CHOSONMINHANG** (ie, Civil Aviation Administration of Korea – CAAK) operated at least four IL-18s. By 1994 the airline had been renamed **AIR KORYO** [JS/KOR] – and it's just as well, because the old name was a real tongue-twister. ('Koryo' is apparently how 'Korea' is pronounced in the native language; the North Koreans themselves, however, call their country by an even less pronounceable phrase which translates as 'the land of the morning freshness'.)

However, virtually all of North Korea's aviation is of a near-military (or *very* military) nature and some of the aircraft nominally owned by Chosonminhang/Air Koryo are in fact operated by the **NORTH KOREAN AIR FORCE**. The IL-18s are known to have been used on government VIP flights.

Registration	*Version*	*C/n*	*Notes*
525	?	?	Seen 6-70. Reportedly ex-825 but not confirmed
825	?	?	Seen 4-68. Reportedly reregistered 525
834	IL-18D	188010604	D/D 13-2-68, no titles. Sold to Air Guinée 8-8-68 as 3X-GAT
835	IL-18D	188011205	D/D 20-3-69.Reregistered P-835 ?-80; to Air Koryo
836	IL-18V	185008204	D/D 16-5-65.Reregistered P-836 ?-80; to Air Koryo

POLAND

Starting in April 1961, **LOT POLISH AIRLINES** [LO/LOT] (Polskie Linie Lotnicze LOT; 'lot' is the Polish word for 'flight') acquired ten IL-18s with a seating capacity ranging from 84 to 102.[5]

Registration	*Version*	*C/n*	*Notes*
SP-LSA	IL-18V	180002403	Regd 21-4-61; named 'Warszawa'. Converted to IL-18Gr; last flight 20-10-88, WFU Warsaw-Okęcie and scrapped 2-90
SP-LSB	IL-18V	180002404	Regd 21-4-61; named 'Westerplatte'. Last flight 2-11-88; WFU Warsaw-Okęcie and scrapped 2-90
SP-LSC	IL-18V	181002805	Regd 21-4-61; named 'Narwik'. Sold to Balkan 18-4-88 as LZ-BEI
SP-LSD	IL-18V	184007102	Regd 5-4-64; named 'Tobruk'. WFU and preserved near Podłaski, later to Wikloń near Kruszyna, still later to Kościelec near Częstochowa
SP-LSE	IL-18V	180002504	Ex-Polish Air Force IL-18V 'Salon' '101 Red' #1, regd 11-9-65; named 'Lenino'. WFU Warsaw-Okęcie ?-89, preserved Warsaw 1991
SP-LSF	IL-18E	185008601	Regd 25-11-65; named 'Falaise'. Sold to Balkan before 8-91 as LZ-BEW
SP-LSG	IL-18E	185008603	Regd 1-12-65; named 'Monte Casino'. Sold to Balkan before 7-91 as LZ-BEZ
SP-LSH	IL-18V	181002701	Ex-Polish Air Force IL-18V 'Salon' '102 Red' #1, regd 16-7-66; named 'Kołobrzeg'. WFU ?-89, preserved as a restaurant in Radomsko, later moved to Strumień
SP-LSI	IL-18E	186008905	Ex-Polish Air Force IL-18E 'Salon' '102 Red' #2, regd 28-3-75. Sold to Balkan as LZ-BEH, date unknown
SP-LSK	IL-18E	185008503	Ex-Polish Air Force IL-18E 'Salon' '101 Red' #3, regd 6-1-88. Cancelled from Polish register 21-9-90, sold to unknown operator as 75711, later to Kryla as UR-75850

The IL-18 was the first turboprop airliner to enter service in Poland; as a symbolic gesture the type was first put into service on the Warsaw-Moscow route. The Polish aircrews were quick to appreciate the *Coot*'s well-equipped avionics suite, particularly the weather radar with its secondary traffic collision avoidance function. Later the IL-18 had the distinction of being the aircraft with which LOT launched several new services. For instance, the route from Warsaw-Okęcie to London-Heathrow via Amsterdam-Schiphol was inaugurated on 1st April 1962; on 17th July 1962 LOT started the Warsaw-Cairo service.

This busy scene at Moscow-Sheremet'yevo sometime between mid-1966 and mid-1970 shows an unidentified North Korean IL-18 sandwiched between a Bulgarian *Coot* (TABSO IL-18E LZ-BED) and a pair of Aeroflot Tu-104Bs. Yefim Gordon archive

This 36. SPLT IL-18E 'Salon' (c/n 185008503) was the third and final Polish Air Force *Coot* to be serialled '101 Red'. It was sold to LOT Polish Airlines as SP-LSK on 6th January 1988. Yefim Gordon archive

The 36. SPLT operating the Polish Air Force's VIP aircraft had a habit of reusing serials. This is IL-18V 'Salon' '102 Red' No 1 (c/n 181002701). Note the rounded style of the zero; it was replaced by an angular presentation on IL-18E 'Salon' '102 Red' No 2 (c/n 186008905). Yefim Gordon archive

IL-18V YR-IMG displays the red/white livery initially applied to the type by TAROM Romanian Airlines. Yefim Gordon archive

Interestingly, even in the 1950s and 1960s the Polish flag carrier had operated Western types along with Soviet aircraft. (Speaking of which, Poland always stood out among the Eastern Bloc nations with its pro-Western views – and its smouldering anti-Russian sentiment which flares up from time to time; the latter goes back for centuries and is outside the scope of this book. Assessing the Soviet Union's allies, one Western politologist wrote that East Germany was the best and Poland the worst: *'the Soviets can never be sure which side it would take in a conflict'*.) Thus in 1962-67 LOT operated three Vickers Viscount Srs 804s alongside the IL-18s.

After LOT introduced the IL-62 *sans suffixe* in 1972, the IL-18s were gradually relegated to high-density domestic routes. Two aircraft were converted to freighters, carrying perishable goods – or baggage. Interestingly, the Tu-134 *sans suffixe* and Tu-134A short/medium-haul airliners delivered to LOT from 1968 onwards were criticised for their limited baggage space, a direct result of the narrow fuselage with a diameter of only 2.9m (9ft 6¹¹⁄₆₄in). This shortcoming sometimes led to absurd situations when a Tu-134 flying from A to B was followed by an IL-18 carrying the passengers' luggage, among other things!

In 1977 the airline introduced its current livery which was then the height of fashion and still does not look dated today; the reader may be interested to learn that a contest was announced via the *Skrzydlata Polska* (Winged Poland) magazine and a winner was selected from half a dozen liveries suggested by the readers. Thus the IL-18s were repainted along with the rest of the fleet to feature a huge *LOT* logo on the forward fuselage and a blue tail with a stylised flying crane in a white circle. Two aircraft were converted to freighters, gaining additional CARGO titles. The last Polish *Coots* were retired in 1989. It should be noted that, unlike many operators of the type, not a single LOT IL-18 was lost in an accident.

A total of five IL-18s saw service with the **POLISH AIR FORCE** (PWL – *Polskie Wojsko Lotnicze*), although no more than two were in service at any one time. The *Coots* were operated by the 36. SPLT (*Specjalny Pułk Lotnictwa Transportowego* – special transport air regiment) at Warsaw-Okęcie airport. Unlike regular PWL aircraft which have serials based on the construction number,[6] the fixed-wing aircraft of the 36. SPLT had serials which had nothing to do with the c/ns and were frequently reused.

Serial	*Version*	*C/n*	*Notes*
'101 Red' #1	IL-18V 'Salon'	180002504	D/D 2-3-61. Sold to LOT as SP-LSE 11-9-65
'101 Red' #2	IL-18V 'Salon'	185008305	D/D 7-8-65, short-term lease only pending delivery of '101 Red' #3. To CCCP-75593 (Magadan CAD/1st Magadan UAD/185th Flight)
'101 Red' #3	IL-18E 'Salon'	185008503	D/D 12-1-66. Sold to LOT as SP-LSK 6-1-88
'102 Red' #1	IL-18V 'Salon'	181002701	D/D 9-3-61. Sold to LOT as SP-LSH 16-6-66
'102 Red' #2	IL-18E 'Salon'	186008905	D/D 8-4-66. Sold to LOT as SP-LSI 28-3-75

In late 1990 a Polish-Japanese joint venture named **POLNIPPON CARGO** [–/PLN] bought two IL-18s formerly operated by the defunct Interflug. IL-18E D-AOAR (c/n 186009202) purchased on 2nd November was registered SP-FNB and named 'Agata', while IL-18D D-AOAT (c/n 188010903) bought on 7th December became SP-FNC 'Hubert'. Operating from Warsaw-Okęcie, the aircraft retained Interflug's red cheatline with the addition of a grey tail and an emblem representing a cross-breed between the Polish and Japanese flags. The airline ceased operations in March 1996.

After the demise of Polnippon Cargo both of its aircraft were promptly sold to another airline named **POLONIA AIRWAYS** [–/PAW]. In fact, SP-FNB and SP-FNC were reregistered on 1st March 1996, becoming SP-FNW and SP-FNZ respectively. Makes you wonder if the whole affair was merely a change of facade! Yet the new venture was shortlived; in the summer of 1997 both aircraft were sold to Air Cess and placed on the Swazi register soon afterwards as 3D-SBW and 3D-SBZ respectively.

ROMANIA

In 1961-67 the Romanian flag carrier **TAROM** [RO/ROT] (Transporturile Aeriene Române – Romanian Air Transport) took delivery of 12 IL-18s. Unfortunately very little is known about the airline's IL-18 operations, but TAROM's *Coots* have been leased abroad to places as far apart as China and Cuba.

Registration	*Version*	*C/n*	*Notes*
YR-IMA	IL-18V	181003602	Registered 15-1-61; was leased to CAAC as B-230 ?-9-85 to 30-3-87. WFU Bucharest-Otopeni 7-8-91, used as a fire trainer
YR-IMB	IL-18V	181003702	Registered 20-10-61. Damaged in belly landing Paphos 24-2-62; repaired and returned to the USSR 11-11-62, used as instructional airframe in Omsk
YR-IMC	IL-18V	182004802	Registered 29-5-62. Preserved Banasti 90km north of Bucharest by 1992
YR-IMD	IL-18V	182004804	Registered 11-6-62. WFU 26-6-91; sold to Air Cess ?-98 as 3D-ALQ
YR-IME	IL-18V	183006205	Registered 7-6-63; was leased to CAAC as B-232 in 1985. WFU 26-6-91; sold to Southern Cross as 3D-AHO or Air Cess as EL-AHO ?-98
YR-IMF	IL-18V	184007105	Registered 30-4-64; was leased to CAAC as B-234 28-3-85 to 21-3-86. Converted to IL-18Gr. Sold to GVG Air Company 3-98 as UN 75111
YR-IMG	IL-18V	184007301	Registered 30-4-64; sold to Renan ?-98 as ER-ICG
YR-IMH	IL-18V	185008301	Registered 3-7-65. Crashed Carpathian Mountains 13-8-91
YR-IMI	IL-18V	185008302	Registered 6-7-65. Crashed Bucharest-Otopeni 21-4-77
YR-IMJ	IL-18E	186009102	Registered 25-6-66; sold to Renan ?-97 as ER-ICJ
YR-IMK	IL-18E	186009104	Registered 9-7-66. Leased to Egyptair; crashed into Red Sea 9-12-74 while on lease
YR-IML	IL-18D	187009903	Registered 22-5-67; sold to Renan ?-98 as ER-ICL

The **ROMANIAN GOVERNMENT FLIGHT** operated a total of three IL-18s which originally wore TAROM livery. In 1990 the two remaining aircraft (which were either delivered as 105-seaters or converted to this configuration) received the titles and logo of the newly-established airline **ROMAVIA** [VQ/RMV].

Registration	*Version*	*C/n*	*Notes*
YR-IMM	IL-18D ('Salon'?)	187009904	Registered 29-5-67. Transferred to Romavia; sold to Yuzhnaya Aircompany 9-02 or 10-02 as UN 75001
YR-IMZ #1	IL-18E 'Salon'	185008602	Registered 26-3-66; returned to the USSR ?-66 as CCCP-75445 (Urals CAD/1st Sverdlovsk UAD/120th Flight)
YR-IMZ #2	IL-18D ('Salon'?)	187009802	Registered 16-6-67. Transferred to Romavia Converted to IL-18GrM [IL-18D/F (SCD)] 1 2-94/27-3-95. Sold to Expo Aviation ?-00 as 4R-EXD

In May 1990 IL-18D YR-IMZ No 2 was briefly leased by the charter carrier **LAR ROMANIAN AIRLINES** [QR/RLA] (Liniile Aeriene Române).

ACVILA AIR [WZ/RRM] (that is, 'Eagle Air') leased IL-18s from Romavia and the Moldovan airline Vichi as required. For instance, from August 1995 to November 1997 it operated the latter carrier's demilitarised IL-22M-11 ER-75929 (c/n 187010505). The aircraft retained basic Aeroflot colours and the Moldovan flag but the fuselage was adorned with a large stylised eagle and 'Acvila Air – Romanian Carrier' titles.

SOMALIA

DAMAL AIRLINES, an air carrier based at Sharjah and Dubai (UAE), leases Soviet-built airliners and transports from Air Cess as required. The single known lease involved IL-18E 3C-KKR (c/n 185008603) during 2000-2001.

JUBBA AIRWAYS, another carrier operating from the same locations, leases IL-18Ds from Phoenix Aviation as required. Again, only one instance is known; IL-18D c/n 187010004 was leased first as ST-APZ with Phoenix/Jubba Airways titles in May-August 2000, then as RA-75449 (this time with Jubba Airways titles only) in September-October 2000 and finally as EX-75449 in October 2000/May 2001. By September 2001 the aircraft had been returned to Phoenix Aviation.

STAR AIRLINES – again based at Sharjah and Dubai – operated an ex-Santa Cruz Imperial IL-18V (formerly EL-ADY No 2, c/n 182004804) in 1999-2001. At first the airliner was flown with the non-standard four-digit Kyrghyz registration EX-7504 and then with the Moldovan registration ER-ICM (in the latter case, jointly with the Kyrghyz airline Aerovista). In 2001 Star Airlines reportedly also leased IL-18V EX-75427 (c/n 183005905) from Phoenix Aviation.

SOUTH AFRICA

SHUTTLE AIR CARGO chartered IL-18GrM RA-75554 (c/n 185008404) from Ramair in December 1997. Unfortunately the aircraft was wrecked at Johannesburg International airport on 17th December 1997 while operating for the company.

SUDAN

SUDAN AIRWAYS [SD/SUD], the nation's flag carrier, occasionally leases Soviet-built aircraft for cargo operations, sometimes even going so far as to apply full colours. The three known *Coots* operated by the airline, however, only sported small additional Sudan Airways titles.

Registration	*Version*	*C/n*	*Notes*
EX-75442	IL-18D	187009702	Leased from Phoenix Aviation by 4-00, carried both titles.
T9-ABB	IL-18V	184007405	Leased from Phoenix Aviation late 1999 to 2-00
3C-KKL	IL-18D	187010204	Leased from Air Cess by 11-00, carried both titles.

SWAZILAND

AIR PASS, a sister company of Air Cess (Liberia),[7] operated five IL-18s, one of which had been transferred from Air Cess (Liberia) in 1997. 'Pass' is an acronym for Pietersburg Aviation Services & Systems; this was because, though nominally a Swazi company, the airline was based at Pietersburg-Gateway International airport, South Africa. Air Pass suspended operations in late 1998; most of the IL-18s were transferred to yet another Air Cess, this time in Equatorial Guinea (which see).

Registration	*Version*	*C/n*	*Notes*
3D-ALQ	IL-18V	182004804	Ex-YR-IMD, bought ?-98; basic TAROM c/s, no titles. Sold to Santa Cruz Imperial 11-98 as EL-ADY #2
3D-SBC	IL-18V	184006903	Ex-Air Zory LZ-AZC, bought 7-97 and originally operated with this registration! Reregistered 3D-SBC by 3-98. To Air Cess as 3C-KKJ
3D-SBQ	IL-18D	187010204	Ex-Air Cess (Liberia) EL-AKQ, named 'Zlatoust'; transferred 1-98. To Air Cess as 3C-KKL
3D-SBW	IL-18E	186009202	Ex-Polonia Airways SP-FNW, bought 10-97 and originally operated with this registration! Reregistered 3D-SBW 11-97. To Air Cess as 3C-KKK
3D-SBZ	IL-18D	188010903	Ex-Polonia Airways SP-FNZ, bought 10-97 and originally operated with this registration! Reregistered 3D-SBZ 11-97; subsequent fate unknown

In 1998 an ex-TAROM IL-18V (YR-IME, c/n 183006205) and an ex-Balkan IL-18E (LZ-BEW, c/n 185008601) were registered to **SOUTHERN CROSS** as 3D-AHO and 3D-ALD respectively. They are believed never to have flown with these registrations and were probably resold immediately to Air Cess (Liberia) and Santa Cruz Imperial respectively as EL-AHO and EL-ALD.

Sometime between January 2002 and March 2003 IL-18V D2-FAM (c/n 184007401) belonging to the Angolan carrier ALADA was leased to an unknown airline, gaining the Swazi registration 3D-SEP – which was already in use since 1999 for a Sud Aviation SE210 Caravelle 11R (c/n 251) belonging to Gabon Express.

VIETNAM

HÁNG KHÔNG VIÊT NAM [VN/HVN] operated a total of five IL-18s. Two of them remained in service long enough to see the carrier renamed **VIETNAM AIRLINES** in 1990.

Registration	*Version*	*C/n*	*Notes*
CCCP-75424	IL-18V 'Salon'	181003305	Ex-Magadan CAD/1st Magadan UFD/185th Flight. Reregistered as, see next line
BH-195			Regd 23-1-73; not VN-195 as reported! Rereg, see below
VN-B195			Reregistered before 8-76. Returned to the Soviet Union as CCCP-75424 (again to Magadan CAD/1st Magadan UFD/185th Flight) after 5-77
VN-B196	IL-18D	187010402	Ex-CCCP-75465, D/D 17-4-79. WFU Hanoi by 1990
198	IL-18D	189011304	Ex-Air Mauritanie 5T-CJL, bought 1971. Rereg, see below
VN-B198			WFU Hanoi by 1996
CCCP-75535	IL-18D	180002405	Leased from Aeroflot 10-72, returned (date unknown)
?	IL-18D	188010703	

The first *Coot* delivered to Vietnam was a special one. Firstly, it had had quite a career by then; delivered to Ghana Airways on 21st April 1961 as 9G-AAM, it was returned to the USSR on 17th June 1965 to become CCCP-75424. Secondly, the aircraft was a gift from the Soviet government to the Vietnamese leader Ho Chi Minh and, of course, fitted with a VIP interior. Originally the idea was to present it on his 80th birthday (19th May 1970), but the Vietnamese urgently needed a VIP aircraft with adequate range so they could fly to Europe to hold peace talks; hence the Soviet government decided to make the gift in advance.

Originally the aircraft retained its Soviet registration but on 23rd January 1973 it became BH-195. Contrary to Western reports the unusual BH- prefix was *not* the Vietnamese VN- prefix written in Cyrillic letters by mistake; it stood for *Bâc Ho* ('Granddad Ho', as Ho Chi Minh was popularly known among his people) and '195' denoted 19th May. The registration was applied in full on the nose but only as 195 on the aft fuselage. Because of the special interior BH-195 had a non-standard window arrangement (see Chapter 2/IL-18V 'Salon').

YEMEN ARAB REPUBLIC (NORTH YEMEN)

On 24th November 1963 the **YEMEN ARAB REPUBLIC AIR FORCE** (YARAF) government flight took delivery of an IL-18V 'Salon' which had been only briefly operated by Aeroflot as CCCP-75870 (c/n 183005905). The aircraft was reregistered YE-AYE and retained basic pre-1973 Aeroflot IL-18 colours, with 'Yemen Arab Republic Aviation' titles.

Like China, Vietnam initially applied simple three-digit numbers by way of registrations (without any nationality prefix), as IL-18D '198' seen on overhaul at Moscow-Bykovo testifies. It was later reregistered VN-B198. Yefim Gordon archive

The two Yugoslav Government IL-18D 'Salons' originally had civil registrations, as illustrated by YU-AIB. Sergey and Dmitriy Komissarov archive

After the sale of YU-AIB the remaining Yugoslav *Coot* (YU-AIA) received military markings – first as 7501 and then as 73201. Yefim Gordon archive

In 1970 North Yemen changed its nationality prefix to 4W and the aircraft was reregistered again, becoming 4W-ABO. Besides VIP transportation, it was occasionally operated by **YEMEN AIRWAYS** on regular passenger flights. Finally, on 24th September 1984 the aircraft was sold to Balkan Bulgarian Airlines as LZ-BEU.

YUGOSLAVIA

In the spring of 1967 the **YUGOSLAV AIR FORCE** (JRV – *Jugoslovensko Ratno Vazduhoplovstvo*) acquired two IL-18D 'Salon' VIP aircraft. Both were initially quasi-civil, being registered YU-AIA (c/n 187009805) and YU-AIB (c/n 187009803). By January 1968 YU-AIA had received overt military markings and the serial 7501; this was changed to 73201 by May 1970 in keeping with the current JRV serial system.

In March 1978 YU-AIB was sold to Air Guinée and reregistered 3X-GAX. The other *Coot* soldiered on until replaced by two ex-Sterling Airways Boeing 727-2L8 Advanced trijets – YU-AKD/74302 and YU-AKH/74301 (the latter was the presidential aircraft). After that, 73201 was sold to the Soviet Union and converted to passenger configuration, becoming CCCP-75780 No 2 with Aeroflot (Krasnoyarsk CAD/1st Krasnoyarsk UAD/214th Flight); it was eventually converted to an IL-22M-11, changing its identity to CCCP-75919.

ZAÏRE (NOW DEMOCRATIC REPUBLIC OF CONGO OR CONGO-KINSHASA)

Back in Zaïrean days when Mobutu Sésé Séko was running (or ruining?) the country, **ATO – AIR TRANSPORT OFFICE**, a cargo airline based at Kinshasa-N'djili, operated two IL-18Ds – RA-75466 (c/n 187010403) leased from GosNII GA from October 1993 to July 1995 and SP-FNC 'Hubert' (c/n 188010903) leased from Polnippon Airways in November 1994 – March 1995.

Later, when the country had been renamed the Democratic Republic of Congo after a fierce civil war and a change of government, **COMPAGNIE AFRICAINE D'AVIATION** [–/CAA] based at Kinshasa-N'dolo purchased IL-18V RA-75431 (c/n 180002003) from Elf Air in 2000. After an overhaul at Moscow-Bykovo the aircraft was delivered in September 2000, still as RA-75431 but already in the new owner's colours with an all-red tail, a red cheatline and stepped 'CAA' titles on the nose and tail; on arrival it was reregistered 9Q-CHB. Another IL-18 registered 9Q-CAA (c/n unknown) was added to the fleet by September 2003.

OPERATORS FROM UNKNOWN NATIONS

Tyumen' Airlines IL-18D RA-74267 (c/n 188011105) was briefly leased by **AFRICAN AIRLINES** in January 2000. This may have been the Kenyan carrier **AFRICAN AIRLINES INTERNATIONAL LTD** [–/CAA] based at Nairobi-Jomo Kenyatta International airport.

A certain **ALFA LINE** leased IL-18V YR-IMF (c/n 184007105) and IL-18D YR-IML (c/n 187009903) from TAROM in late 1997/early 1998. This might be the Turkish airline Alfa Hava Yollari [H7/LFA].

Balkan Bulgarian Airlines IL-18E LZ-BEZ (c/n 185008603) was leased by an organisation called **COMCO** in the autumn of 1993, returning to the owner by December 1994.

EXCELSIOR AIRLINES briefly leased IL-18V SP-LSA (c/n 180002403) from LOT Polish Airlines in 1987.

INTER TROPIC AIRLINES (not to be confused with the Surinamese carrier Inter Tropical Aviation) bought IL-18V LZ-BFU (c/n 183005905) from Bulgarian Flying Cargo sometime before September 2000. However, this airline operated the aircraft only briefly; in October 2000 LZ-BFU was sold to Phoenix Aviation end registered in Kyrghyzstan as EX-75427.

Production List

IL-18 family production is presented in construction number/ fuselage number order, with all identities worn consecutively by each aircraft. The split presentation of c/ns is used for the sake of convenience to show how the numbers accrue. The symbol § means the aircraft was reconverted to standard configuration from a special-mission variant. 'Deceased' (ie, crashed) examples are duly marked with † (RIP cross) followed by the date of the accident. IL-18 registrations followed by * indicate this is a quasi-civil aircraft, while ** indicates the aircraft was military at first but was later sold to a civil operator with the same registration. Registrations given in italics were allocated but not taken up; those in square brackets are ATC callsigns (not worn visibly) of aircraft in overt military markings – eg, East German AF '493 Black' [DM-VAX].

Construction number	*Version*	*Registration/ tactical code/serial*	*Manufacture date*
000.01	IL-18 *sans suffixe*	CCCP-Л5811	
18.7.0.000.02?	IL-18 *sans suffixe*	CCCP-Л5812?	
	IL-18I	CCCP-75888	
18.7.0.001.01	IL-18A	CCCP-Л5818?, CCCP-75634	
18.7.0.001.02	IL-18A	CCCP-Л5819; CCCP-75635?	2-10-1957
18.7.0.001.03	IL-18A	CCCP-Л5820, CCCP-75820	
18.8.0.001.04	IL-18A	CCCP-Л58...?, CCCP-75636?	
18.8.0.001.05	IL-18A	CCCP-Л58...?, CCCP-75748	
18.8.0.002.01	IL-18A, IL-18LL	CCCP-Л5821, CCCP-75637?	
18.8.0.002.02	IL-18A	CCCP-75638?	
18.8.0.002.03	IL-18A	CCCP-75639?	
18.8.0.002.04	IL-18A	CCCP-75640?	
18.8.0.002.05	IL-18A	CCCP-75641?	
18.8.0.003.01	IL-18A	CCCP-75642	
18.8.0.003.02	IL-18A, IL-18SL	CCCP-75643	
18.8.0.003.03	IL-18A	CCCP-75644?	
18.8.0.003.04	IL-18A	CCCP-75645	
18.8.0.003.05	IL-18A	CCCP-75646?	
18.8.0.004.01	IL-18A	CCCP-75647	
	IL-18SIP	CCCP-27220	
18.8.0.004.02	IL-18A	CCCP-75648? † 27-4-1960	
18.8.0.004.03	IL-18A	CCCP-75649?	
18.8.0.004.04	IL-18A	CCCP-75650?	
18.8.0.004.05	IL-18A	CCCP-75651? † 26-12-1960	
18.8.0.005.01	IL-18B	*CCCP-75652?*, CCCP-75473	
18.8.0.005.02	IL-18B	CCCP-75653? † 13-8-1961	
18.8.0.005.03	IL-18B	CCCP-75654? † 17-12-1961	
18.8.0.005.04	IL-18B	CCCP-75655?	
18.8.0.005.05	IL-18B	CCCP-75656	
18.8.0.006.01	IL-18B	CCCP-75657?	
18.8.0.006.02	IL-18B	CCCP-75658?	
18.8.0.006.03	IL-18B	CCCP-75659	
18.8.0.006.04	IL-18B	CCCP-75660?	
18.8.0.006.05	IL-18B	CCCP-75661? † 2-7-1964	
18.8.0.007.01	IL-18B	CCCP-75662	
18.8.0.007.02	IL-18B	CCCP-75663 † 26-8-1972	
18.8.0.007.03	IL-18B	CCCP-75664?	
18.8.0.007.04	IL-18B	CCCP-75665? † 22-11-1966	
18.8.0.007.05	IL-18B (ABCP)	CCCP-75666*	
18.8.0.008.01	IL-18S	*CCCP-75667?*, Soviet AF '001 Red' [CCCP-33569], CCCP-33569*, CCCP-75479*	
18.8.0.008.02	IL-18B	CCCP-75668?	
18.8.0.008.03	IL-18B	CCCP-75669?	
18.8.0.008.04	IL-18B	CCCP-75670?	
18.8.0.008.05	IL-18B	CCCP-75671	
18.8.0.009.01	IL-18B	CCCP-75672 † 22-6-1961	
18.8.0.009.02	IL-18B	CCCP-75673	
18.8.0.009.03	IL-18B	CCCP-75674	
18.9.0.009.04	IL-18B	CCCP-75675	
18.9.0.009.05	IL-18S?	CCCP-75601*?	
18.9.0.010.01	IL-18B	CCCP-75676* No 1? (see c/n 185008605) † 2-9-1959	
18.9.0.010.02	IL-18B	CCCP-75677	
18.9.0.010.03	IL-18B	CCCP-75678	
18.9.0.010.04	IL-18B	CCCP-75679	
18.9.0.010.05	IL-18B	CCCP-75680?	

Construction number	*Version*	*Registration/ tactical code/serial*	*Manufacture date*
18.9.0.011.01	IL-18B	CCCP-75681	
18.9.0.011.02	IL-18B	CCCP-75682	
18.9.0.011.03	IL-18B	CCCP-75683	
18.9.0.011.04	IL-18B	CCCP-75684	
18.9.0.011.05	IL-18B	CCCP-75685 † 3-1-1965	
18.9.0.012.01	IL-18B	CCCP-75686? † 10-11-1963	
18.9.0.012.02	IL-18B	CCCP-75687 † 11-5-1973	
18.9.0.012.03	IL-18B	CCCP-75688 † 23-12-1965	
18.9.0.012.04	IL-18B	CCCP-75689	
18.9.0.012.05	IL-18B	CCCP-75690?	
18.9.0.013.01	IL-18B	CCCP-75691	
18.9.0.013.02	IL-18V [proto]	*CCCP-75692?*, CCCP-75710	
18.9.0.013.03	IL-18B	CCCP-75693?	
18.9.0.013.04	IL-18B	CCCP-75694	
18.9.0.013.05	IL-18B	CCCP-75695	
18.9.0.014.01	IL-18B	B-202 † 24-12-82	
18.9.0.014.02	IL-18B	CCCP-75699 † 11-12-1969	31-8-1959
18.9.0.014.03	IL-18B	CCCP-75696?	
18.9.0.014.04	IL-18B	CCCP-75697?	
18.9.0.014.05	IL-18B	CCCP-75698	
18.9.0.015.01	IL-18B	CCCP-75700?	
18.9.0.015.02	IL-18B	CCCP-75701?	
18.9.0.015.03	IL-18B	CCCP-75702	
18.9.0.015.04	IL-18B	B-200?	
18.9.0.015.05	IL-18B	CCCP-75703, RA-75703	30-10-1959
18.9.0.016.01	IL-18B	XT-806, 206	
18.9.0.016.02	IL-18B	XT-804?, B-204 † ?-2-1977	
18.9.0.016.03	IL-18S?	CCCP-75749*?	
18.9.0.016.04	IL-18B	OK-NAA 'Piešťany' † 2-1-1977	
18.9.0.016.05	IL-18B	OK-NAB 'Košice' † 28-7-1976	
18.9.0.017.01	IL-18B	CCCP-75704?	
18.9.0.017.02	IL-18S	CCCP-75705 † 17-8-1960	
18.9.0.017.03	IL-18B	CCCP-75706	
18.9.0.017.04	IL-18B	CCCP-75707	
18.9.0.017.05	IL-18B	CCCP-75708?	
18.9.0.018.01	IL-18B	CCCP-75709	
18.9.0.018.02	IL-18V	CCCP-75711 (see c/n 181008503)	
18.*.0.018.03	IL-18V	CCCP-75712? † 24-2-1973	
18.0.0.018.04	IL-18V	CCCP-75713 No 1 (see c/n 186009403)	
18.0.0.018.05	IL-18V	CCCP-75714?	
18.0.0.019.01	IL-18V	CCCP-75715	
18.0.0.019.02	IL-18V 'Salon', IL-18V 'Meteor' §	CCCP-75716	
18.0.0.019.03	IL-18V, IL-18Gr	HA-MOA	22-2-1960
18.0.0.019.04	IL-18V 'Salon'	CCCP-75717	
18.0.0.019.05	IL-18V	DM-STA, DDR-STA	
18.0.0.020.01	IL-18V	DM-STB, DDR-STB	
18.0.0.020.02	IL-18V	*HA-MOB*, HA-MOD † 23-11-1962	
18.0.0.020.03	IL-18V, IL-18Gr	CCCP-04...?, CCCP-75431 *geo research* §, RA-75431	
18.0.0.020.04	IL-18V	CCCP-75719?, 3X-NZE No 1?	
18.0.0.020.05	IL-18V	CCCP-75718	
18.0.0.021.01	IL-18V	OK-OAC 'Šliačkupe'	
18.0.0.021.02	IL-18V	OK-OAD	
18.0.0.021.03	IL-18V	CCCP-04356, CCCP-75422	
18.0.0.021.04	IL-18V	CCCP-75720	
18.0.0.021.05	IL-18V	CCCP-75721	

Construction number	Version	Registration/ tactical code/serial	Manufacture date
18.0.0.022.01	IL-18V	CCCP-75722	
18.0.0.022.02	IL-18V	DM-STC, DDR-STC	
	IL-18LL	OK-018	
18.0.0.022.03	IL-18V	CCCP-75723	
18.0.0.022.04	IL-18V	CCCP-75724	
18.0.0.022.05	IL-18V	CCCP-75725	
18.0.0.023.01	IL-18V		
18.0.0.023.02	IL-18V 'Salon' §	CCCP-75726, East German AF '493 Black' [DM-VAX], DM-STD**, DDR-STD	
18.0.0.023.03	IL-18V	CCCP-75727 † 21-1-1971	
18.0.0.023.04	IL-18V	CCCP-75728	
18.0.0.023.05	IL-18V	CCCP-75729	
18.0.0.024.01	IL-18V	CCCP-75730	
18.0.0.024.02	IL-18V	9G-AAI; CCCP-75531? † 2-9-1964	
18.0.0.024.03	IL-18V, IL-18Gr	SP-LSA 'Warszawa'	
18.0.0.024.04	IL-18V	SP-LSB 'Westerplatte'	
18.0.0.024.05	IL-18V	9G-AAJ, CCCP-75535	
18.0.0.025.01	IL-18V	9G-AAK; CCCP-75533? † 5-6-1970	
18.0.0.025.02	IL-18V	9G-AAL; CCCP-75534?	
18.0.0.025.03	IL-18V	CCCP-75731?	
18.0.0.025.04	IL-18V 'Salon' §	Polish AF '101 Red' No 1, SP-LSE 'Lenino'	
18.*.0.025.05	IL-18V		
18.*.0.026.01	IL-18V	CCCP-75732? † 26-2-1963	
18.*.0.026.02	IL-18V	CCCP-75733?	
18.*.0.026.03	IL-18V	CCCP-75734	
18.*.0.026.04	IL-18V	CCCP-75735	
18.1.0.026.05	IL-18V	CCCP-75736	
18.1.0.027.01	IL-18V 'Salon' §	Polish AF '102 Red' No 1, SP-LSH 'Kołobrzeg'	
18.1.0.027.02	IL-18V	CCCP-75737	
18.1.0.027.03	IL-18V	CCCP-75738	
18.1.0.027.04	IL-18V	CCCP-75739	
18.1.0.027.05	IL-18V	CCCP-75740	
18.1.0.028.01	IL-18V	CCCP-75741?	
18.1.0.028.02	IL-18V 'Salon'	CCCP-75742	
18.1.0.028.03	IL-18V	9G-AAX, CCCP-75426	
18.1.0.028.04	IL-18V	9G-AAY	
18.1.0.028.05	IL-18V, IL-18Gr	SP-LSC 'Narwik', LZ-BEI	
18.1.0.029.01	IL-18V (Polar)	CCCP-75743	
18.1.0.029.02	IL-18V	OK-PAE 'Karlový Vary'	
18.1.0.029.03	IL-18V	HA-MOC † 28-8-1971	
18.1.0.029.04	IL-18V	OK-PAF † 12-7-1962	
18.1.0.029.05	IL-18V	CCCP-75744	
18.1.0.030.01	IL-18V	CCCP-75745	18-4-1961
18.1.0.030.02	IL-18V	CCCP-75746?	
18.1.0.030.03	IL-18V	CCCP-75747	
18.1.0.030.04	IL-18V	CCCP-75750	
18.1.0.030.05	IL-18V	CCCP-75751	
18.1.0.031.01	IL-18V	CCCP-75752	
18.1.0.031.02	IL-18V	CCCP-75753?	
18.1.0.031.03	IL-18V	CCCP-75754	
18.1.0.031.04	IL-18V	CCCP-75755	
18.1.0.031.05	IL-18V 'Salon' §	OK-BYP, OK-PAI	
18.1.0.032.01	IL-18V		
18.1.0.032.02	IL-18V	CCCP-75756?	
18.1.0.032.03	IL-18V	CCCP-75757? † 31-12-1961	
18.1.0.032.04	IL-18V	CCCP-75758?	
18.1.0.032.05	IL-18V	CCCP-75759?	
18.1.0.033.01	IL-18V	CCCP-75760?	
18.1.0.033.02	IL-18V	CCCP-75761?	
18.1.0.033.03	IL-18V	CCCP-75762, TZ-ABD, CCCP-75762	
18.1.0.033.04	IL-18V	CCCP-75763, TZ-ABE † 11-8-1974	
18.1.0.033.05	IL-18V, IL-18V 'Salon' §	9G-AAM, CCCP-75424, BH-195, VN-B195, CCCP-75424	
18.1.0.034.01	IL-18V		
18.1.0.034.02	IL-18V 'Salon'	CCCP-75764, 7T-VRA	
18.1.0.034.03	IL-18V	9G-AAN, CCCP-75425 † 9-5-1974	27-7-1961
18.1.0.034.04	IL-18V	CCCP-75765? † 5-3-1963	
18.1.0.034.05	IL-18V	CCCP-75766	
18.1.0.035.01	IL-18V	CCCP-75767?	
18.1.0.035.02	IL-18V	CCCP-75768? (also quoted as c/n 189001201!)	
18.1.0.035.03	IL-18V	CCCP-75769	
18.1.0.035.04	IL-18V	CCCP-75770?	
18.1.0.035.05	IL-18V	CCCP-75771	
18.1.0.036.01	IL-18V	CCCP-75772	
18.1.0.036.02	IL-18V	YR-IMA, B-230, YR-IMA	
18.1.0.036.03	IL-18V	CCCP-75773 † 31-12-1970	
18.1.0.036.04	IL-18V 'Salon'	CCCP-75774*	
18.1.0.036.05	IL-18V	CCCP-75775	
18.1.0.037.01	IL-18V	CCCP-75776	
18.1.0.037.02	IL-18V	YR-IMB	
18.1.0.037.03	IL-18V	3X-KKN, 3X-GAB † 9-7-1967	
18.1.0.037.04	IL-18V	3X-NZE No 2, 3X-GAC, CCCP-75428	
18.1.0.037.05	IL-18V	3X-LBE, 3X-GAA, CCCP-74299 30-9-1961	
18.1.0.038.01	IL-18V	CCCP-75777?	
18.1.0.038.02	IL-18V	CCCP-75778?	
18.1.0.038.03	IL-18V	CCCP-75779?	
18.1.0.038.04	IL-18V	CCCP-75780 No 1? (see c/n 187009805)	
18.1.0.038.05	IL-18V	CCCP-75781	
18.1.0.039.01	IL-18V	CCCP-75782?	
18.1.0.039.02	IL-18V	CCCP-75783?	
18.1.0.039.03	IL-18V	CCCP-75784?	
18.1.0.039.04	IL-18V, IL-18Gr	CCCP-75785	
18.1.0.039.05	IL-18V, SL-18V	CCCP-75786, RA-75786, 75786	
18.1.0.040.01	IL-18V	CCCP-75787	
18.1.0.040.02	IL-18V	CCCP-75788	
18.1.0.040.03	IL-18V	CCCP-75789	
18.1.0.040.04	IL-18V	CCCP-75790	
18.1.0.040.05	IL-18V	CCCP-75791 † 10-9-1969	
18.1.0.041.01	IL-18V	CCCP-75792	
18.1.0.041.02	IL-18V	CCCP-75793	
18.1.0.041.03	IL-18V	CCCP-75794, CCCP-78732, RA-78732	
18.1.0.041.04	IL-18V	CCCP-75797	
18.1.0.041.05	IL-18V	CCCP-75796, DM-STF No 1 (see c/n 184007405), CCCP-75475 (see c/n 184007401)	
18.1.0.042.01	IL-18V	*CCCP-75795?*, OK-PAG 'Vysoké Tatry'	
18.1.0.042.02	IL-18V	OK-PAH 'Marianské Lazne'	
18.2.0.042.03	IL-18V (ABCP)	CCCP-75602*, RA-75602*	
18.2.0.042.04	IL-18V	CCCP-75799	
18.2.0.042.05	IL-18V	75800, CCCP-75800?	
18.2.0.043.01	IL-18V	CCCP-75801 † 12-2-1975	...-1-1962
18.2.0.043.02	IL-18V	CCCP-75802?	
18.2.0.043.03	IL-18V		
18.2.0.043.04	IL-18V	75803, CCCP-75803	
18.2.0.043.05	IL-18V, SL-18I §	CCCP-75804, RA-75804 §	
18.2.0.044.01	IL-18V	CCCP-75805?	
18.2.0.044.02	IL-18V	DM-STG, DDR-STG	
18.2.0.044.03	IL-18V	CCCP-75806?	
18.2.0.044.04	IL-18V	CCCP-75807?	
18.2.0.044.05	IL-18V (ABCP)	CCCP-75606*, RA-75606*	
18.2.0.045.01	IL-18V	CCCP-75808?	
18.2.0.045.02	IL-18V	CCCP-75809?	
18.2.0.045.03	IL-18V	CCCP-75810?	
18.2.0.045.04	IL-18V, IL-18REO §	CCCP-75811, RA-75811 §	
18.2.0.045.05	IL-18V	TZ-ABY, CCCP-?????	
18.2.0.046.01	IL-18V	LZ-BEL	
18.2.0.046.02	IL-18V	CCCP-75812	
18.2.0.046.03	IL-18V	LZ-BEK	
18.2.0.046.04	IL-18V	CCCP-75813?	
18.2.0.046.05	IL-18V	CCCP-75814?	
18.2.0.047.01	IL-18V	CCCP-75815	
18.2.0.047.02	IL-18V	CCCP-75816	
18.2.0.047.03	IL-18V	CCCP-75817	
18.2.0.047.04	IL-18V	CCCP-75818	
18.2.0.047.05	IL-18V	CCCP-75819	
18.2.0.048.01	IL-18V 'Salon', IL-18V	CCCP-75894, RA-75894	
18.2.0.048.02	IL-18V	YR-IMC	
18.2.0.048.03	IL-18V	CCCP-75821	
18.2.0.048.04	IL-18V	YR-IMD, 3D-ALQ, EL-ADY No 2 (see c/n 184007405!), EX-7504, ER-ICM; EX-011?	...-5-1962
18.2.0.048.05	IL-18V		
18.2.0.049.01	IL-18V	CCCP-75822	
18.2.0.049.02	IL-18V 'Salon' §	CCCP-75823 † 23-8-1970	
18.2.0.049.03	IL-18V	CCCP-75824 † 3-8-1964	
18.2.0.049.04	IL-18V 'Salon' §	CCCP-75825*, RA-75825, EL-ALW, RA-75825, EX-75825, EX-904 † 15-9-2002?	28-6-1962
18.2.0.049.05	IL-18V	CCCP-75826 No 1?, CU-T830 † 10-7-1966	
18.2.0.050.01	IL-18V	CCCP-75827	
18.2.0.050.02	IL-18V	CCCP-75828	
18.2.0.050.03	IL-18V	CCCP-75829?	
18.2.0.050.04	IL-18V	CCCP-75830	
18.2.0.050.05	IL-18V	CCCP-75831?	
18.2.0.051.01	IL-18V 'Salon' §	East German AF '499 Black' [DM-VAY], DM-STE, DDR-STE	
18.2.0.051.02	IL-18V	CCCP-75832	
18.2.0.051.03	IL-18V	CCCP-75833?	
18.2.0.051.04	IL-18V, IL-18E [proto]	CCCP-75834, RA-75834	1-8-1962
18.2.0.051.05	IL-18V	CCCP-75835	

Construction number	Version	Registration/ tactical code/serial	Manufacture date
18.2.0.052.01	IL-18V	CCCP-75842	
18.2.0.052.02	IL-18V	CCCP-75836?, CU-T831	
18.2.0.052.03	IL-18V	CCCP-75837	
18.2.0.052.04	IL-18V	CCCP-75838	
18.2.0.052.05	IL-18V	CCCP-75839	
18.2.0.053.01	IL-18V, IL-18RT §	CCCP-75840, RA-75840 § † 19-11-2001	25-9-1962
18.2.0.053.02	IL-18V	75841, CCCP-75841	
18.2.0.053.03	IL-18V	CCCP-75843	
18.2.0.053.04	IL-18V-26A, IL-18V	CCCP-75844	
18.2.0.053.05	IL-18V-26A, IL-18V	CCCP-75845	
18.2.0.054.01	IL-18V-26A, IL-18V	CCCP-75846	
18.2.0.054.02	IL-18V-26A, IL-18V	CCCP-75847	
18.2.0.054.03	IL-18V-26A, IL-18V	CCCP-75848	
18.2.0.054.04	IL-18V	CCCP-75849?	
18.2.0.054.05	IL-18V	CCCP-75850? (see c/n 185008503)	
18.2.0.055.01	IL-18V, IL-18Gr	CCCP-75851, CU-T832, CCCP-75851 *testbed* §, RA-75851	28-11-1962
18.2.0.055.02	IL-18V	CCCP-75852	
18.2.0.055.03	IL-18V	CCCP-75853	
18.2.0.055.04	IL-18V	CCCP-75854	
18.2.0.055.05	IL-18V, IL-18Gr	HA-MOE	14-11-1962
18.2.0.056.01	IL-18V	CCCP-75855?, CCCP-75423 *eco research*, RA-75423 §	14-12-1962
18.2.0.056.02	IL-18V	LZ-BEM † 3-3-1973	
18.2.0.056.03	IL-18V	CCCP-75856	
18.2.0.056.04	IL-18V	CCCP-75857	
18.2.0.056.05	IL-18V	CCCP-75858?	
18.2.0.057.01	IL-18V		
18.*.0.057.02	IL-18V		
18.3.0.057.03	IL-18V	CCCP-75859	
18.3.0.057.04	IL-18V		
18.3.0.057.05	IL-18V	CCCP-75860	
18.3.0.058.01	IL-18V	CCCP-75861?	
18.3.0.058.02	IL-18V	CCCP-75862?	
18.3.0.058.03	IL-18V	CCCP-75863	
18.3.0.058.04	IL-18V	CCCP-75864?	
18.3.0.058.05	IL-18V	CCCP-75865	
18.3.0.059.01	IL-18V	CCCP-75866? † 4-4-1963	
18.3.0.059.02	IL-18V	CCCP-75867?	
18.3.0.059.03	IL-18V	CCCP-75868	
18.3.0.059.04	IL-18V	CCCP-75869?	
18.3.0.059.05	IL-18V 'Salon' §	CCCP-75870, YE-AYE, 4W-ABO, LZ-BEU, LZ-BFU, EX-75427	...-2-1963
18.3.0.060.01	IL-18V	CCCP-75871?	
18.3.0.060.02	IL-18V	CCCP-75872?	
18.3.0.060.03	IL-18V 'Salon'	CCCP-75873	
18.3.0.060.04	IL-18V	CCCP-75874	
18.3.0.060.05	IL-18V	CCCP-75875?	
18.3.0.061.01	IL-18V	CCCP-75876	
18.3.0.061.02	IL-18V	CCCP-75879	
18.3.0.061.03	IL-18V	CCCP-75880	
18.3.0.061.04	IL-18V	CCCP-75881	24-4-1963
18.3.0.061.05	IL-18V	CCCP-75882	
18.3.0.062.01	IL-18V	CCCP-75883?	
18.3.0.062.02	IL-18V	CCCP-75884	
18.3.0.062.03	IL-18V	CCCP-75885?	
18.3.0.062.04	IL-18V	CCCP-75886	
18.3.0.062.05	IL-18V	YR-IME, B-232, YR-IME, EL-AHO	15-3-1963
18.3.0.063.01	IL-18V	HA-MOF † 23-11-1977	
18.3.0.063.02	IL-18V	CCCP-75887	
18.3.0.063.03	IL-18V	CCCP-75500*; RA-75500*?	
18.3.0.063.04	IL-18V	CCCP-75501?	
18.3.0.063.05	IL-18V	CCCP-75502	
18.3.0.064.01	IL-18V	CCCP-75503	
18.3.0.064.02	IL-18V	CCCP-75504	
18.3.0.064.03	IL-18V	CCCP-75505?	
18.3.0.064.04	IL-18V	CCCP-75506	31-7-1963
18.3.0.064.05	IL-18V	CCCP-75507? † 1-10-1972	
18.3.0.065.01	IL-18V	CCCP-75508	
18.3.0.065.02	IL-18V	CCCP-75509?	
18.3.0.065.03	IL-18V	CCCP-75510?	
18.3.0.065.04	IL-18V	CCCP-75511	
18.3.0.065.05	IL-18V	CCCP-75512	
18.3.0.066.01	IL-18V	CCCP-75513?	
18.3.0.066.02	IL-18V	CCCP-75514	3-10-1963?[ii]
18.3.0.066.03	IL-18V	CCCP-75515?	
18.3.0.066.04	IL-18V (ABCP)	CCCP-75516*, RA-75516*	
18.3.0.066.05	IL-18V	CCCP-75517	

Construction number	Version	Registration/ tactical code/serial	Manufacture date
18.3.0.067.01	IL-18V	CCCP-75518	
18.3.0.067.02	IL-18V	CCCP-75519? † 9-1-1968	
18.3.0.067.03	IL-18V	CCCP-75520 † 15-2-1977	19-10-1963
18.3.0.067.04	IL-18V	CCCP-75521	
18.3.0.067.05	IL-18V	CCCP-75522	
18.3.0.068.01	IL-18V, IL-18USh §	CCCP-75523	30-12-1963
18.3.0.068.02	IL-18V	CCCP-75524	
18.3.0.068.03	IL-18V	CCCP-75525	
18.3.0.068.04	IL-18V	CCCP-75526? † 22-4-1968	
18.3.0.068.05	IL-18V	CCCP-75527	
18.3.0.069.01	IL-18V, IL-18RT	CCCP-75528	
18.4.0.069.02	IL-18V	CCCP-75529	
18.4.0.069.03	IL-18V 'Salon' §	CCCP-75530*, LZ-AZC, 3D-SBC, 3C-KKJ, UN 7500...	12-12-1963
18.4.0.069.04	IL-18V	CCCP-75532?	
18.4.0.069.05	IL-18V	CCCP-75536	
18.4.0.070.01	IL-18V	CCCP-75537	
18.4.0.070.02	IL-18V	CCCP-75538 † 16-11-1967	
18.4.0.070.03	IL-18V	CCCP-75539	
18.4.0.070.04	IL-18V	CCCP-75540	
18.4.0.070.05	IL-18V	CCCP-75541	
18.4.0.071.01	IL-18V	*CCCP-75542?*, LZ-BEN † 24-11-66	
18.4.0.071.02	IL-18V	SP-LSD 'Tobruk'	
18.4.0.071.03	IL-18V, IL-18Gr	HA-MOG	...-1-1964
18.4.0.071.04	IL-18V	HA-MOH † 15-1-1975	...-1-1964
18.4.0.071.05	IL-18V, IL-18Gr	YR-IMF, B-234, YR-IMF, UN 75111	...-2-1964
18.4.0.072.01	IL-18V	CCCP-75543	
18.4.0.072.02	IL-18V	CCCP-75544	
18.4.0.072.03	IL-18V	LZ-BER	
18.4.0.072.04	IL-18V	CCCP-75545	
18.4.0.072.05	IL-18V	CCCP-75546	
18.4.0.073.01	IL-18V	YR-IMG, ER-ICG	3-4-1964
18.4.0.073.02	IL-18V	CCCP-75547	
18.4.0.073.03	IL-18V	CCCP-75548	
18.4.0.073.04	IL-18V	CCCP-75549	
18.4.0.073.05	IL-18V 'Salon' §	DM-STH**, DDR-STH	
18.4.0.074.01	IL-18V 'Salon' IL-18V/calibrator §	East German AF '497 Black' [DM-VAZ],DM-STP**, DDR-STP, D-AOAQ §, UR-75475 (see c/n 181004105), 3D-SEP, D2-FAM	4-6-1964
18.4.0.074.02	IL-18V	CCCP-75550	
18.4.0.074.03	IL-18V	CCCP-75551	
18.4.0.074.04	IL-18V	CCCP-75552 † 27-8-1966	
18.4.0.074.05	IL-18V	CCCP-75553, DM-STF No 2 (see c/n 181004105), DDR-STF, D-AOAO, RA-75553, LZ-AZO, EL-ADY No 1 (see c/n 182004804), T9-ABB, EX-405	16-6-1964
18.4.0.075.01	IL-18V	CCCP-75554 (see c/n 185008404)	
18.4.0.075.02	IL-18V	CCCP-75555	
18.4.0.075.03	IL-18V	CCCP-75556	
18.4.0.075.04	IL-18V	CCCP-75557	
18.4.0.075.05	IL-18V	CCCP-75558 † 30-1-1976	26-8-1964
18.4.0.076.01	IL-18V	B-218	
18.4.0.076.02	IL-18V	B-210	
18.4.0.076.03	IL-18V	CCCP-75564	
18.4.0.076.04	IL-18V 'Salon' §	PLAAF '50850 Red', B-240	
18.4.0.076.05	IL-18V 'Salon'	PLAAF '50851 Red', 'B-230 No 2' (fake registration!)	
18.4.0.077.01	IL-18V	CCCP-75569	
18.4.0.077.02	IL-18V	B-212	
18.4.0.077.03	IL-18V	CCCP-75559 † 27-4-1974	15-10-1964
18.4.0.077.04	IL-18V	CCCP-75560 † 24-2-1968	
18.4.0.077.05	IL-18V	CCCP-75561	
18.4.0.078.01	IL-18V	CCCP-75562, CU-T1269	
18.4.0.078.02	IL-18V 'Salon'	CCCP-75563? † 6-4-1967	
18.4.0.078.03	IL-18D [proto]	CCCP-75581	
18.4.0.078.04	IL-18V	CCCP-75565	
18.4.0.078.05	IL-18V 'Salon'	PLAAF '50854 Red', B-224 No 1	
18.4.0.079.01	IL-18V	CCCP-75566	
18.4.0.079.02	IL-18V	CCCP-75567	
18.4.0.079.03	IL-18V	CCCP-75568	
18.5.0.079.04	IL-18V	CCCP-75570	
18.5.0.079.05	IL-18V	CCCP-75571	
18.5.0.080.01	IL-18D [proto]	CCCP-75572	
18.5.0.080.02	IL-18V	CCCP-75573	
18.5.0.080.03	IL-18V 'Salon'	*CCCP-75574?*, CCCP-75591*, RA-75591*	
18.5.0.080.04	IL-18V	CCCP-75575 † 30-10-1976	20-2-1965
18.5.0.080.05	IL-18V	CCCP-75576	
18.5.0.081.01	IL-18V	CCCP-75577	
18.5.0.081.02	IL-18V	CCCP-75578? † 16-10-1970	
18.5.0.081.03	IL-18V	CCCP-75579	
18.5.0.081.04	IL-18V	LZ-BES † 21-12-1971	
18.5.0.081.05	IL-18V	LZ-BEP † 16-6-1984	

Construction number	Version	Registration/ tactical code/serial	Manufacture date
18.5.0.082.01	IL-18V	LZ-BEV	
18.5.0.082.02	IL-18V	CCCP-75580	
18.5.0.082.03	IL-18V	CCCP-75582	
18.5.0.082.04	IL-18V	836, P-836	
18.5.0.082.05	IL-18V	CCCP-75583	
18.5.0.083.01	IL-18V	YR-IMH † 14-8-1991	
18.5.0.083.02	IL-18V	YR-IMI † 21-4-1977	
18.5.0.083.03	IL-18V	CCCP-75584	
18.5.0.083.04	IL-18V	CCCP-75585	
18.5.0.083.05	IL-18V 'Salon' §	Polish AF '101 Red' No 2, CCCP-75593	
18.5.0.084.01	IL-18V	CCCP-75586	
18.5.0.084.02	IL-18V	CCCP-75587	
18.5.0.084.03	IL-18V	CCCP-75588	
18.5.0.084.04	IL-18V 'Salon' §	DM-STI, DDR-STI	3-9-1965
	IL-18GrM	D-AOAP IL-18V/F (SCD), RA-75554 (see c/n 184007501) † 17-12-1997	
18.5.0.084.05	IL-18V	CCCP-75589	
18.5.0.085.01	IL-18V	CCCP-75590	
18.5.0.085.02	IL-18E	CCCP-75592	
18.5.0.085.03	IL-18E 'Salon' §	Polish AF '101 Red' No 3, SP-LSK, 75711 (No 2 – see c/n 189001803), UR-75850* (see c/n 182005405)	
18.5.0.085.04	IL-18E	CCCP-75594	
18.5.0.085.05	IL-18E	B-214	
18.5.0.086.01	IL-18E	SP-LSF 'Falaise', LZ-BEW, *3D-ALD*, EL-ALD, EX-601	
18.5.0.086.02	IL-18E 'Salon' §	YR-IMZ No 1 (see c/n 186009802), CCCP-75445	
18.5.0.086.03	IL-18E	SP-LSG 'Monte Casino', LZ-BEZ, EL-ARK, 3C-KKR, UN 75002	8-10-1965
18.5.0.086.04	IL-18E 'Salon'	PLAAF '50852 Red'	
18.5.0.086.05	IL-18E	CCCP-75676 No 2* (see c/n 189001001), RA-75676*	11-12-1965
18.5.0.087.01	IL-18E 'Salon'	208, B-208	
18.5.0.087.02	IL-18E 'Salon'	PLAAF '50855 Red', B-228*	
18.5.0.087.03	IL-18E		
18.5.0.087.04	IL-18E	B-216	
18.5.0.087.05	IL-18E	CCCP-75595	
18.5.0.088.01	IL-18E	CCCP-75596	
18.6.0.088.02	IL-18E, IL-18E 'Meteor' §	CCCP-75598, RA-75598 §	18-7-1967!
18.6.0.088.03	IL-18E	CCCP-75597	
18.6.0.088.04	IL-18E		
18.6.0.088.05	IL-18E	CCCP-75599	
18.6.0.089.01	IL-18E	CCCP-75400	
18.6.0.089.02	IL-18E	CCCP-75401	
18.6.0.089.03	IL-18E	CCCP-75402	
18.6.0.089.04	IL-18E	LZ-BET	
18.6.0.089.05	IL-18E 'Salon' §, IL-18E, IL-18Gr	Polish AF '102 Red' No 2, SP-LSI, LZ-BEH, LZ-ZAH, EX-75905 (see IL-22)	
18.6.0.090.01	IL-18E	CCCP-75403	
18.6.0.090.02	IL-18E	LZ-BED † 18-1-1971	
18.6.0.090.03	IL-18E	CCCP-75404	
18.6.0.090.04	IL-18E 'Salon' §	OK-BYZ, OK-VAF	
18.6.0.090.05	IL-18E	CCCP-75405 † 24-6-1974	21-5-1966
18.6.0.091.01	IL-18E	LZ-BEG † 3-9-68	
18.6.0.091.02	IL-18E	YR-IMJ, ER-ICJ, EL-ALY, ER-ICJ, UR-TMD, ER-ICJ	...-5-1966
18.6.0.091.03	IL-18E	CCCP-75406	
18.6.0.091.04	IL-18E	YR-IMK † 9-12-1974	
18.6.0.091.05	IL-18E	CCCP-75407	
18.6.0.092.01	IL-18E	CCCP-75408 † 6-3-1976	29-6-1966
18.6.0.092.02	IL-18E, IL-18Gr	DM-STK, DDR-STK, D-AOAR, SP-FNB 'Agata', SP-FNW, 3D-SBW, 3C-KKK, UN 7500...	29-6-1966
18.6.0.092.03	IL-18E	CCCP-75409	
18.6.0.092.04	IL-18E	CCCP-75410	
18.6.0.092.05	IL-18E, SL-18P §	CCCP-75411, RA-75411 §	
18.6.0.093.01	IL-18D 'Salon'	CCCP-75412	
18.6.0.093.02	IL-18D	CCCP-75413	
18.6.0.093.03	IL-18D	CCCP-75414 † 10-5-1979	19-9-1966
18.6.0.093.04	IL-18D	CCCP-75415	
18.6.0.093.05	IL-18D	CCCP-75416	
18.6.0.094.01	IL-18D	CCCP-75417	
18.6.0.094.02	IL-18D	DM-STL † 26-3-1979	
18.6.0.094.03	IL-18D, SL-18D §	CCCP-75713 No 2 (see c/n 180001804), 75713, RA-75713 §, 75713	
18.6.0.094.04	IL-18D	CCCP-75418	
18.6.0.094.05	IL-18D	CCCP-75419	
18.6.0.095.01	IL-18D	CCCP-75432	
18.6.0.095.02	IL-18D	CCCP-75433	
18.6.0.095.03	IL-18D	CCCP-75434	
18.6.0.095.04	IL-18D	CCCP-75435	
18.6.0.095.05	IL-18D	CCCP-75436 † 20-10-1968	
18.6.0.096.01	IL-18D	CCCP-75437	
18.6.0.096.02	IL-18D	CCCP-75438, LZ-BEO	
18.6.0.096.03	IL-18D	CCCP-75439	
18.7.0.096.04	IL-18D	CCCP-75440	
18.7.0.096.05	IL-18D	B-220	
18.7.0.097.01	IL-18D	CCCP-75441	
18.7.0.097.02	IL-18D, IL-18DTs §	CCCP-75442, RA-75442 §, EX-75442	
18.7.0.097.03	IL-18D	B-224 No 2, '208 No 2' (fake registration, see c/n 185008701!)	
18.7.0.097.04	IL-18D 'Salon'	PLAAF '50853 Red'	
18.7.0.097.05	IL-18D	*CCCP-75443?*, OK-WAI † 5-9-1967	
18.7.0.098.01	IL-18D	CCCP-75444	
18.7.0.098.02	IL-18D 'Salon', IL-18GrM	YR-IMZ No 2 (see c/n 185006802), 4R-EXD	
18.7.0.098.03	IL-18D 'Salon'	YU-AIB*, 3X-GAX † 3-9-1978	
18.7.0.098.04	IL-18D 'Salon'	CCCP-75498*, RA-75498*	
18.7.0.098.05			
296.40.098.05	IL-18D 'Salon' §	YU-AIA*, Yugoslav AF '7501', Yugoslav AF '73201', CCCP-75780 No 2 (see c/n 181003804)	
	IL-22M-11	CCCP-75919	
18.7.0.099.01	IL-18D	222, B-222 † 18-1-1988	
18.7.0.099.02	IL-18D 'Salon' §	PLAAF '50856 Red', B-226	
18.7.0.099.03	IL-18D	YR-IML, ER-ICL	
18.7.0.099.04	IL-18D 'Salon'	YR-IMM, UN 75001	
18.7.0.099.05	IL-18D	CCCP-75446, RA-75446	
18.7.0.100.01	IL-18D	CCCP-75447	
18.7.0.100.02	IL-18D, IL-18Gr	HA-MOI	...-5-1967
18.7.0.100.03	IL-18D	CCCP-75448,	
	IL-22M-11	CCCP-75928	
18.7.0.100.04	IL-24N §, IL-18D	CCCP-75449, RA-75449 §, ST-APZ, EX-75449	30-6-1967
18.7.0.100.05	IL-18D	CCCP-75450	
18.7.0.101.01	IL-18D	*CCCP-75451?*, OK-WAJ 'Podébrady'	
18.7.0.101.02	IL-18D	CCCP-75452	
18.7.0.101.03	IL-18D 'Salon', IL-18D/relay	CCCP-75453, RA-75453	26-7-1967
18.7.0.101.04	IL-18D 'Salon', IL-18D/relay	CCCP-75454, RA-75454	
18.7.0.101.05	IL-18D 'Salon' §	Afghan AF T 001 'Kandahar' §, CCCP-75451	
	IL-22M-11	CCCP-75917, RA-75917	
18.7.0.102.01	IL-18D; IL-18Gr?	CCCP-75455	22-8-1967
18.7.0.102.02	IL-18D	CCCP-75456	
18.7.0.102.03	IL-18D	CCCP-75457	
18.7.0.102.04	IL-18D 'Salon' §	CCCP-75497*, RA-75497, EL-AKQ 'Zlatoust', 3D-SBQ 'Zlatoust', 3C-KKL 'Zlatoust', UN 75005	19-8-1967
18.7.0.102.05	IL-18D	CCCP-75458	
18.7.0.103.01	IL-18D	CCCP-75459, TZ-ADF, CCCP-75459, CU-T1270 † 15-11-1992	
18.7.0.103.02	IL-18D	CCCP-75460	
18.7.0.103.03	IL-18D	CCCP-75461	
18.7.0.103.04	IL-18D, IL-18DORR §	CCCP-75462, RA-75462 §	
18.7.0.103.05	IL-18D,	CCCP-75463	
	IL-22M-11	CCCP-75923, RA-75923 No 1 (see IL-20M c/n 173011501)	
18.7.0.104.01	IL-18D 'Salon', IL-18D/relay	CCCP-75464, RA-75464	
18.7.0.104.02	IL-18D	CCCP-75465, VN-B196	
18.7.0.104.03	IL-24N §, IL-18D	CCCP-75466, RA-75466 §	
	IL-18GrM	EX-75466	
18.7.0.104.04	IL-18D	CCCP-75467	
18.7.0.104.05	IL-18D	CCCP-75468	
18.7.0.105.01	IL-18D	CCCP-75469	21-11-1967
18.7.0.105.02	IL-18D	CCCP-75470	
18.7.0.105.03	IL-18D	CCCP-75471	
18.7.0.105.04	IL-18D	CCCP-74250	
18.7.0.105.05	IL-18D	CCCP-74251	
	IL-22M-11 §	CCCP-75929, ER-75929 §	
18.7.0.106.01	IL-18D	CCCP-74252 † 29-2-1968	
18.7.0.106.02	IL-18D	CCCP-75472, CCCP-74295*, RA-74295* † 25-10-2000	17-1-1968
18.8.0.106.03	IL-18TD §, IL-18D	*CCCP-75473?*, CCCP-74296, RA-74296	16-1-1968
18.8.0.106.04	IL-18D	[P-] 834, 3X-GAT 'Conakry'	
18.8.0.106.05	IL-18D	CCCP-74253	
18.8.0.107.01	IL-18D	CCCP-74254	
18.8.0.107.02	IL-18D	CCCP-74255	
18.8.0.107.03	IL-18D		
18.8.0.107.04	IL-18D	CCCP-74256, CU-T1268, CU-T1517	
18.8.0.107.05	IL-18D	CCCP-74257	
18.8.0.108.01	IL-18D	CCCP-74258	
18.8.0.108.02	IL-18D 'Salon' §, IL-18D, IL-18Gr	LZ-BEA**	
18.8.0.108.03	IL-18D	CCCP-74259	
18.8.0.108.04	IL-18D	CCCP-74260	
18.8.0.108.05	IL-18D 'Salon'	DM-STM**, DDR-STM	
	IL-18GrM	D-AOAS IL-18D/F (SCD), LZ-AZZ, CU-T132, CU-C132	5-5-1968

Construction number	Version	Registration/ tactical code/serial	Manufacture date
18.8.0.109.01	IL-18D	CCCP-74261 RA-74261	
18.8.0.109.02	IL-18D	CCCP-74262 RA-74262	
18.8.0.109.03	IL-18D, IL-18Gr	DM-STN, DDR-STN, D-AOAT, SP-FNC 'Hubert', SP-FNZ, 3D-SBZ	15-5-1968
18.8.0.109.04	IL-18D	DM-STO, DDR-STO, D-AOAU, LZ-AZR, CU-T131, CU-T1532	30-5-1968
18.8.0.109.05	IL-18D	CCCP-74263	
18.8.0.110.01	IL-18D	CCCP-74264	
18.8.0.110.02	IL-18D	CCCP-74265?	
18.8.0.110.03	IL-18D	SU-AOV † 29-1-1973	
18.8.0.110.04	IL-18D 'Salon' §	CCCP-75499*, RA-75499*; EX-75499?	30-8-1968
18.8.0.110.05	IL-18D	SU-AOX, CCCP-75430	
18.8.0.111.01	IL-18D	SU-AOY, CCCP-75429	
18.8.0.111.02	IL-18D	CU-T899 † 19-1-1985	
18.8.0.111.03	IL-18D	CCCP-74266	
18.8.0.111.04	IL-18D, IL-18Gr	CU-T900 'Fernando Alvaréz Peréz', CU-C900	
18.8.0.111.05 [iii]	IL-18D, IL-18D-GAL §	CCCP-74267, RA-74267, EX-74267, EX-105	
18.8.0.112.01	IL-18D, IL-18DORR §	CCCP-74268, RA-74268 §; EX-201?	9-12-1968
18.8.0.112.02	IL-18D	CCCP-74269	
18.8.0.112.03	IL-18D	CCCP-74270	
	IL-22M-11	CCCP-75926	
18.8.0.112.04	IL-18D		
18.8.0.112.05	IL-18D	835, P-835	
18.8.0.113.01	IL-18D	SU-APC? † 20-3-1969	
18.9.0.113.02	IL-18D (ABCP)	CCCP-75478*, RA-75478*	28-2-1969
18.9.0.113.03 [iv]	IL-18D (ABCP)	CCCP-75496*, RA-75496*	26-3-1969
18.9.0.113.04	IL-18D	5T-CJL, VN-B198 28-3-1969	

Construction number	Version	Registration/ tactical code	Manufacture date
17.2.0.114.01	IL-20M	no code, Ukrainian AF '07 Red'?	
	'IL-18D'	UR-BXD	
17.2.0.114.02	'IL-18D' (ABCP?)	CCCP-75903* No 2 (see IL-22 c/n 0393610235), RA-75903* No 1!	
17.*.0.114.03			
17.3.0.114.04	IL-20M	Soviet/Russian AF '88 White'	
17.3.0.114.05	IL-20RT	CCCP-75480	
17.3.0.115.01	IL-20M	Soviet/Russian AF '90 Red', RA-75923 No 2 (see IL-18D/IL-22M-11 c/n 187010305)	
17.3.0.115.02	IL-20M	Soviet/Russian AF '20 Red'	
17.3.0.115.03	IL-20RT	CCCP-75481	
17.3.0.115.04	IL-20M	Soviet/Russian AF '21 Red'	
17.3.0.115.05	IL-20RT	CCCP-75482, RA-75482	
17.3.0.116.01	IL-20RT	CCCP-75483, RA-75483	
17.3.0.116.02?	IL-20M	no code	
17.4.0.116.03	IL-20M	no code	
17.4.0.116.04	IL-20M	no code	
17.*.0.116.05			
17.5.0.117.01	IL-20M	no code	
17.*.0.117.02			
17.*.0.117.03			
17.*.0.117.04			
17.*.0.117.05			
17.*.0.117.06			
17.6.0.117.07?	IL-20M	Soviet/Russian AF '34 Red'	

Construction number	Version	Registration	Manufacture date
039.36.07.150	IL-22 (IL-18D-36) §	CCCP-75896, 75896, UR-75896 §, D2-FFR	
039.36.07.430	IL-22 (IL-18D-36)	CCCP-75897, RA-75897	
039.36.07.850	IL-22 (IL-18D-36)	CCCP-75895, RA-75895	
039.36.07.950	IL-22 (IL-18D-36)	CCCP-75898	
	IL-22 (IL-18D-36)	CCCP-75899, RA-75899	
039.36.0*.*40?	IL-22 (IL-18D-36)	CCCP-75900	
039.36.09.935	IL-22 (IL-18D-36)	CCCP-75901, RA-75901	
039.36.10.226	IL-22 (IL-18D-36)	CCCP-75902, 75902	
039.36.10.235	IL-22 (IL-18D-36) §	CCCP-75903 No 1 (see IL-20 c/n 172011402), RA-75903 No 2! §	
	IL-22 (IL-18D-36)	CCCP-75904	
	IL-22 (IL-18D-36)	CCCP-75905, RA-75905 (see IL-18 c/n 186008905)	
039.36.10.501	IL-22 (IL-18D-36)	CCCP-75906	25-6-1979
	IL-22 (IL-18D-36)?	CCCP-75907	
039.40.11.091	IL-22M-11, ?	CCCP-75908, RA-75908	
039.40.11.092	IL-22M-11, ?	CCCP-75909, RA-75909	
039.40.11.094	IL-22M-11	CCCP-75910	
039.40.11.096	IL-22M-11	CCCP-75911, RA-75911	
039.40.11.097	IL-22M-11	CCCP-75912, RA-75912	
039.40.11.098	IL-22M-11	CCCP-75913, RA-75913	
039.40.11.099?	IL-22M-11	CCCP-75914	
296.40.17.101	IL-22M-11 §	CCCP-75915, UN-75915 § † ...-1-1995	
296.40.17.102	IL-22M-11 §	CCCP-75916 §, YL-LAO, CU-T1539	27-9-1983
	IL-22M-11	CCCP-75918, 75918 (Ukrainian AF)	
296.40.17.551	IL-22M-11	CCCP-75920, RA-75920	8-6-1984
296.40.17.552	IL-22M-11	CCCP-75922, RA-75922	
296.40.17.553?	IL-22M-11	CCCP-75921	
296.40.17.554	IL-22M-11	CCCP-75924, RA-75924	
296.40.17.557	IL-22M-11	CCCP-75925, RA-75925	19-11-1985
296.40.17.558	IL-22M-11	CCCP-75927, RA-75927	

Construction number	Version	Tactical Code	Manufacture date
08.7.0.101.06	IL-38	no code; Soviet Navy '10 Red'	...-12-1967
08.*.0.101.07	IL-38		
08.8.0.101.08	IL-38		
08.8.0.101.09	IL-38		
08.8.0.101.10	IL-38		
08.8.0.102.06	IL-38		
08.8.0.102.07	IL-38		
08.8.0.102.08	IL-38		
08.8.0.102.09	IL-38		
08.8.0.102.10	IL-38		
08.*.0.103.06	IL-38		
08.*.0.103.07	IL-38		
08.*.0.103.08	IL-38		
08.*.0.103.09	IL-38		
08.*.0.103.10	IL-38		
08.*.0.104.06	IL-38		
08.*.0.104.07	IL-38		
08.*.0.104.08	IL-38		
08.*.0.104.09	IL-38		
08.*.0.104.10	IL-38		
08.*.0.105.06	IL-38		
08.*.0.105.07	IL-38		
08.*.0.105.08	IL-38		
08.*.0.105.09	IL-38		
08.*.0.105.10	IL-38		
08.*.0.106.06	IL-38		
08.*.0.106.07	IL-38		
08.*.0.106.08	IL-38		
08.*.0.106.09	IL-38		
08.*.0.106.10	IL-38		
08.*.0.107.06	IL-38		
08.*.0.107.07	IL-38		
08.*.0.107.08	IL-38		
08.*.0.107.09	IL-38		
08.*.0.107.10	IL-38		
08.*.0.108.06	IL-38		
08.*.0.108.07	IL-38		
08.*.0.108.08	IL-38		
08.*.0.108.09	IL-38		
08.*.0.108.10	IL-38		
08.*.0.109.06	IL-38		
08.1.0.109.07	IL-38	not known; Russian Navy '25 Red'	
08.1.0.109.08	IL-38		
08.1.0.109.09	IL-38		
08.1.0.109.10	IL-38	not known; Russian Navy '21 Red'	
08.*.0.110.06	IL-38	not known; Russian Navy '22 Red'	
08.*.0.110.07	IL-38		
08.*.0.110.08	IL-38		
08.*.0.110.09	IL-38		
08.*.0.110.10	IL-38		
08.2.0.111.06	IL-38	not known; Russian Navy '79 Red'	
08.2.0.111.07	IL-38		
08.2.0.111.08	IL-38		
08.2.0.111.09	IL-38		
08.2.0.111.10	IL-38		
08.2.0.112.06	IL-38		
08.2.0.112.07	IL-38	not known; Russian Navy '71 Red'	
08.2.0.112.08	IL-38		

i Chronologically the manufacture dates do not always follow the c/n order; this is explained by the need to eliminate defects discovered during the first flight (if any) and by the higher complexity of outfitting jobs on VIP and special mission aircraft as compared to ordinary passenger aircraft. Sometimes different sources state different manufacture dates for the same aircraft. For instance, the manufacture date of IL-18D CCCP-75472 was quoted as 21st December 1967, whereas two official documents for RA-74295 (which is the same aircraft!) state the manufacture date as 17th January 1968! The manufacture date of IL-18D CCCP-74296 was also quoted as 6th March 1968.

ii The manufacture date of CCCP-75514 has been reported as 3rd January 1963, which is impossible. This is obviously a typing error – either a digit being omitted in the month or, less likely, the wrong year being quoted (ie, should be 3rd January 1964?).

iii The c/n was applied in error as 187011105.

iv The c/n was applied in error as 185011303.

Accident Attrition

The following is a brief run-down of known IL-18 accidents as of 1st January 2003. Only fatal and non-fatal accidents of varying seriousness are listed here; flight incidents and cases of ground damage to parked aircraft are not dealt with because with most aircraft types the number of such incidents per year runs into the hundreds![1]

Within a 43-year time frame (1959-2002), 96 fatal and non-fatal accidents involving IL-18s have been reported in which 85 aircraft have been written off. The worst period was 1966-1971, with five aircraft being written off in crashes each year; but then, these were the years when IL-18 utilisation reached its peak. On the other hand, there were 'lucky' years when not a single accident occurred with the type – for instance, 1981, 1983 and 1986-87, and this was a time when the *Coot* was still around in substantial numbers. (The latter comment is in case someone should feel like cracking a joke that 'they did not crash because there weren't any left to crash'!)

Unfortunately very little information is available on most of the accidents, so what follows is basically a simple list in chronological order.

The first total hull loss occurred on 2nd September 1959 when Aeroflot IL-18B CCCP-75676 No 1 (c/n 189001001?) was damaged beyond repair after suffering a quadruple engine failure at 10,000m (32,810ft); the location is unknown but it **is** known that no-one was killed. It can only be conjectured that fuel contamination was the cause and the damage was incurred in the ensuing off-field emergency landing.

The first fatal crash of an IL-18 was on 27th April 1960 when IL-18A CCCP-75648 (c/n 188000402?) operated by Aeroflot's Urals CAD crashed on final approach to Sverdlovsk-Kol'tsovo airport.

On 17th August 1960 IL-18S CCCP-75705 (c/n 189001702) operated by the 235th Independent Air Detachment crashed near Kiev on the way back from Cairo to Moscow-Vnukovo after suffering an engine fire, killing all 35 occupants. Of course all IL-18s were grounded and a huge investigation began. It was established that the combustor nozzle flame in one of the affected engine's burner cans had been impinging on the wall of the can, causing it to burn out, whereupon the fire had spread to the wing structure. Apparently corrective measures were developed very rapidly, as the IL-18 was back in service in October 1960.

On 26th December 1960 IL-18A CCCP-75651 (c/n 188000405?) belonging to the Ul'yanovsk Higher Civil Aviation Flying School crashed fatally on take-off from its home base, probably due to icing.

On 28th March 1961 ČSA Czechoslovak Airlines IL-18V OK-OAD (c/n 180002102) crashed near Forchheim 27km (17 miles) north of Nürnberg, West Germany; there were no survivors among the eight crew and 44 passengers. The aircraft was en route from Prague to Zürich and the accident occurred about 25 minutes before OK-OAD was due to land at Zürich-Kloten. An in-flight explosion was suspected at an early stage of the investigation, since one of the wings (complete with both engines) was located more than a mile from the main wreckage. Eventually, however, it was established that the aircraft had broken up in severe turbulence.

On 22nd June 1961 IL-18B CCCP-75672 (c/n 188000901) of the Moscow Territorial CAD was written off after force-landing in a field near Tula, central Russia, when an engine-driven generator caught fire.

On 13th August 1961 another Ul'yanovsk HFS IL-18B, CCCP-75653 (c/n 188000502?), was damaged beyond repair when it overran the runway at Riga-Spilve in fog.

On 17th December 1961 IL-18B CCCP-75654 (c/n 188000503?) crashed fatally near the small town of Millerovo, Russia.

Two weeks later the 1962 New Year's Day celebrations were marred by a tragic accident at Mineral'nyye Vody airport when IL-18V CCCP-75757 (c/n 181003203?) of the Armenian CAD/Yerevan UAD/279th Flight crashed fatally on 31st December 1961.

Sometime before February 1962 an Aeroflot IL-18 (identity unknown) belonging to the Moscow Territorial CAD/Vnukovo UAD/65th Flight was en route from Moscow-Vnukovo to Adler with 80 passengers when an engine fire occurred. The captain, Boris Yevgen'yevich Gratsianov, executed a wheels-up emergency landing in an oat field, the aircraft ploughing a 450-m (1,480-ft) furrow in the soft earth before coming to a standstill. Reportedly nobody was hurt and the crew managed to extinguish the fire. For this performance the entire crew received government awards, Gratsianov being awarded the Order of the Red Banner of Labour. The ultimate fate of the aircraft is not known.

On 24th February 1962 sloppy ground support procedures at Bucharest-Otopeni nearly resulted in disaster. En route from Bucharest to Nicosia, Cyprus, all four engines of TAROM Romanian Airlines IL-18V YR-IMB (c/n 181003702) flamed out consecutively at high altitude. The crew managed to glide to a perfect belly landing in a corn field on Paphos Island; no-one was hurt and the aircraft sustained only superficial damage. It turned out that fuel contamination was the cause, the fuel filters being clogged with ice and mud. The *Coot* was transported to Yeroskipou, a disused wartime airfield, for temporary repairs; thence it was flown to Nicosia and subsequently returned to the Soviet Union on 11th November 1962. The aircraft was destined to fly no more, becoming an instructional airframe at the Omsk Technical School. Again the crew received government awards for saving the passengers' lives.

On 12th July 1962 ČSA Czechoslovak Airlines IL-18V OK-PAF (c/n 181002904) hit power cables while performing a go-around after a missed approach to Casablanca, Morocco, and crashed, killing 72.

On 23rd November 1962 MALÉV Hungarian Airlines IL-18V HA-MOD (c/n 180002002) crashed near Roissy-en-France whilst on approach to Paris-le Bourget; all 13 passengers and eight crew perished. The cause was eventually traced to a stabiliser stall which rendered the aircraft uncontrollable.

On 26th February 1963 Polar Aviation IL-18V CCCP-75732 (c/n 180002601 or 181002601?) force-landed on ice near Cape Schmidt after suffering a double engine failure en route from Anadyr' to Magadan, broke up and sank. Several of the occupants lost their lives.

On 5th March 1963 IL-18V CCCP-75765 (c/n 181003404?) of the Turkmen CAD/Ashkhabad UAD (369th Flight?) crashed fatally while landing at its home airport in a dust storm.

On 4th April 1963 IL-18V CCCP-75866 (c/n 183005901) of the Krasnoyarsk CAD/1st Krasnoyarsk UAD/214th Flight crashed fatally near Kazan' after encountering technical problems.

On 10th November 1963 IL-18B CCCP-75686 (c/n 189001201?) of the Uzbek CAD/Tashkent UAD/202nd Flight was damaged beyond repair at Kuibyshev-Kurumoch.

On 2nd July 1964 IL-18B CCCP-75661 (c/n 188000605?) was written off in an accident at Krasnodar-Pashkovskiy airport.

One month and one day later (3rd August 1964) IL-18V CCCP-75824 (c/n 182004903) of the Far Eastern CAD/1st Khabarovsk UAD was damaged beyond repair in an accident at Magadan-Sokol.

On 2nd September 1964 IL-18V CCCP-75531 (c/n 180002402?) slammed into a hill on approach to Yuzhno-Sakhalinsk/Khomutovo airport. There were no survivors. (It should be noted that the region is often plagued by bad weather, and several other controlled flight into terrain (CFIT) accidents have taken place there under similar circumstances.)

On 19th October 1964 a quasi-civil Soviet Air Force/8th Special Mission Air Division IL-18 'Salon' (exact version and identity unknown) crashed into Mount Avala on approach to Belgrade, inbound from Moscow with a Soviet military delegation. The victims of the crash included Marshal Sergey S Biryuzov. (Interestingly, the identity of the *Coot* which crashed on this day has been reported in error as YR-IMB!)

On 3rd January 1965 Kazakh CAD IL-18B CCCP-75685 (c/n 189001105) crashed fatally while attempting to land at Alma-Ata, Kazakhstan, in below-minima weather conditions.

On 23rd December 1965 another IL-18 was totalled at Magadan-Sokol – fortunately again with no fatalities. This time the aircraft in question was IL-18B CCCP-75688 (c/n 189001203).

On 10th July 1966 Cubana IL-18V CU-T830 (c/n 182004905) crashed while attempting to land at Cienfuegos, Cuba; two people were killed.

On 27th August 1966 IL-18V CCCP-75552 (c/n 184007404) of the Latvian CAD/Riga UAD (280th Flight?) attempted to take off at Arkhangel'sk-Talagi with the gust locks still engaged. The take-off was aborted but the aircraft could not be stopped in time, overran and was destroyed – fortunately with no loss of life.

On 22nd November 1966 IL-18B CCCP-75665 (c/n 188000704?) suffered a fatal crash at Alma-Ata.

Two days later TABSO Bulgarian Air Transport IL-18V LZ-BEN (c/n 184007101) crashed into hills about 13km (8 miles) from Bratislava-Ivanka airport shortly after take-off, killing 82.

In January 1967 Interflug IL-18V DM-STF No 1 (c/n 181004105) was extensively damaged at Moscow-Bykovo when a fire broke out in the work hangar of ARZ No 402 while the aircraft was in mid-refurbishment. The prospects of repairing it looked pretty nebulous, so another IL-18V (CCCP-75553, c/n 184007405) was requisitioned from Aeroflot and supplied to Interflug as a replacement, becoming DM-STF No 2.

Meanwhile, despite the extent of the damage, the plant managed to rebuild the airliner after all. The 'undead' aircraft was registered CCCP-75475 and delivered to the Krasnoyarsk CAD/1st Krasnoyarsk UAD/214th Flight where it served on until finally retired in 1979 as time-expired.

On 6th April 1967 IL-18V 'Salon' CCCP-75563 (c/n 184007802?) of the 235th IAD crashed at Moscow-Domodedovo during a night take-off. Accounts of this tragedy vary; some sources indicate the artificial horizon failed, causing pilot disorientation and CFIT, while others say the No 4 engine failed and the propeller could not be feathered, creating strong drag and causing loss of control.

On 9th July 1967 Air Guinée IL-18V 3X-GAB (c/n 181003703) struck a building with the port wingtip on landing at Casablanca and slewed off the runway onto rough ground. Nobody was killed but the aircraft was a complete write-off.

On 5th September 1967 ČSA Czechoslovak Airlines IL-18D OK-WAI (c/n 187009705) crashed immediately after take-off at Gander, Newfoundland, Canada. Thirty-seven of the occupants died in the crash.

On 16th November 1967 IL-18V CCCP-75538 (c/n 184007002) of the Urals CAD (probably 1st Sverdlovsk UAD/120th Flight) hit high ground shortly after take off from Sverdlovsk-Kol'tsovo. There were no survivors among the 130 occupants.

On 9th January 1968 IL-18V CCCP-75519 (c/n 183006702?) of the Northern Territorial CAD was damaged beyond repair when it landed 700m (2,300ft) short at Karaganda, Kazakhstan.

On 24th February 1968 IL-18V CCCP-75560 (c/n 184007704) of the Urals CAD was written off when it overran a poorly cleaned runway during an aborted take-off at Donetsk, the Ukraine.

Five days later IL-18D CCCP-74252 (c/n 187010601) broke up in mid-air in severe turbulence, killing all on board

On 3rd September 1968 TABSO Bulgarian Air Transport IL-18E LZ-BEG (c/n 186009101) crashed on approach to Bourgas, killing 47.

On 20th October 1968 IL-18D CCCP-75436 (c/n 186009505) of the West Siberian CAD/Tolmachovo UAD (384th Flight?) crashed near Krasnoyarsk while attempting an emergency landing in poor weather.

On 20th March 1969, only two weeks after delivery, United Arab Airlines IL-18D SU-APC (c/n 188011301?) crashed near Aswan, Egypt, on its third landing attempt. One hundred people lost their lives in the accident.

22nd April is the birthday of Vladimir I Lenin, the founder of the Soviet state, and it was customary for Soviet enterprises to celebrate this semi-official holiday by making more or less notable accomplishments in their line of work. On 22nd April 1969, however, the Moscow Territorial CAD got a most unwelcome 'gift' when IL-18V CCCP-75526 (c/n 183006804?) of the Domodedovo UAD/212th Flight struck a power line during a training flight and crashed; fortunately there were no fatalities but the aircraft was totalled.

On 26th August 1969 an Aeroflot IL-18 (identity unknown) reportedly belly-landed at Moscow-Vnukovo on a flight from Sochi and was damaged beyond repair. However, no evidence of this has been found in Russian documents.

On 10th September 1969 IL-18V CCCP-75791 (c/n 181004005) of the Krasnoyarsk CAD/1st Krasnoyarsk UAD/214th Flight was written off after colliding with an ambulance on landing at Yakutsk.

On 11th December 1969 IL-18B CCCP-75699 (c/n 189001402) was declared a write-off when inadmissible structural deformation was discovered after a flight through turbulence.

On 6th February 1970 IL-18 CCCP-75798 (version and c/n unknown) crashed fatally 32km (19.8 miles) from Samarkand.

On 5th June 1970 IL-18V CCCP-75533 (c/n 180002501?) crashed on take-off at Samarkand when the crew forgot to unlock the controls.

On 23rd August 1970 IL-18V CCCP-75823 (c/n 182004902), a former VIP machine which by then had been transferred to the Far Eastern CAD/1st Khabarovsk UAD, was damaged beyond repair after a hard landing at Yuzhno-Sakhalinsk/Khomutovo.

On 16th October 1970 IL-18V CCCP-75578 (c/n 185008102?) of the Armenian CAD/Yerevan UAD/279th Flight was written off in a non-fatal accident at Simferopol', the Ukraine, caused by a hydraulics failure.

31st December 1970 was an uncanny replay of the 1961 Mineral'nyye Vody tragedy when IL-18V CCCP-75773 (c/n 181003603), another Armenian CAD/Yerevan UAD/279th Flight aircraft, crashed shortly after taking off from Leningrad-Pulkovo bound for Yerevan. Ninety-three people died in the crash.

On 18th January 1971 TABSO Bulgarian Air Transport IL-18E LZ-BED (c/n 186009002) crashed on approach to Zürich-Kloten where it was to make a refuelling stop en route from Paris to Sofia. The aircraft had strayed to the right from the runway centreline on short finals in thick fog. Trying to take corrective action, the pilots banked the aircraft excessively and the port wingtip struck the ground; the airliner groundlooped and exploded. Only three of the 37 occupants (the captain and two of the passengers) were found alive on the crash site but one of the two passengers later died in hospital.

On 21st January 1971 IL-18V CCCP-75727 (c/n 180002303) was damaged beyond repair at Rostov-on-Don.

On 28th August 1971 MALÉV Hungarian Airlines IL-18V HA-MOC (c/n 181002903) dived into Øresund Strait on approach to Copenhagen-Kastrup, breaking up on impact and sinking; 32 people aboard were killed. The wreckage was quickly salvaged but the flight data recorder could not provide any clues... because there wasn't one; HA-MOC was one of the last two IL-18s where an FDR was not yet installed. Wind shear was cited as a possible cause. (Incidentally, runway 27, which the aircraft was due to land on, is no longer extant (only a small section of it is left which functions as a taxiway); Kastrup's runways are now 04L/22R, 04R/22L and 12/30.)

On 21st December 1971 TABSO Bulgarian Air Transport IL-18V LZ-BES (c/n 185008104) crashed immediately after taking off from Sofia-Vrazhdebna on an unscheduled flight to Algiers. The aircraft was carrying a Bulgarian folk dance group which was due to make a tour of Algeria. Of the 73 people aboard, 28 died in the crash.

The following day MALÉV Hungarian Airlines IL-18D HA-MOI (c/n 187010002) had a narrow escape on approach to Damascus when the landing gear brushed a hill located 16km (10 miles) from the runway threshold and the aircraft ran 65m (213ft) across the hilltop before becoming airborne again! Luckily no one was hurt and the aircraft was undamaged, landing safely. (ČSA Czechoslovak Airlines IL-62 *sans suffixe* OK-DBF was not so lucky: on 20th August 1975 it went smack into the same hill and was destroyed.)

On 26th August 1972 IL-18B CCCP-75663 (c/n 188000702) of the Northern Territorial CAD was damaged beyond repair in an unsuccessful landing at Arkhangel'sk-Talagi in ground fog.

Five days later IL-18 CCCP-74298 (version and c/n unknown) of the Kirghiz CAD/Frunze UAD (250th Flight?) crashed fatally near Magnitogorsk, Kirghizia.

On 1st October 1972 IL-18V CCCP-75507 (c/n 183006405?) of the Moscow Territorial CAD/Vnukovo UAD/65th Flight crashed into the Black Sea shortly after take-off from Sochi-Adler airport, killing 100.[2]

On 29th January 1973 Egyptair IL-18D SU-AOV (c/n 188011003) crashed into mountains on approach to Nicosia, inbound from Cairo; all 37 occupants were killed.

On 24th February 1973 IL-18V CCCP-75712 (c/n 189001803 or 180001803?) of the Tajik CAD/Dushanbe UAD lost speed, stalled and spun in near Leninabad, Tajikistan.

On 3rd March 1973 TABSO Bulgarian Air Transport IL-18V LZ-BEM (c/n 182005602) hit the ground in a steep dive on final approach to Moscow-Sheremet'yevo, killing 25 of the occupants.

On 11th May 1973 IL-18B CCCP-75687 (c/n 189001202) of the Azerbaijan CAD/Baku UFD/107th Flight crashed fatally near Semipalatinsk, Kazakhstan.

On 27th April 1974 IL-18V CCCP-75559 (c/n 184007703) of the Leningrad Territorial Civil Aviation Production Association, as the Leningrad CAD was known at the time, crashed 3.2km (2 miles) from Leningrad-Pulkovo after suffering an engine fire on take-off for Krasnodar.

Accidents have a way of showing complete disregard for public holidays – or perhaps it's people in a bit too relaxed state of mind on occasion of the holiday that make the accidents happen. On 9th May 1974 IL-18V CCCP-75425 (c/n 181003403) of the Urals CAD/1st Sverdlovsk UFD/120th Flight was on approach to Ivano-Frankovsk, the Ukraine, when the crew mistook an agricultural airstrip only 500m (1,640ft) long used by An-2SKh biplanes for the airport's runway which was still several miles ahead. Landing long on the airstrip which was all too short for the *Coot* as it was, the airliner had no chances to stop the in time and overran into a ravine. No one was killed but the aircraft broke its back aft of the wings and was a total loss.

On 24th June 1974 IL-18E CCCP-75405 (c/n 186009005) of the Uzbek CAD/Tashkent UAD/202nd Flight crashed at Tashkent-Yuzhnyy. Birdstrike was the cause of the accident this time.

On 11th August 1974 Air Mali IL-18V TZ-ABE (c/n 181003304) crashed near Lingomin, Upper Volta, after running out of fuel. Forty-seven of the occupants died in the crash.

On 9th December 1974 TAROM Romanian Airlines IL-18E YR-IMK (c/n 186009104) leased by Egyptair crashed into the Red Sea en route from Jeddah to Cairo, killing nine of the occupants.

On 12th February 1975 IL-18V CCCP-75801 (c/n 182004301) of the Moscow Territorial CAD/Domodedovo UFD/212th Flight undershot 250m (820ft) when landing at Krasnoyarsk-Yemel'yanovo in bad weather, then ran onto the runway and veered off onto the left shoulder, collapsing the nose and port main gear units. No one was hurt but the aircraft suffered major damage to the port wing and fuselage and was declared a write-off.

On 15th January 1975 MALÉV Hungarian Airlines IL-18V HA-MOH (c/n 184007104) impacted the ground 1,360m (4,460ft) beyond the runway threshold and 120m (390ft) to the left of the runway centreline and exploded while trying to land at Budapest-Férihegy in thick fog. Analysis of the cockpit voice recorder showed that the crew changed their mind several times as to whether they should land or go around!

On 30th January 1976 IL-18V CCCP-75558 (c/n 184007505) of the Kirghiz CAD/Frunze UFD crashed near Frunze-Manas airport on a training flight. The crew was practicing flying techniques with a simulated dual engine failure on one side – an extremely dangerous exercise which nevertheless was standard operational procedure in those days. With both starboard engines shut down, the airliner became uncontrollable when the No 3 engine failed to relight, the propeller causing strong drag. The aircraft rolled right and impacted with 55-60° bank, killing all on board.

On 6th March 1976 IL-18E CCCP-75408 (c/n 186009201) of the Armenian CAD/Yerevan UFD/279th Flight crashed near the town of Verkhnyaya Khava 50km (31 mile) east-north-east of Voronezh en route from Moscow-Vnukovo to Yerevan-Zvartnots after an electrics failure. There were no survivors among the 120 passengers and crew.

On 28th July 1976 ČSA Czechoslovak Airlines IL-18B OK-NAB (c/n 189001605) crashed on approach to Bratislava-Ivanka after two engines caught fire, killing 76.

On 30th October 1976 IL-18V CCCP-75575 (c/n 185008004) of the Uzbek CAD/Tashkent UFD/202nd Flight was damaged beyond repair when it overran at Tashkent-Yuzhnyy in poor weather.

In the late afternoon of 2nd January 1977 ČSA Czechoslovak Airlines Tu-134A OK-CFD (c/n 2351505) failed to maintain adequate horizontal separation while landing at Prague-Ruzyne. Seconds after touchdown the jet's port wing struck the tail of another ČSA aircraft, IL-18B OK-NAA (c/n 189001604), just as the latter was vacating the runway. Fortuitously, there were no fatalities on either aircraft but both were struck off charge. After serving as a fire trainer at Prague-Ruzyne for eight months OK-CFD was finally broken up on 30th August. The *Coot* had a happier fate – on 8th March 1979 it went to the Military Museum at Prague-Kbely where it resides to this day.

Civil Aviation Administration of China IL-18B B-204 (c/n 189001602) was written off after an unspecified accident in Shenyang in February 1977 (the exact date is unknown).

On 15th February 1977 IL-18V CCCP-75520 (c/n 183006703) of the Uzbek CAD/Tashkent UFD crashed at Mineral'nyye Vody airport while making a go-around in bad weather.

Death of a re-export aircraft. IL-18V CCCP-75425 (formerly Ghana Airways 9G-AAN) broke its back when it overran into a ravine after landing on the wrong runway near Ivano-Frankovsk on 9th May 1974 – luckily with no fatalities.
Courtesy CIS Interstate Aviation Committee

Malév Hungarian Airlines IL-18V HA-MOH caught by the camera on short finals to Moscow-Sheremet'yevo. On 15th January 1975 this aircraft crashed while trying to land at Budapest-Ferihegy in thick fog. Yefim Gordon archive

On 21st April 1977 TAROM Romanian Airlines IL-18V YR-IMI (c/n 185008302) crashed while making a touch-and-go at Bucharest-Otopeni.

In another case of sloppy ground services management at Bucharest-Otopeni, on 23rd November 1977 an airport truck collided with MALÉV Hungarian Airlines IL-18V HA-MOF (c/n 183006301) as the latter moved along a cleared taxiway after landing. The aircraft was declared beyond economical repair.

On 3rd September 1978 Air Guinée IL-18D 3X-GAX (c/n 187009803) crashed into marshland on approach to Conakry on a flight from Moscow, killing 15.

On 26th March 1979 Interflug IL-18D DM-STL (c/n 186009402), which was temporarily converted to cargo configuration for famine relief flights in Africa, suffered an engine failure on take-off at Luanda where it had made a refuelling stop. The take-off was aborted but the heavily loaded aircraft overran, nose-diving into a ravine and bursting into flames; the ten crew and passengers accompanying the cargo died in the blaze.

On 10th May 1979 IL-18D CCCP-75414 (c/n 186009303) of the Urals CAD overran the runway at Sochi-Adler after an aborted take-off, shearing off the landing gear and suffering massive damage to the wing and fuselage underside. There were no fatalities but the aircraft was a complete write-off.

An Aeroflot IL-18 reportedly crashed near Moscow on 6th June 1980; no details are known and no proof of this accident has been found in civil aviation archives. A possible explanation is that if the crash did indeed occur, the aircraft was quasi-civil.

On 24th December 1982 Civil Aviation Administration of China IL-18B B-202 (c/n 189001401) crashed near Canton while attempting an emergency landing with a fire on board, killing 25. Thus the first export IL-18 also became the last IL-18B to crash.

On 16th June 1984 Balkan Bulgarian Airlines IL-18V LZ-BEP (c/n 185008105) landed long at Sana'a, North Yemen, on a flight from Cairo, overran the runway and was damaged beyond repair.

On 19th January 1985 Cubana suffered its last *Coot* loss when IL-18D CU-T899 (c/n 188011102) crashed at San José de las Lajas shortly after taking off from Havana-José Marti on a flight to Managua, Nicaragua. Forty people lost their lives.

On 18th January 1988 Civil Aviation Administration of China IL-18D B-222 (c/n 187009901) hit power cables and crashed on farmhouses 8km (5 miles) out on approach to Chongqing. The total death toll in the aircraft and on the ground was 108.

On 14th August 1991 TAROM Romanian Airlines IL-18V YR-IMH (c/n 185008301) crashed into the Carpathian Mountains 18km (11 miles) west of Uricani, Romania, on a positioning flight.

On 15th November 1992 Aero Caribbean IL-18D CU-T1270 (c/n 187010301) crashed into Isabel de Torres hill, Dominican Republic, and was destroyed.

In January 1995 (the exact date is not known) Kazakstan Government IL-22M-11 UN-75915 (c/n 2964017101, f/n unknown), which had just been converted from an ABCP to a luxurious VIP aircraft, was destroyed by a wayward An-12 which ran into the parked aircraft.

On 17th December 1997 Ramair IL-18GrM, aka IL-18V/F (SCD), RA-75554 (c/n 185008404) chartered by the South African company Shuttle Air Cargo would not become airborne due to overloading when taking off at Johannesburg International airport, bound for Boujoumboura, Burundi. Realising they were not going to make it, the crew aborted the take-off and resorted to emergency braking; still, the aircraft overran and collided with obstacles at the end of the runway, collapsing the nose and port landing gear units, shedding the No 1 engine and No 2 propeller. A small fire broke out but was rapidly extinguished. The crew and sole passenger were unhurt, but the aircraft was declared a write-off; when it was finally broken up in April 2000 the cargo door assembly was carefully salvaged and installed in another IL-18 freighter!

In a much-publicised accident on 25th October 2000, Russian Air Force/223rd Flight Unit State Airline IL-18D RA-74295 (c/n 187010602) deviated considerably from the designated landing pattern and slammed into the wooded slope of Mount Mtirava during a late night approach to Batumi-Chorokh airport, Georgia, inbound from Chkalovskaya AB. The aircraft was carrying relief personnel for the Russian military installations in Georgia, as well as the servicemen's families and mail for the Russian military doing their tour of duty. The aircraft was totally destroyed by the impact and post-crash fire, killing all 86 persons on board.

On 19th November 2001 at 18:19 UTC, IRS Aero IL-18V RA-75840 (c/n 182005301) dive into the ground unexpectedly from high altitude en route from Khatanga to Moscow-Domodedovo, crashing between the villages of Zakharovka and Overkovo 15km (9.3 miles) south-east of Kalyazin, Tver' Region (57°12'50" N, 38°07'00" E). The big turboprop disintegrated utterly, killing all 27 aboard (four flight crew, two cabin crew, two technicians, an IRS Aero manager and 18 paying passengers). A terrorist attack was suspected at first, but eventually it came to light that an autopilot failure had resulted in uncommanded elevator deflection, causing the aircraft to dive steeply.

Finally, on 15th September 2002 Phoenix Aviation IL-18V EX-904 (c/n 182004904) was reportedly damaged beyond repair when it overran at Neghazi, Angola.

END NOTES

Chapter One

1 In 1957 MAP lost its ministerial status together with several other ministries and was 'demoted' to the State Committee for Aviation Hardware (GKAT – *Gosoo**dar**stvennyy komi**tet** po aviatsi**onnoy tekh**nike*) because of the Soviet leader Nikita S Khruschchov's disdainful attitude. In 1965, however, GKAT regained its original name and 'rank'.

Chapter Two

1 *Izdeliye* – product (number such and such), a common designation for Soviet military hardware items.

2 There were also about a dozen registration blocks reserved for Aeroflot's Polar division (04xxx), the Ministry of Aircraft Industry (29xxx, 48xxx, 69xxx, 93xxx, 98xxx etc), the Ministry of Defence and so on. They do not correspond to any specific type and are a mixed bag of assorted aircraft.

3 Now the site of a major aircraft factory, the Lookhovitsy Machinery Plant (LMZ – *Lookho**vits**kiy ma**shino**stroi**tel**'nyy za**vod***) which is part of the Moscow Aircraft Production Association (MAPO – *Mos**kov**skoye aviatsi**on**noye proiz**vod**stvennoye obyedi**nen**iye*).

4 In some cases the c/n is known with certainty but the registration to c/n tie-up is unconfirmed, hence the c/n is given first.

5 It has to be said that Aeroflot's organisation closely resembled an air arm's order of battle – which is hardly surprising, considering that the Soviet civil air fleet constituted an immediately available military reserve (and considering the militarisation of the Soviet economy at large). There was a number of Civil Aviation Directorates (UGA – *Oopravleniye grazh**dahn**skoy aviahtsii*), several of which were in the Russian Federation and one in each of the other Soviet republics. These were broadly equivalent to the air forces of the USAF or the air armies of the Soviet Air Force. Each CAD consisted of several United Air Detachments (OAO – *obyedi**nyon**nyy avia**otryad***) based in major cities or airports; these were equivalent to an air group (USAF) or an air division (SovAF). Each UAD had several Flights (LO – ***lyot**nyy ot**ryad***) similar to an air wing (USAF) or an air regiment (SovAF). Finally, a Flight comprised up to four, or maybe more, squadrons (yes, squadrons – *aviaeskad**ril'**ya*!); not infrequently different squadrons of the same Flight operated different aircraft types.

6 The transport aircraft division of (ex-) GK NII VVS is still located at Chkalovskaya AB; combat aircraft are tested at the main facility in Akhtoobinsk near Saratov in southern Russia. *Krasnozna**myon**nyy* means that the institute was awarded the Order of the Red Banner.

7 This aircraft was later reregistered CCCP-75820. The registration may have been reused later for an IL-18V (c/n 182004805) but this is unconfirmed.

8 Logically this aircraft should have been registered CCCP-75652; quite possibly this registration **was** allocated initially to c/n 188000501 but not taken up.

9 Later renamed the State Civil Aviation Research Institute (GosNII GA – *Gosoo**dar**stvennyy na**ooch**no-iss**led**ovatel'skiy insti**toot** grazh**dahn**skoy avi**ah**tsii*).

10 IL-18B CCCP-75688 (c/n 189001203) had the new-style window arrangement but it is not known if the aircraft was built as such or was a conversion.

11 Logically, this aircraft should have been registered CCCP-75696, while CCCP-75699 should have had c/n 189001405.

12 Some sources claim that the first aircraft to receive AI-20 Srs 2 engines was the first IL-18B delivered to China (registration number 202, later amended to B-202, c/n 189001401).

13 Until the mid-1970s, each type operated by the Soviet airline had its own colour scheme.

14 The registration CCCP-33569 was later reused for a Polish-built WSK-Mielec (Antonov) An-2R agricultural aircraft (c/n 1G 23006).

15 Logically this aircraft should have been registered CCCP-75692; this registration may have been allocated initially to c/n 189001302 but not taken up.

16 The first IL-18 to have this registration; it was later reused for IL-18D c/n 186009403 described later in this chapter.

17 This means 20+50+14 seats (ie, two rows five-abreast and one row four-abreast in the rear cabin).

18 The designations of the engine variants were a bit confusing at first. Just like there were two varieties of the AI-20 with no suffix letter (Srs 1 and Srs 2), the AI-20 Srs 3 was followed by an improved version (Srs 4) bearing the same alternative designation AI-20A!

19 Logically, if the registrations had run in sequence, this aircraft would have been CCCP-75836, whereas CCCP-75842 would have had c/n 182005302. However, CCCP-75842 'moved six places up' in the production order, fitting in between CCCP-75835 and -75836, while CCCP-75841 (c/n 182005302) was followed by CCCP-75843 (c/n 182005303).

20 The c/n is sometimes quoted as 18**2**006801, which is impossible; this may have been a mispaint or a typing error.

21 TG = ***toorbogenerahtor*** – lit. 'turbo generator'. This specific reference to the generator is because the TG-16 had no provisions for using bleed air for engine starting, since the engines were started electrically.

22 Now the ***Aviamotor*** Joint-Stock Co.

23 The c/n has also been reported as 18400**75**02 but this is IL-18V CCCP-75555.

24 Test flown as CCCP-75726.

25 RPSN = ***rahdiolokatsionnyy*** *pri**bor** sle**poy** navi**gah**tsii* – blind navigation radar device.

26 Some sources claim the XQ-1120 propeller synchronisation system was tested on IL-18V CCCP-75834 (c/n 182005104) but this may be true as well.

27 Logically, this aircraft (c/n 184007803) should have been CCCP-75564 (which, curiously enough, already existed by then as another out-of-sequence aircraft with the c/n 184007603!), while CCCP-75581 should have had c/n 185008203 – which was instead allocated to CCCP-75582.

28 Total IL-18 production is often quoted as 565 aircraft, which must include the prototype (CCCP-Л5811). This makes the existence of the alleged second prototype (CCCP-Л5812) very doubtful indeed!

29 C/n was erroneously stencilled on the tail as 18**7**011105.

30 YR-IMZ No 1 was IL-18E 'Salon' c/n 185008602 which was returned to the USSR as CCCP-75445.

31 LNPO = *Lenin**grahd**skoye na**ooch**no-proiz**vod**stvennoye obyedi**nen**iye* – Leningrad Scientific & Production Association. LNPO Leninets is now known as the Leninets Holding Co.

32 Later transformed into the Ministry of Electronic Industry (MRP – *Mini**ster**stvo **rah**dioelek**tron**noy pro**mysh**lennosti*).

33 This aircraft should have been registered CCCP-75820 but this registration already existed as IL-18 c/n 187000103.

34 In 1965 the designation IL-18P was used in error by the Polish magazine *Skrzydlata Polska* for the standard IL-18Vs delivered to Poland; the P apparently meant [*samolot*] *pasażerski* – passenger aircraft!

35 The other aircraft in the series were two Tashkent-built Antonov An-12BPs – CCCP-11530 (c/n 6344503) and CCCP-11531 (c/n 6344506), two highly modified Kazan'-built Tupolev Tu-16K-26 *Badger-G Mod* missile strike aircraft – CCCP-42355 No 1 (c/n 6203203) and CCCP-42484 (c/n 6203208), Tu-104A CCCP-42454 and an unidentified An-26 *Curl*.

36 This aircraft should have been registered CCCP-75718 (which was allocated to IL-18V c/n 180002005 instead). A possible explanation is that the aircraft was originally operated by the Polar Aviation and registered in the CCCP-04xxx block.

37 This aircraft should have been CCCP-75855 (this registration was probably allocated but not taken up).

38 This Russian term is used indiscriminately and can denote any kind of testbed (avionics, engine, equipment, weapons etc), an aerodynamics research aircraft or control configured vehicle (CCV), a weather research aircraft, a geophysical survey aircraft etc.

39 The registration had been previously worn by a Lisunov Li-2 *Cab* (a licence-built Douglas DC-3 derivative).

Chapter Three

1 Early aircraft had 900 x 285mm (35.43 x 11.22in) mainwheels.

Chapter Four

1 Now called Bykovo Air Services Company (BASCO).

2 Not to be confused with the city of Mirnyy in the republic of Yakutia, the Soviet (Russian) Far East.

3 Some sources quote a different figure – 25,793km (16,020 miles).

4 Some sources, however, claim the aircraft was flown to Moscow-Bykovo on 14th September 1979 and scrapped!

Chapter Five

1 Some sources say the IL-20M is equipped with the Nit'-S1 (Thread-S1) SLAR, like the IL-24N.

2 These aircraft (CCCP-76452 through -76456) are also known as IL-76SK (*spetsi**ahl'**nyy ko**mahn**dnyy* [*samo**lyot***] – special command aircraft) and are sometimes erroneously reported in the Western press as 'IL-976' or 'Be-976'.

3 The black, blue, red or yellow colouring of spinner tips on Soviet/CIS Air Force turboprop aircraft indicates the squadron operating the aircraft within the regiment – and thus reveals quasi-civil transports. However, this is by no means 100% sure, as some aircraft (notably An-12s) are known to retain these 'squadron markings' long after being sold to truly civil operators!

4 Also reported as IL-18M-11 'Zebra' but this designation is unconfirmed.

5 The same thing happened to MiG-29 c/ns at the same time. Cf. Soviet Air Force *Fulcrum-As* '03 Blue' (c/n 0390502020, f/n 0103) and '52 White' (c/n 2960507683, f/n 0708).

6 The original CCCP-75780 No 1 was IL-18V c/n 181003804. The original CCCP-75451, Il-18D c/n 187010101, was sold to ČSA Czechoslovak Airlines as OK-WAJ.

7 75908 wore the Russian prefix on the fuselage while still retaining the Soviet prefix on the wings.

Chapter Six

1 The name of a river in central Russia, a tributary of the Volga.

2 DISS = ***dopp**lerovskiy izme**ri**tel' **sko**rosti i **sno**sa* – Doppler speed and drift sensor; RSBN = ***rah**diotekh**nee**cheskaya sis**te**ma **bliz**hney navi**gah**tsii* – SHORAN; SP = *sis**te**ma* [*sle**poy***] *po**sah**dki* – [blind] landing system; ARK = *avto**matee**cheskiy **rah**dio**kom**pas* – ADF; RV = ***rah**dio**vy**soto**mer*** – radio altimeter; SPU = *samo**lyot**noye peregovo**r**noye oo**stroy**stvo* – [aircraft] intercom; R = ***rah**dio**stahn**tsiya* – radio [transmitter or receiver]; MS = *mag**nit**nyy samo**pis**ets* – magnetic recorder.

3 PLAB = *protivo**lod**ochnaya avia**bom**ba* – anti-submarine bomb; KAB = *korrek**tee**ruyemaya avia**bom**ba* – guided bomb; RGB = ***rah**dio**ghid**roakoos**tee**cheskiy booy* – radio sonobuoy.

4 During the Second World War this ministry was also tasked with developing ordnance, and this task remained in post-war years.

5 OAPO = ***Om**skoye aviatsee**on**noye proiz**vod**stvennoye obyedi**nen**iye 'Po**lyot**'* – Omsk Aviation Production Association 'Flight'; KnAAPO = *Komso**mol'**skoye-na-A**moo**re aviatsee**on**noye proiz**vod**stvennoye obyedi**nen**iye* – Komsomol'sk-on-Amur Omsk Aviation Production Association.

6 In Soviet/Russian exercises, Red Force are the 'good guys' and Blue Force are the 'bad guys'; in the West, it's vice versa.

7 Soviet Navy ship.

Chapter Eight

1 Contrary to usual Kazakh practice the registration was applied with a dash.

2 Second use of registration; CCCP-75850 was an IL-18V (c/n 182005405).

3 In 1997 the flight code was reallocated to Buffalo Airways Ltd [J4/BFL] of Hay River, NWT, Canada.

4 The name has nothing to do with the French oil company Elf-Aquitaine.The E6 flight code was used only briefly; in 2000 it was passed to a Russian charter carrier, Aviaexpresscruise [E6/BKS].

5 The name has also been rendered as SP Air and Spaero.

6 The name means 'Wings' in Ukrainian.

7 Sometimes reported in error as 175011701.

8 Some sources have reported that '10 Red' is preserved at the Lugansk Technical School as a ground instructional airframe.

Chapter Nine

1 This is how his last name is usually spelled in Western publications (French-style; a more accurate phonetic rendition in English is 'Boot'). JP Airline-Fleets, however (whether willfully or not), misspelled his name as 'Victor *Butt*'.

2 The 'O' may have stood for Ostdeutschland, hinting at the East German origin of the aircraft.

3 Marxwalde reverted to its original name, Neuhardenberg, following German reunification in 1990.

4 The registration EL-AHO was also worn by An-26 c/n 8610.

5 According to the *Skrzydlata Polska* (Winged Poland) magazine, all *Coots* delivered directly to LOT were originally 87-seaters but were converted to 99-seat configuration during overhauls.

6 For instance, a) Mil' Mi-2T '2617 White' is c/n 512617092, ie, transport version (product code '51'), Batch 26, 17th aircraft in the batch out of 50, manufactured in September (09) 1972 (2); b) Mikoyan MiG-29 '70 Red' is c/n 2960526370, ie, Moscow Aircraft Production Association (factory code 296 – cf. IL-22), *izdeliye* 5 (first production version known at the Mikoyan OKB as *izdeliye* 9.12 and in the West as *Fulcrum-A*), the rest is the 'famous last five'.

7 Air Pass was just a trading name; the official name was Air Cess (Swaziland), Ltd.

Appendix Two

1 Russian terminology is used here. Thus, the word 'accident' refers to fatal and non-fatal accidents when the aircraft suffers more or less serious damage to the airframe and/or engines, applying to both total hull losses and cases where the aircraft is eventually repaired. 'Incident' means situations not serious enough to be rated as a non-fatal accident, such as avionics or systems malfunctions not causing catastrophic results, in-flight engine shutdowns, tyre explosions, birdstrikes (unless they result in a crash), lightning strikes, near-misses, departures from designated air routes, landings in below-minima conditions, go-arounds because of obstacles on the runway and so on.

2 This airport serves the resort cities of Sochi and Adler, being located about halfway between them; the same situation as, for example, with Köln-Bonn airport.

IL-18 Family Drawings

A typical early-production IL-18B representative of batches 5-11 with 14 cabin windows to port.

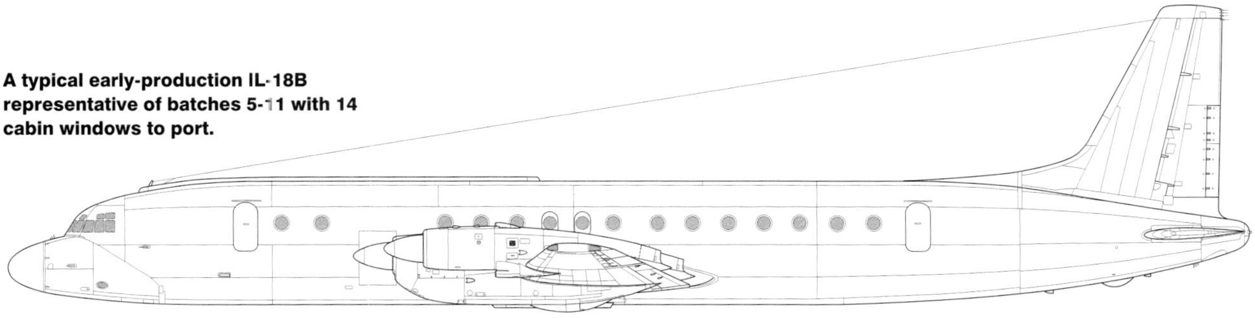

A late-production IL-18B as built from batch 13 onwards with 13 cabin windows to port.

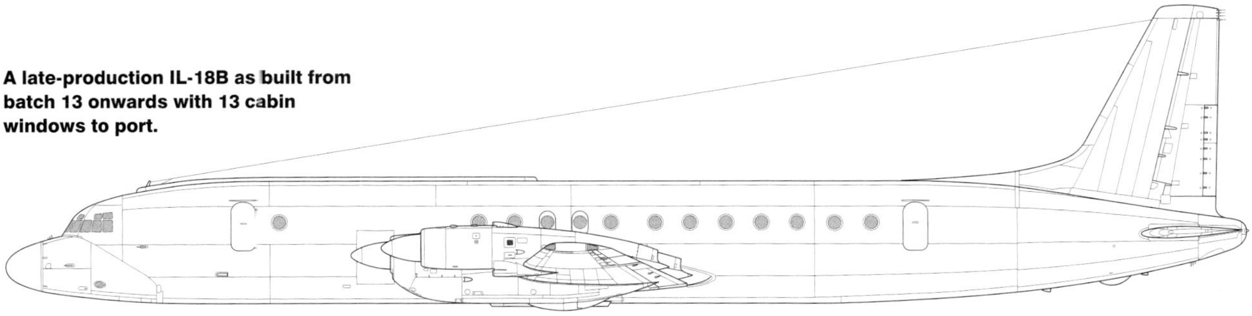

A typical early-production IL-18V with a strake aerial on the forward fuselage and no anti-collision lights, Doppler speed/drift sensor or APU.

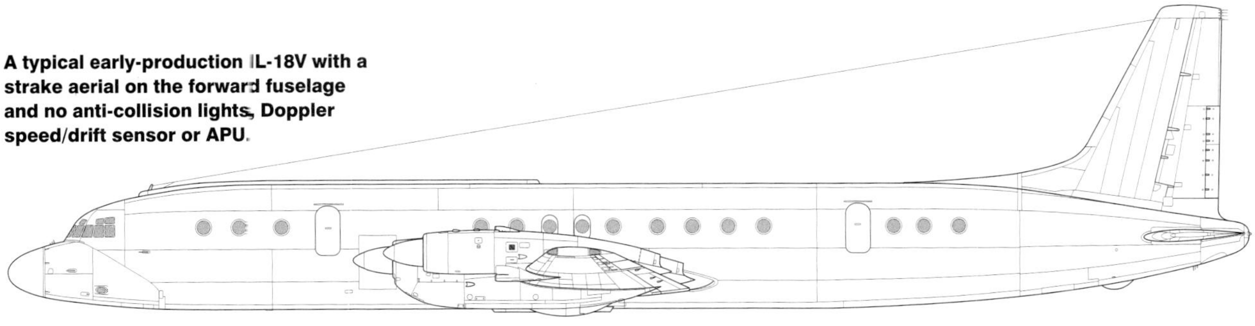

Starboard side view of an early-production IL-18V.

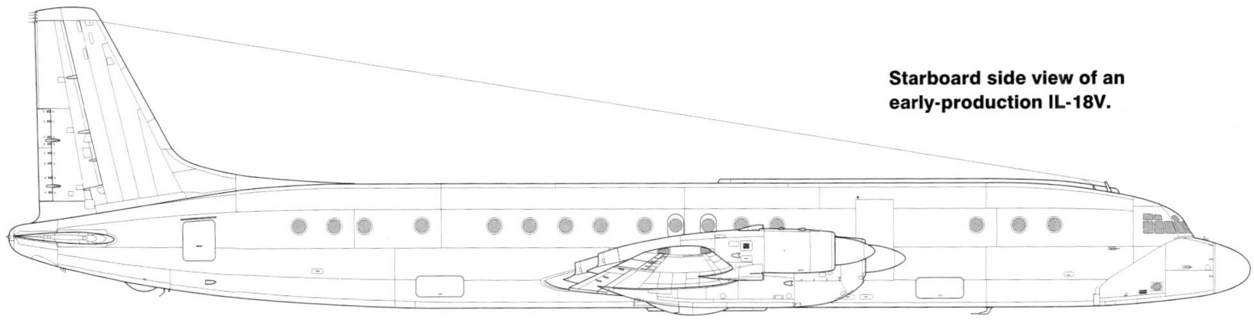

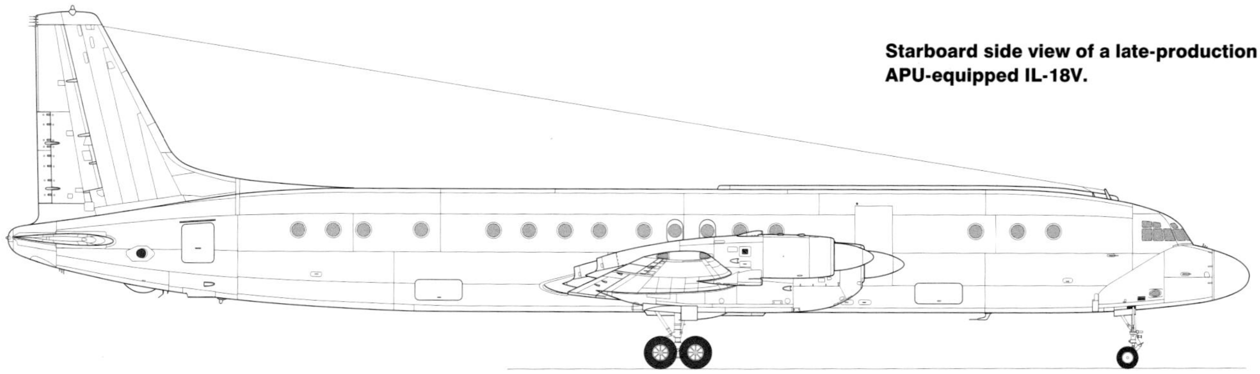

Starboard side view of a late-production APU-equipped IL-18V.

Lower view of a late-production IL-18V.

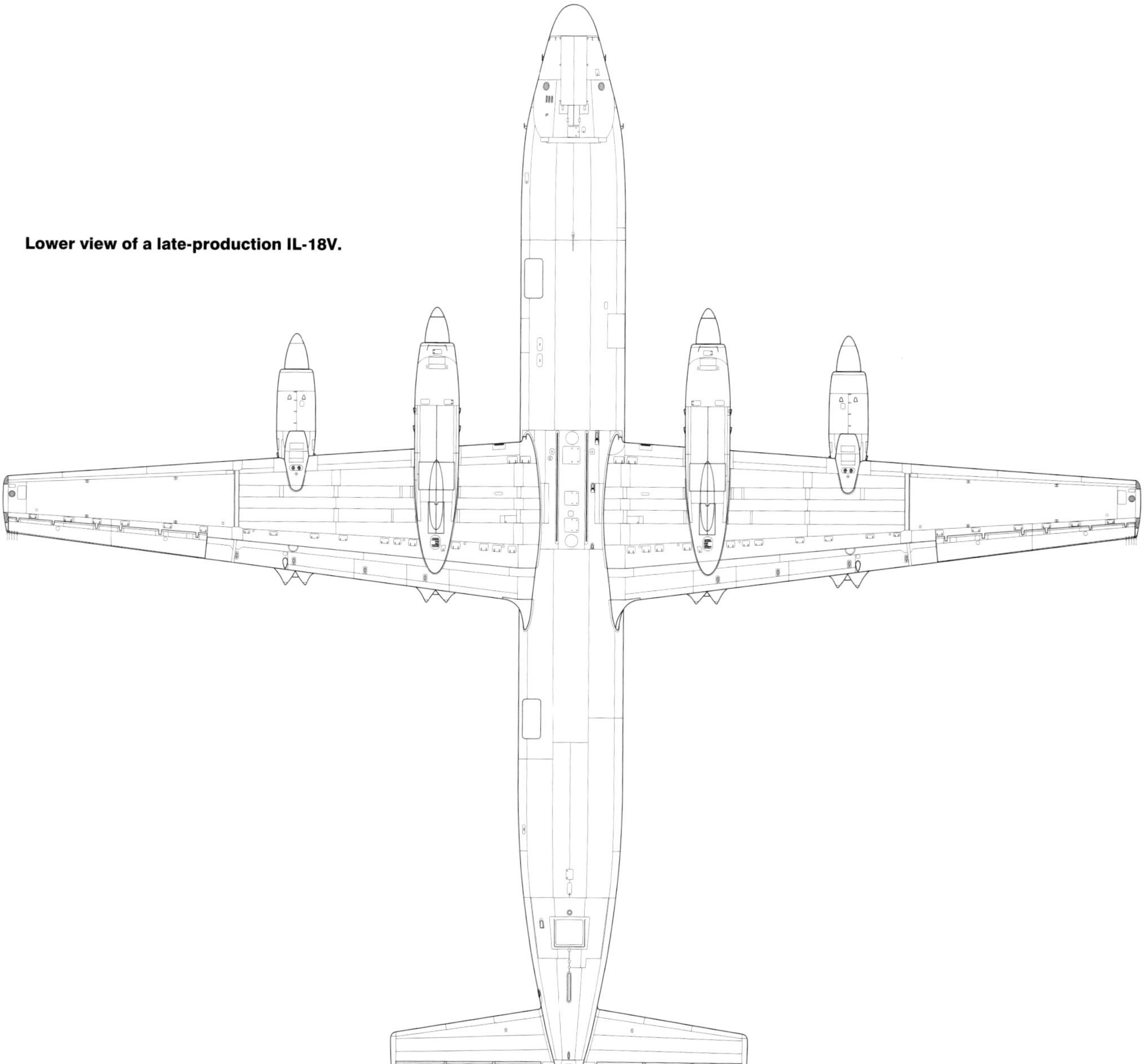

Upper view of a late-production IL-18V.

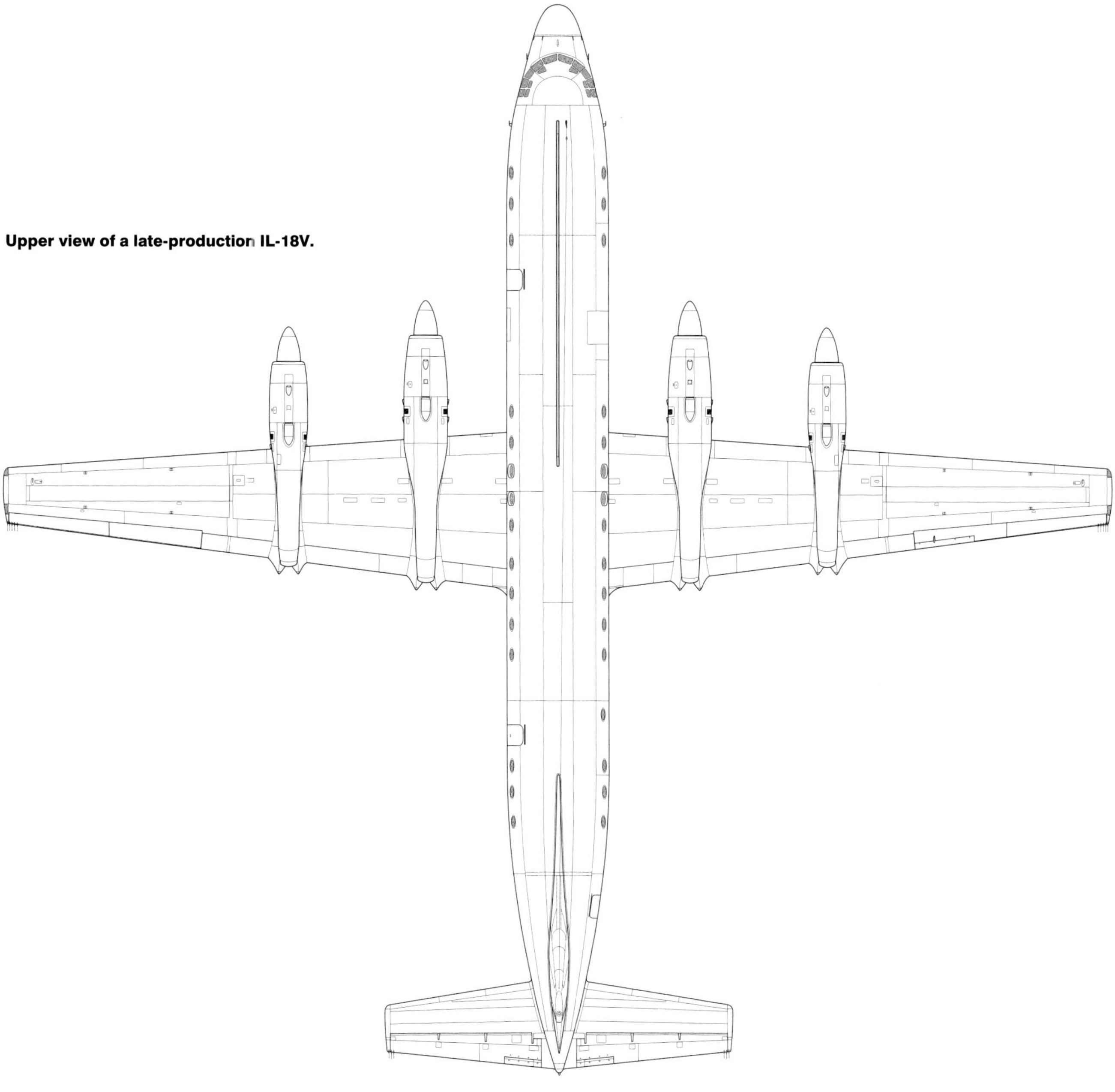

The specially modified IL-18V CCCP-75743 (c/n 181002901) which flew to Antarctica in December 1961/February 1962.

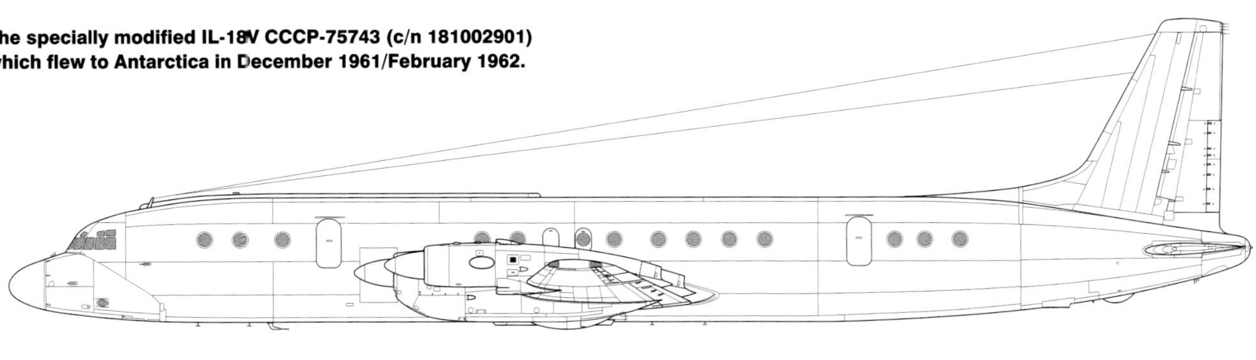

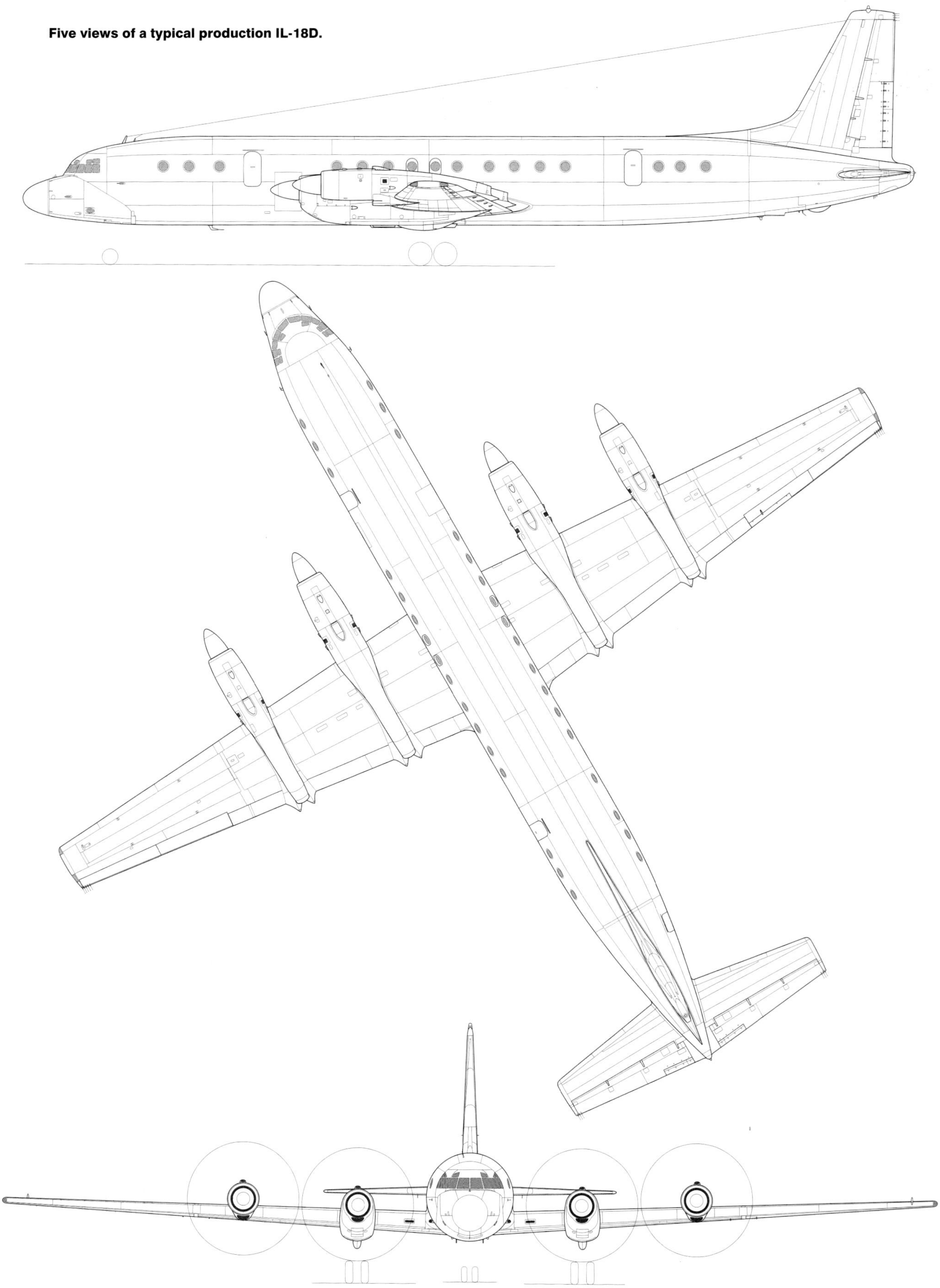

Five views of a typical production IL-18D.

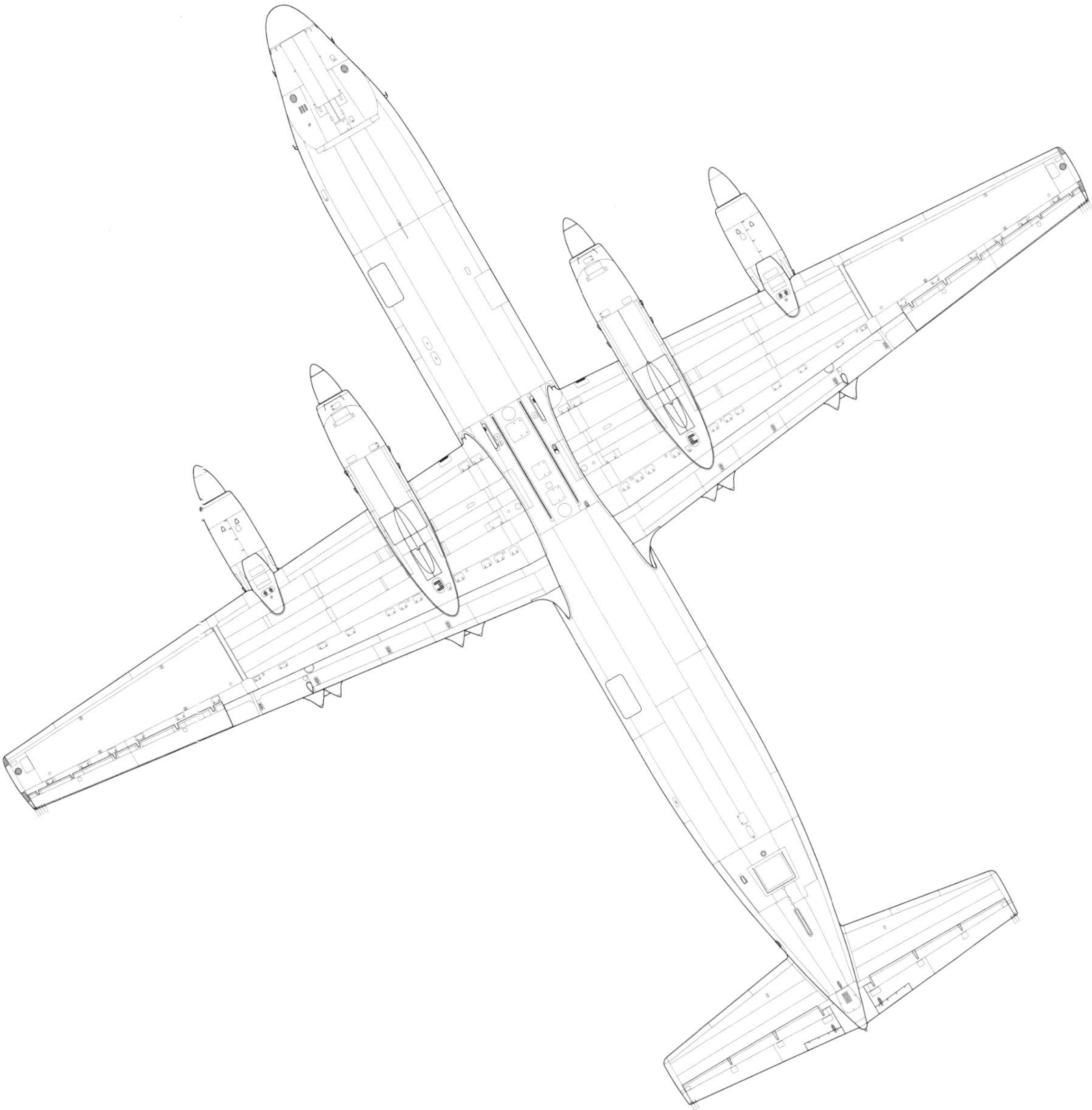

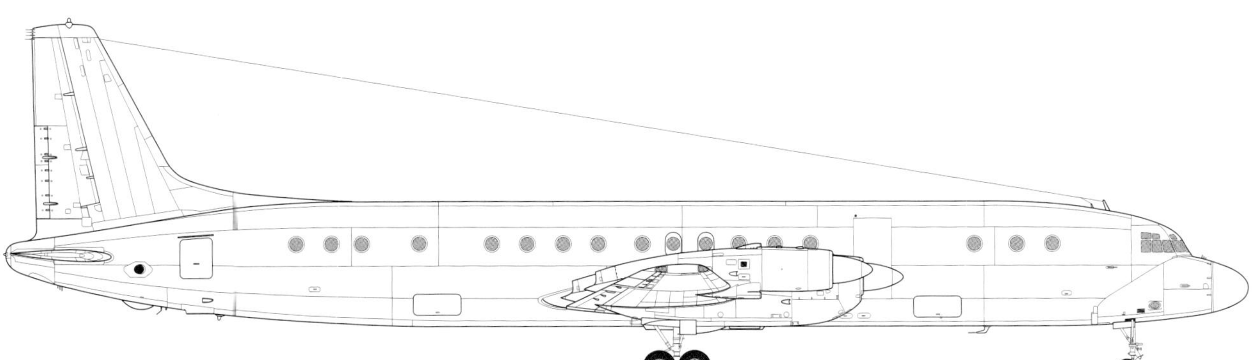

A late-production IL-18V, still with strake aerial but featuring anti-collision lights, Doppler speed/drift sensor and APU.

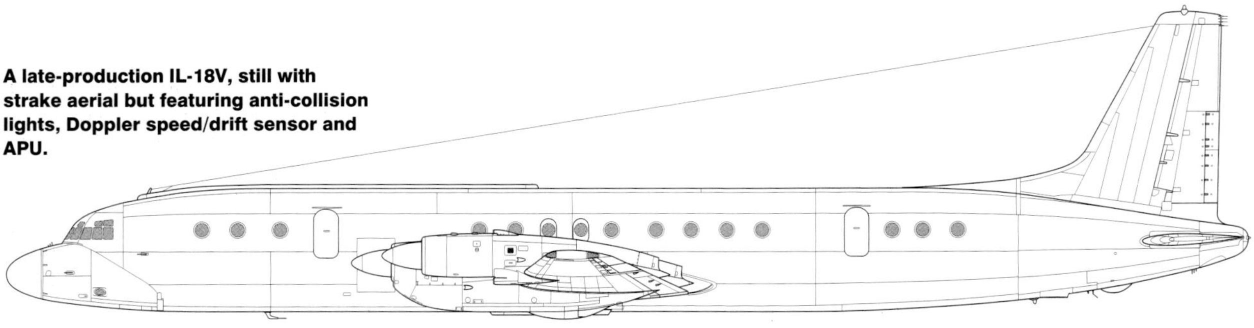

The IL-18V-26A long-range Polar version (CCCP-75844 through -75848, c/ns 182005304, 182005305 and 182005401 through 182005403).

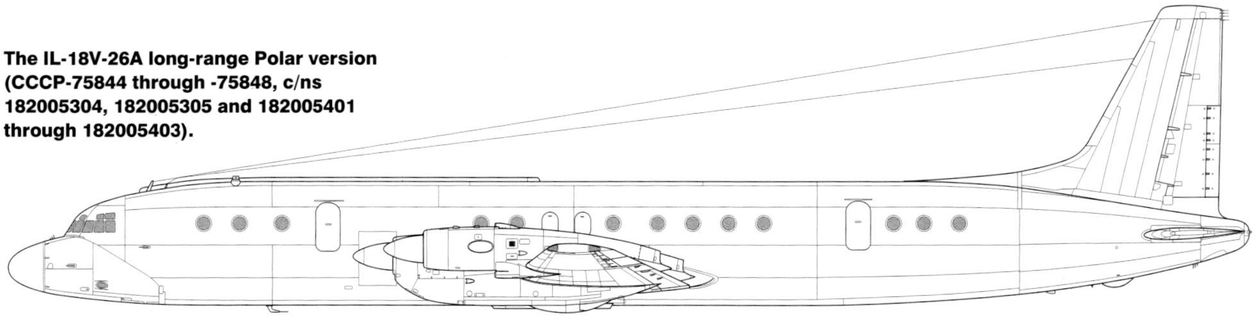

CCCP-75888, the IL-18I development aircraft.

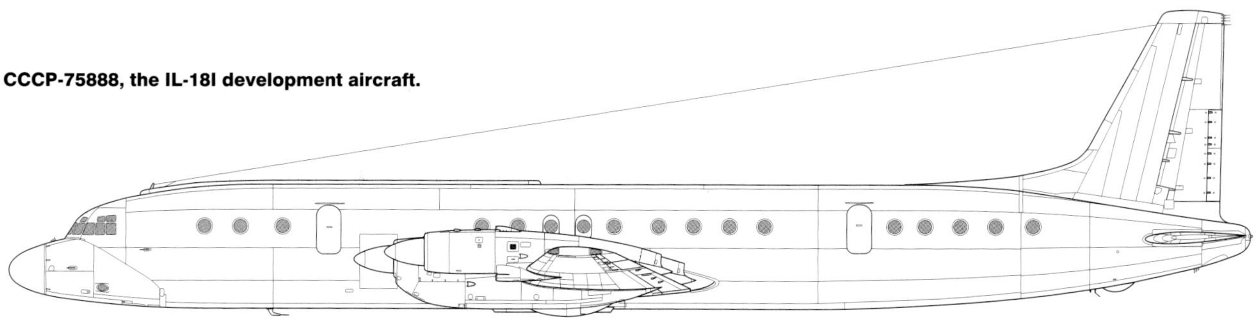

Soviet/Russian Air Force IL-18Ds CCCP-75478 (c/n 189011302) and CCCP-75496 (c/n 185011303) with HF communications gear and a TA-6 APU.

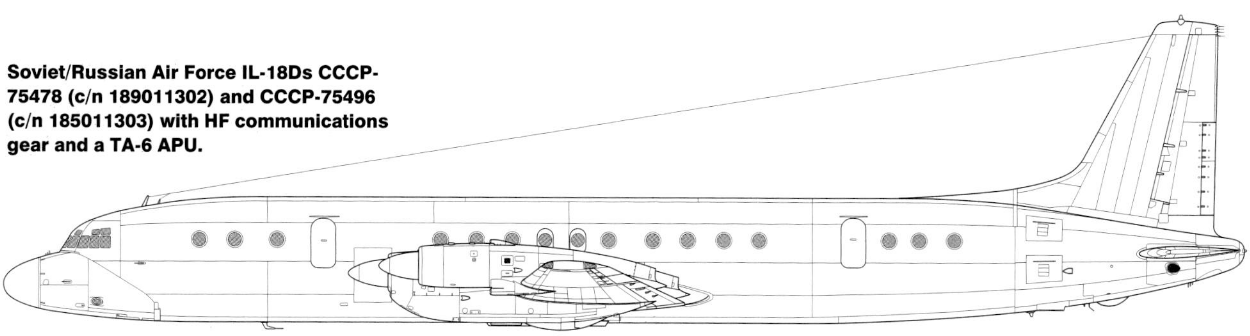

IL-18GrM, alias IL-18D/F (SCD), D-AOAS (c/n 188010805).

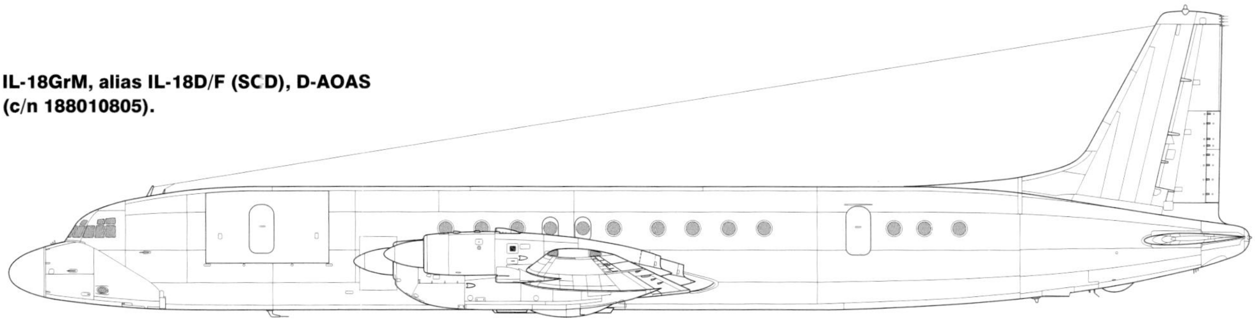

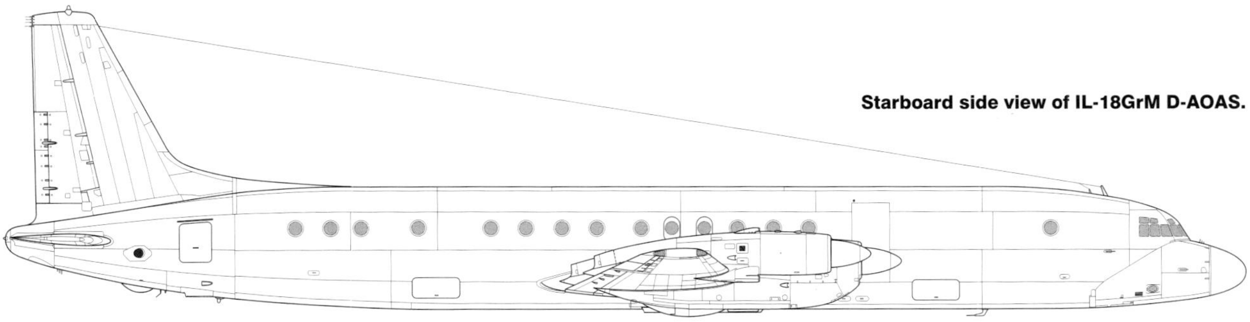

Starboard side view of IL-18GrM D-AOAS.

An IL-18D communications relay aircraft operated by the 235th Independent Flight Detachment (CCCP-75453, -75454 and -75464, c/ns 187010103, 187010104 and 187010401).

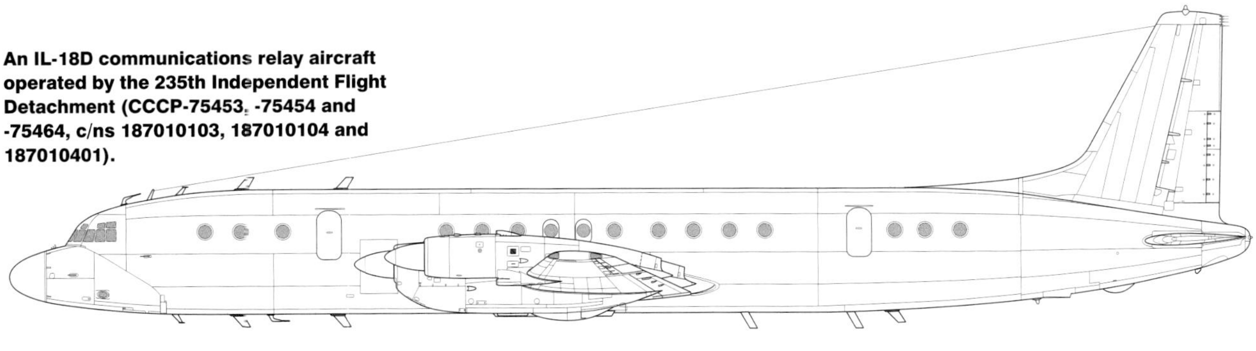

The IL-18DORR long-range ocean fishery reconnaissance aircraft (CCCP-75462, c/n 187010304 and CCCP-74268, c/n 188011201).

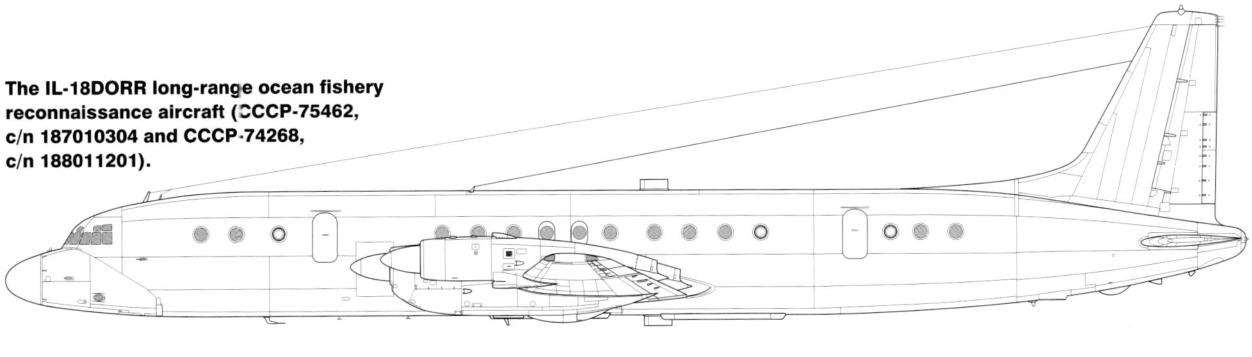

Five views of the IL-20M ELINT aircraft.

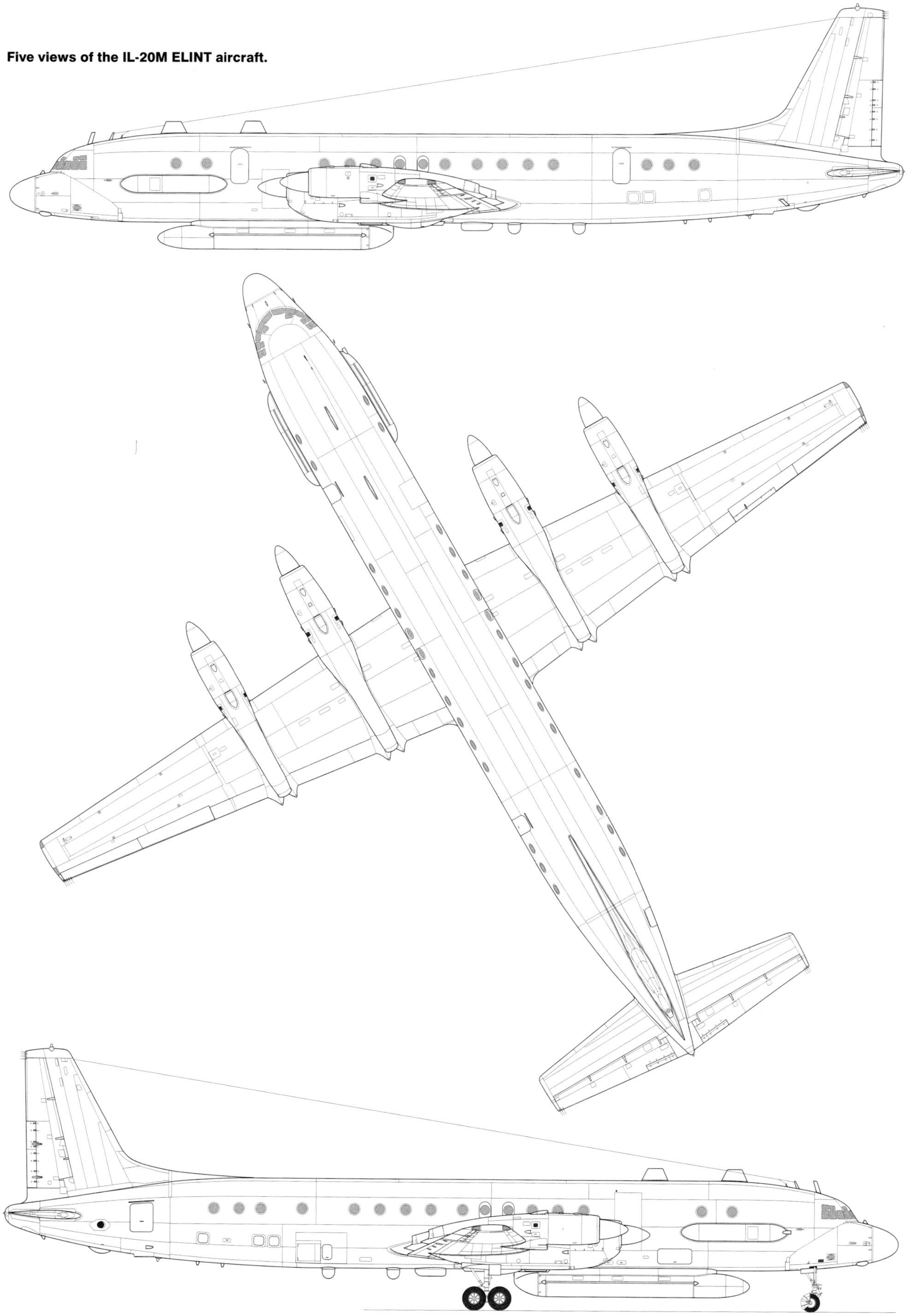

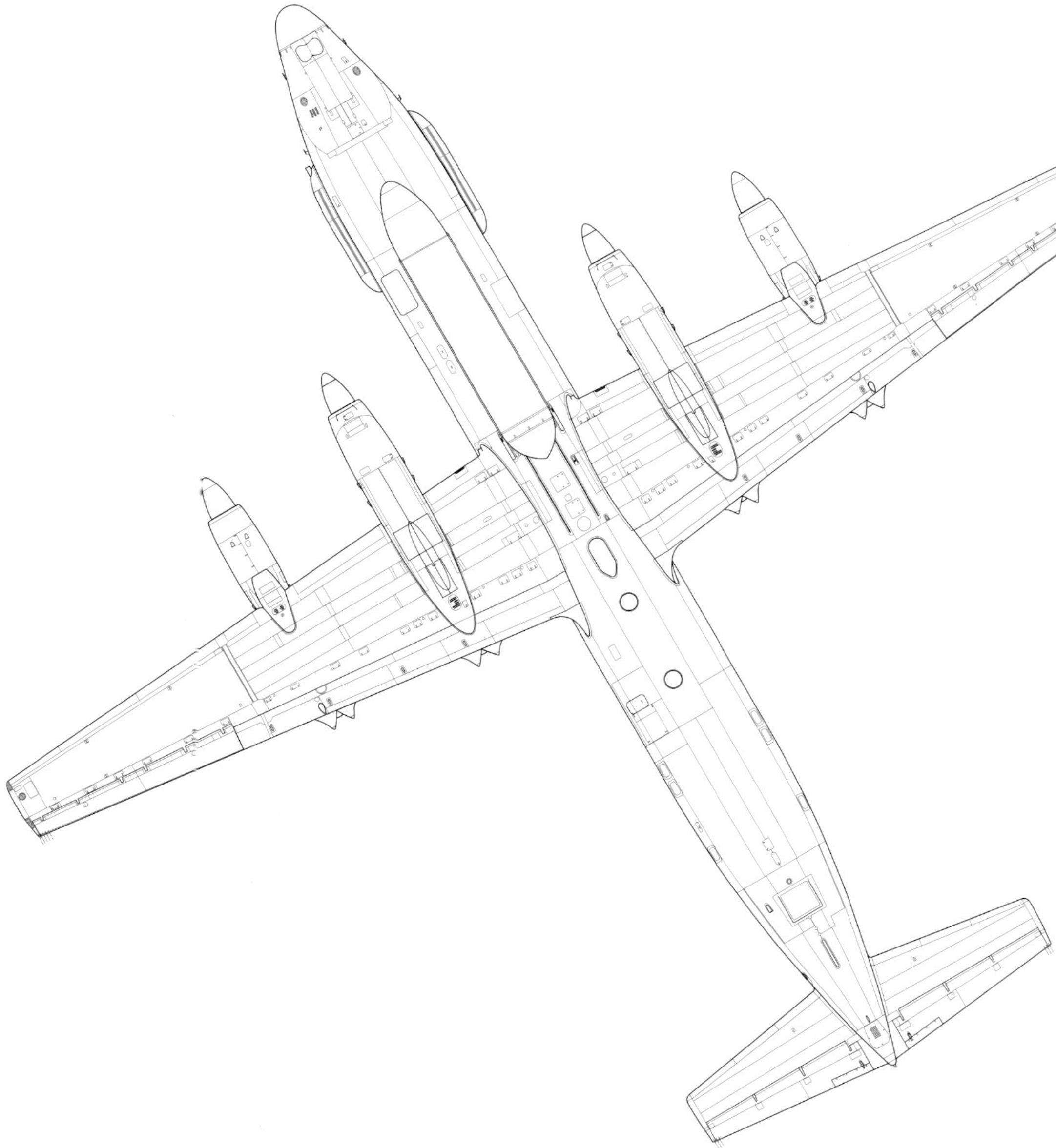

The IL-18USh navigator trainer (CCCP-75523, c/n 183006801).

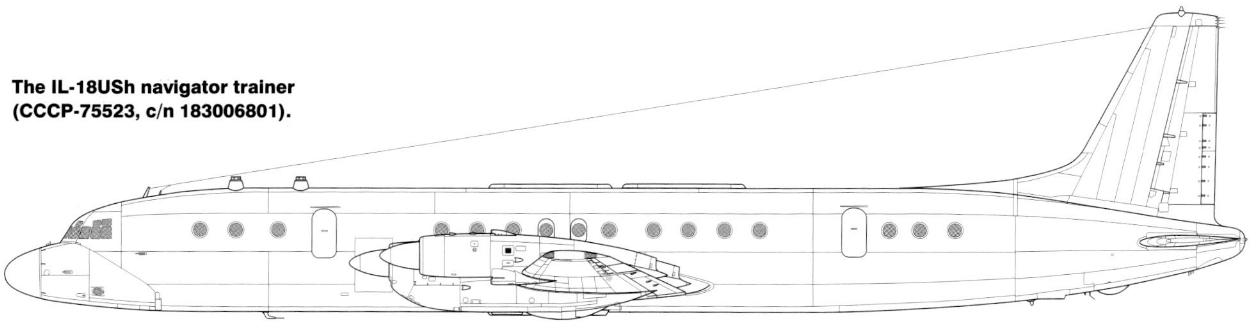

The IL-24N ice reconnaissance aircraft (CCCP-75449, c/n 187010004 and CCCP-75466, c/n 187010403).

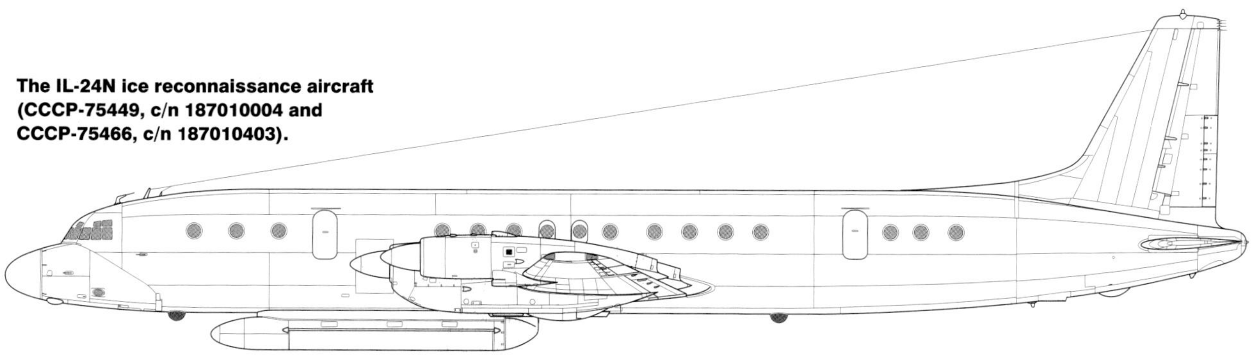

The IL-18D Tsiklon weather research aircraft (CCCP-75442, c/n 187009702).

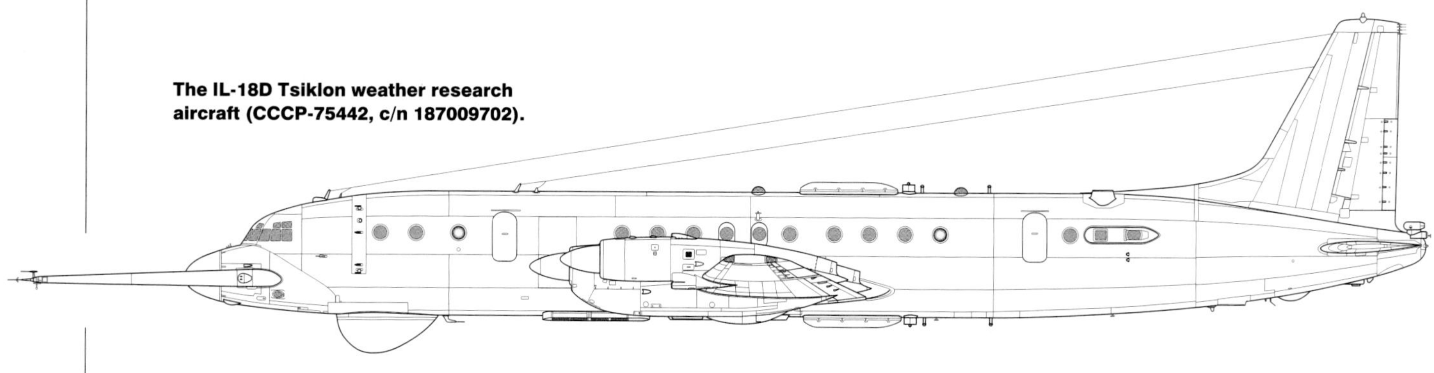

This view and opposite page: four views of a production IL-20RT space tracker/telemetry and communications relay aircraft.

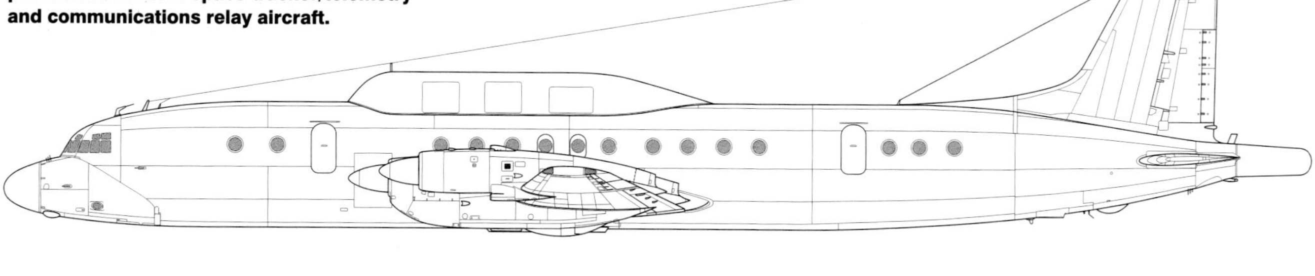

The IL-18E Meteor weather research aircraft (CCCP-75598, c/n 186008802).

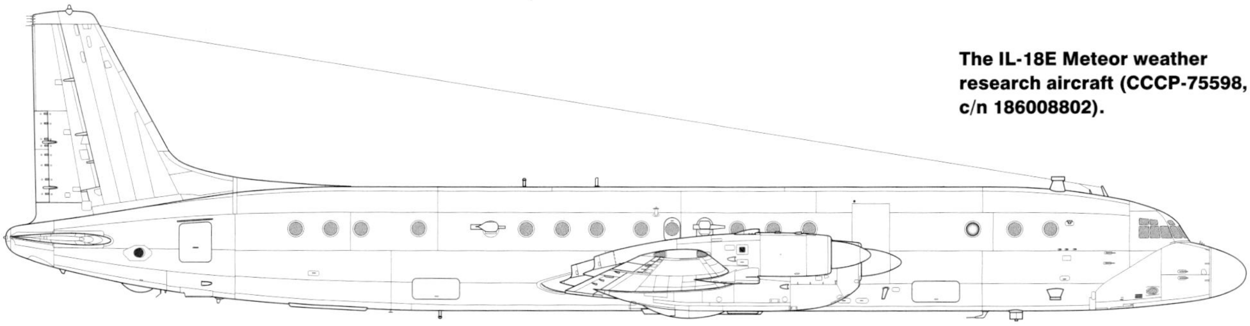

The former IL-18E Meteor following conversion for GosNII GA with a SLAR aft of the port wing.

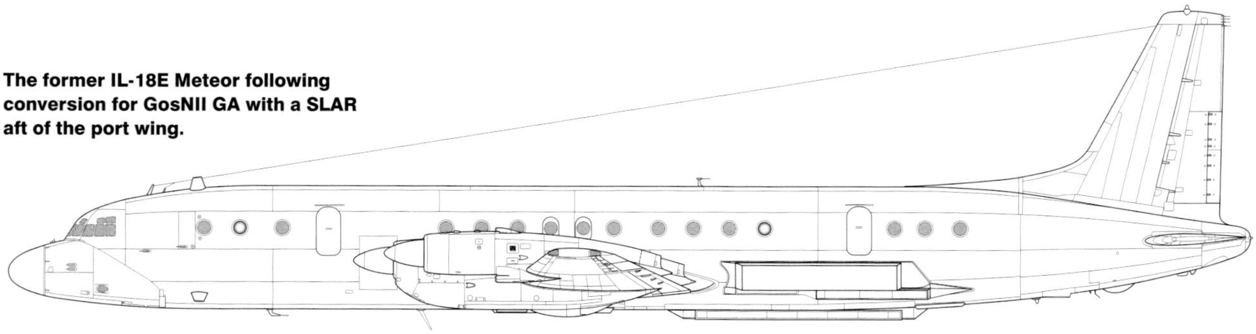

The IL-18RT communications relay aircraft (CCCP-75528, c/n 183006901 and CCCP-75840, c/n 182005301).

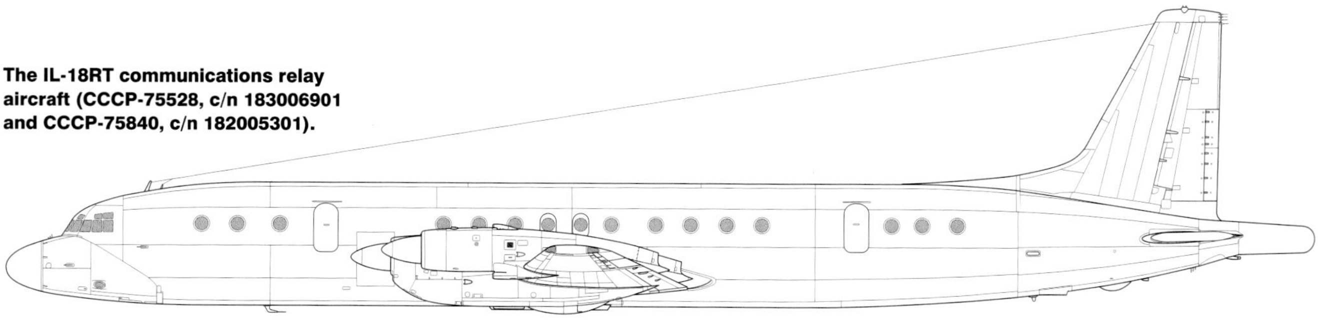

The IL-18SIP (Il-18RTL) space tracker/telemetry and communications relay aircraft (CCCP-27220, c/n 188000401).

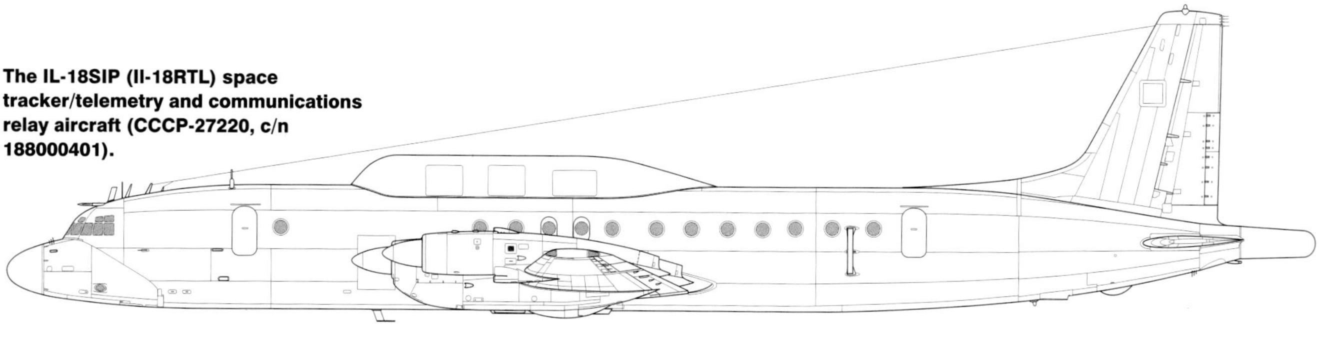

Lower view of an IL-22 (IL-18D-36 Zebra).

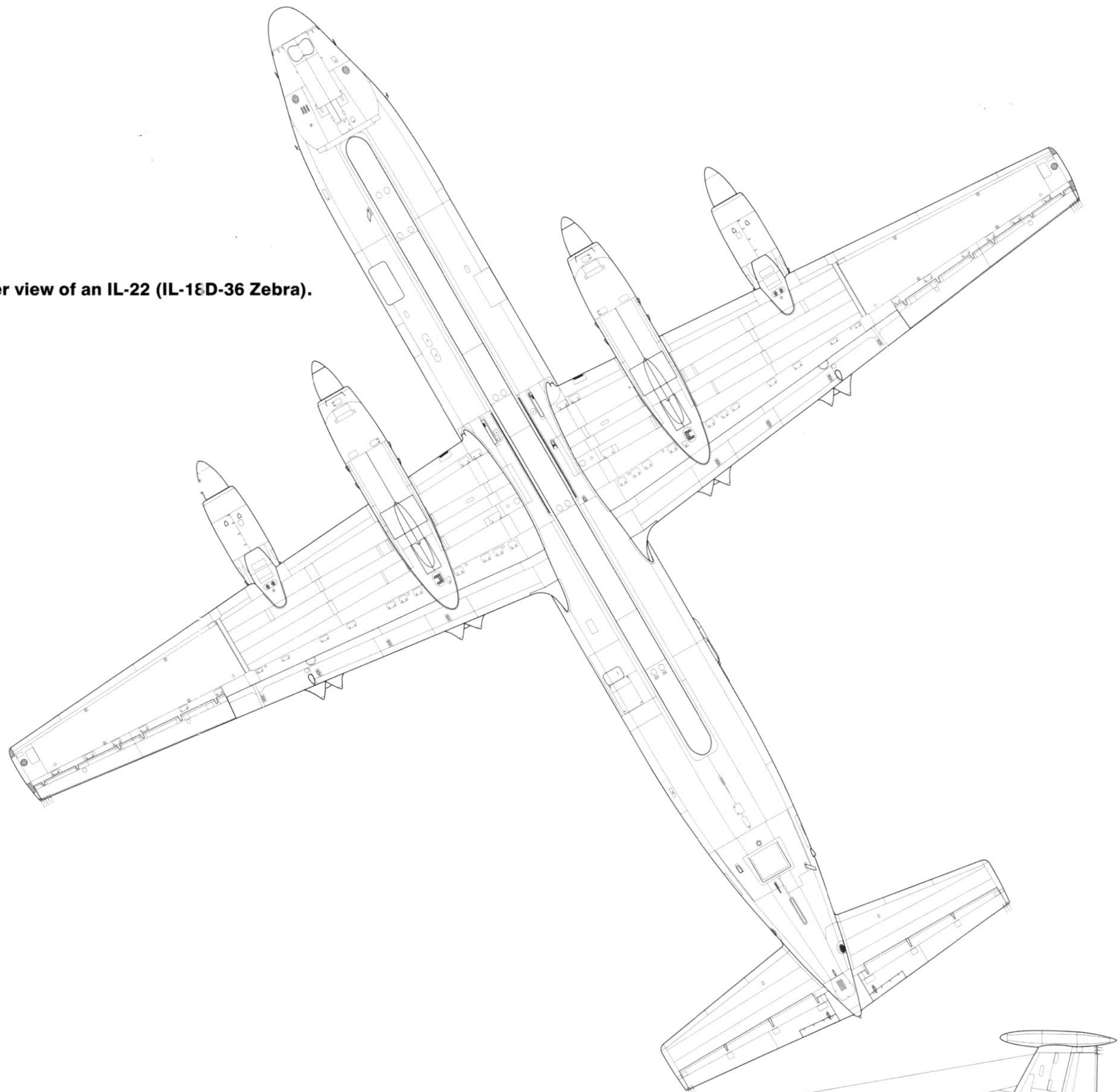

A typical IL-22M (IL-22M-11 Bizon) airborne command post.

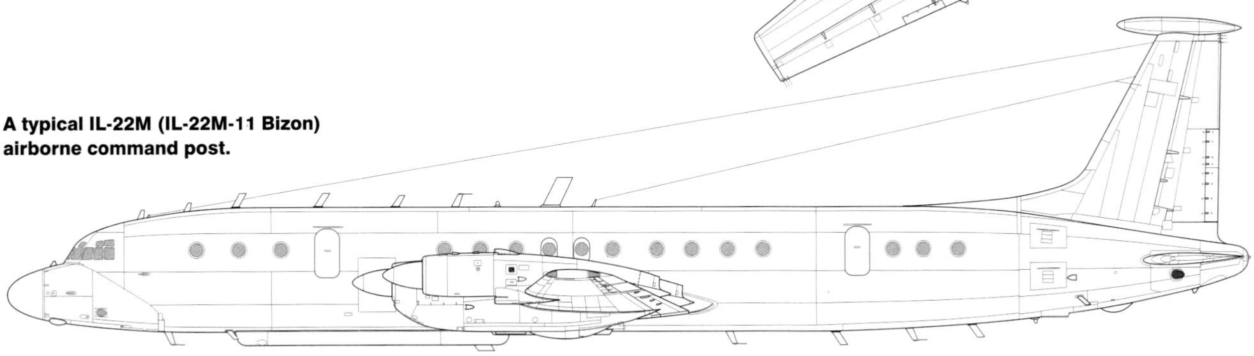

Starboard side view of an IL-22M.

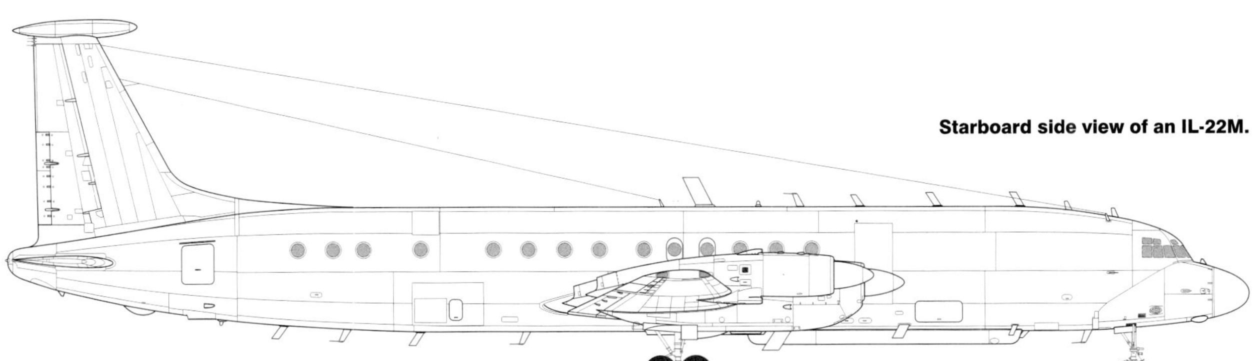

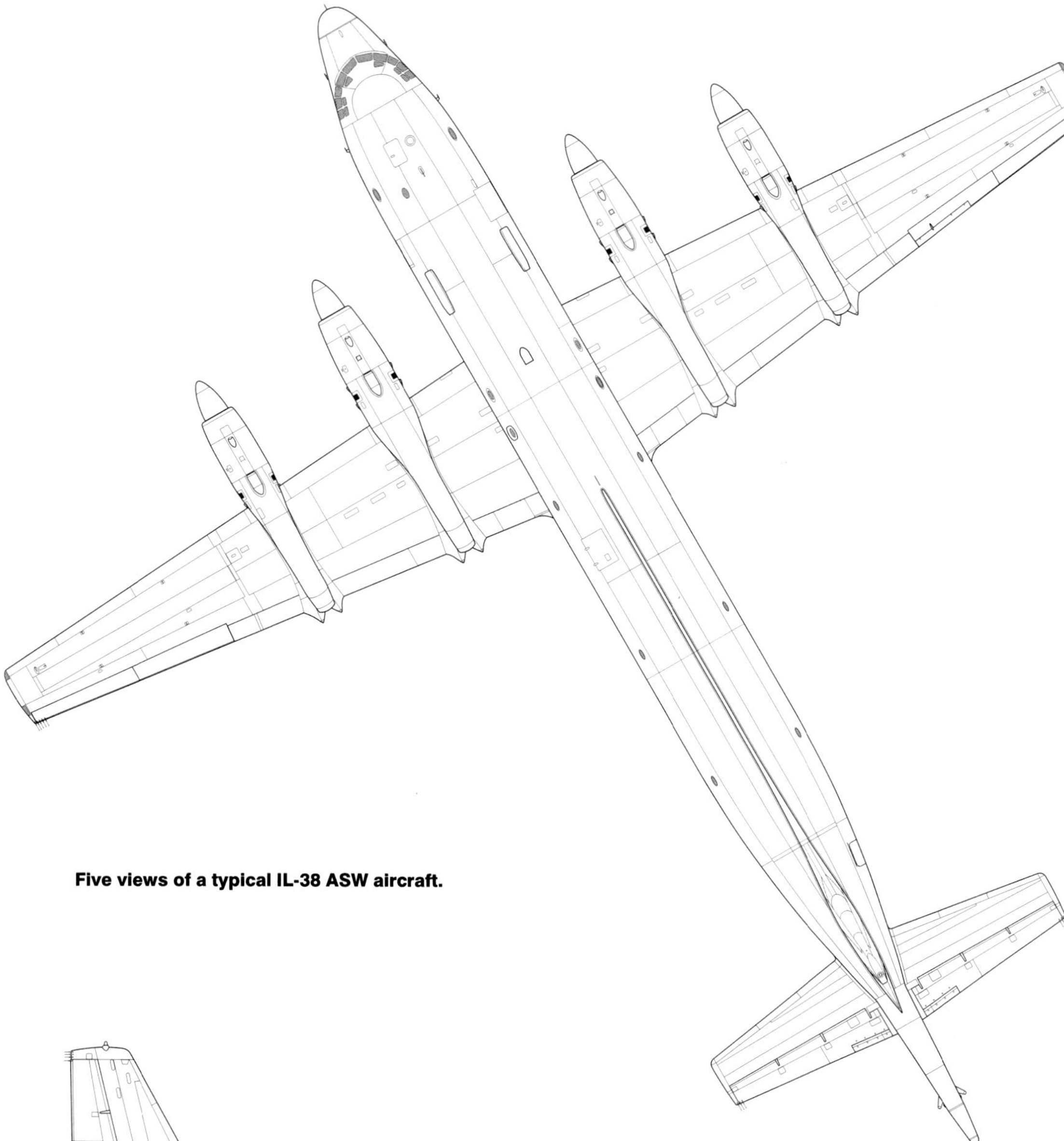

Five views of a typical IL-38 ASW aircraft.

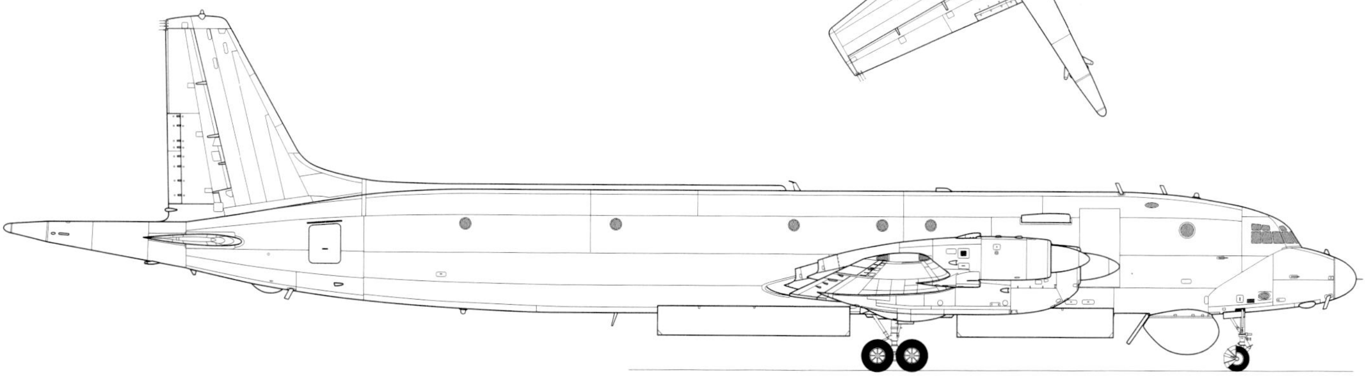

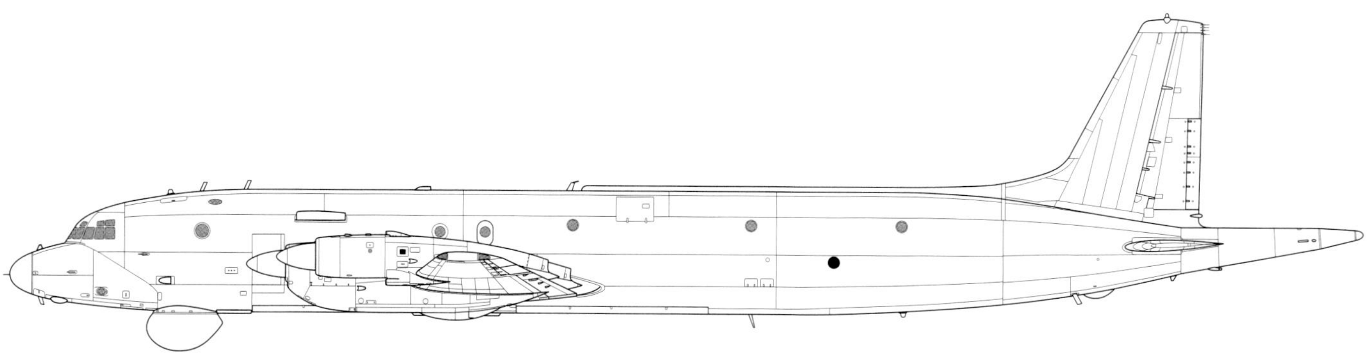

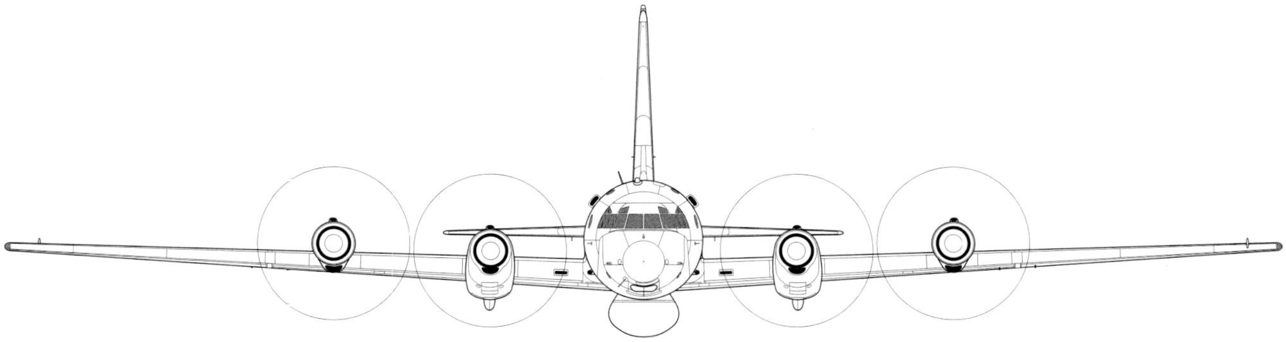

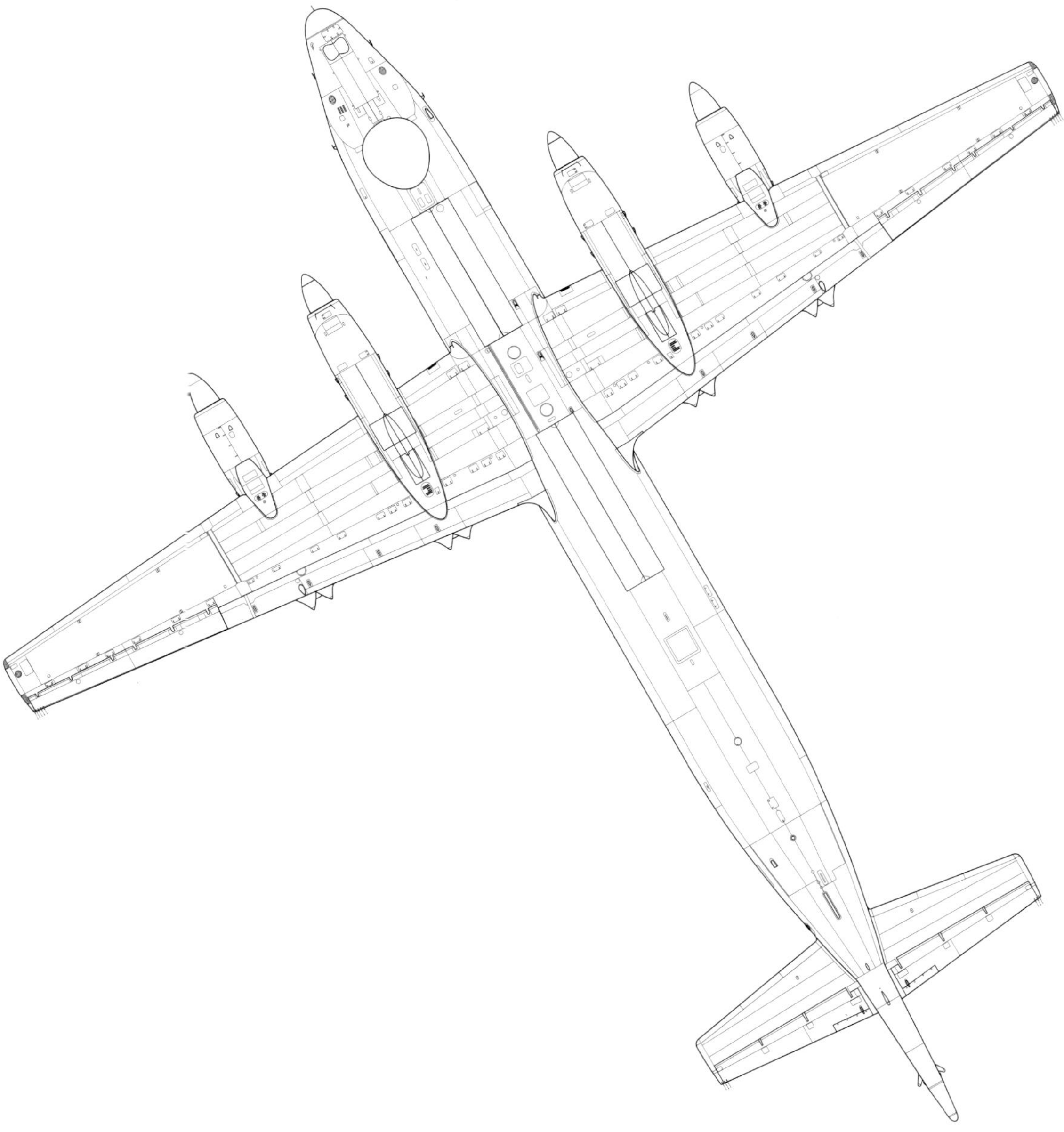

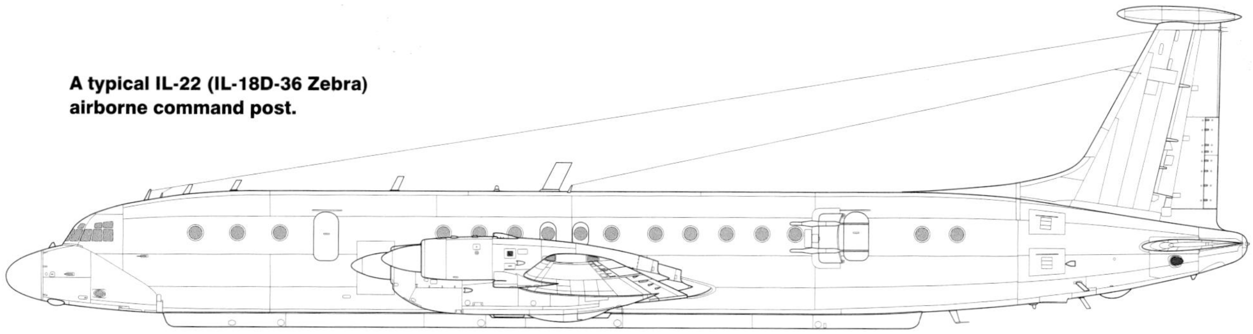

A typical IL-22 (IL-18D-36 Zebra) airborne command post.

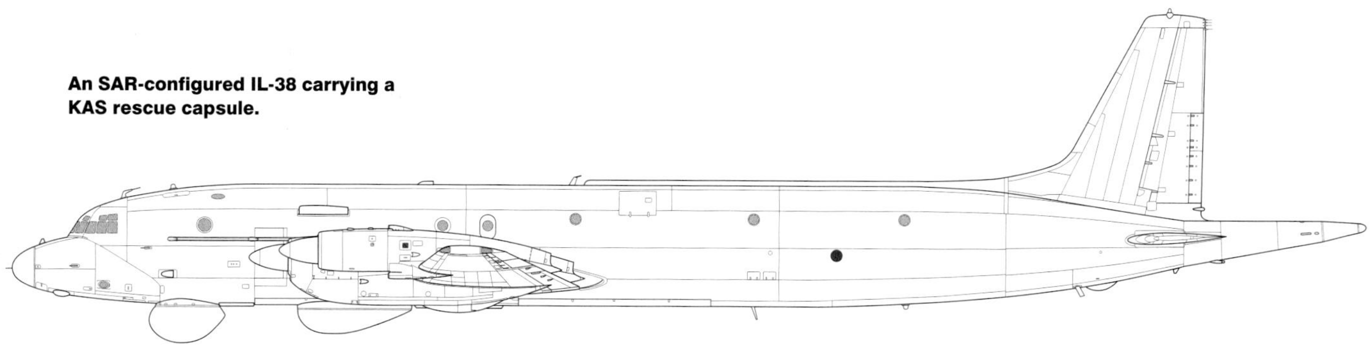

An SAR-configured IL-38 carrying a KAS rescue capsule.

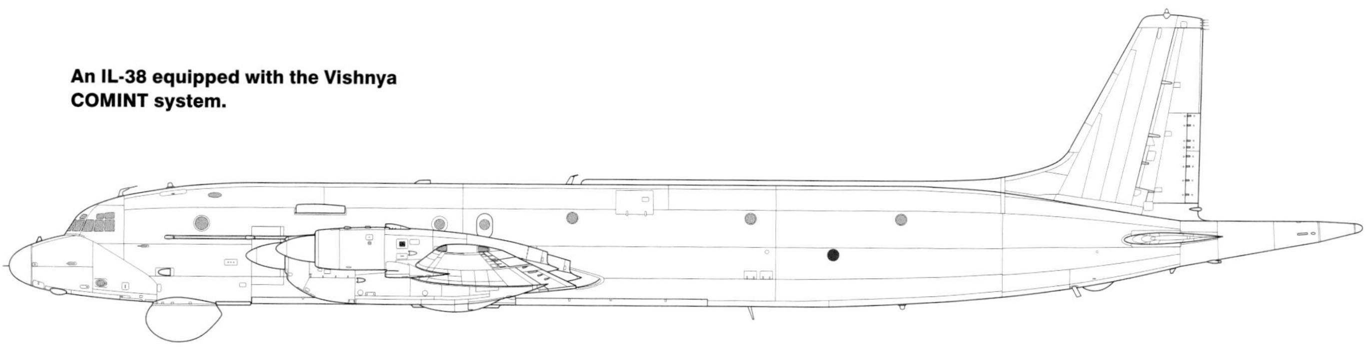

An IL-38 equipped with the Vishnya COMINT system.

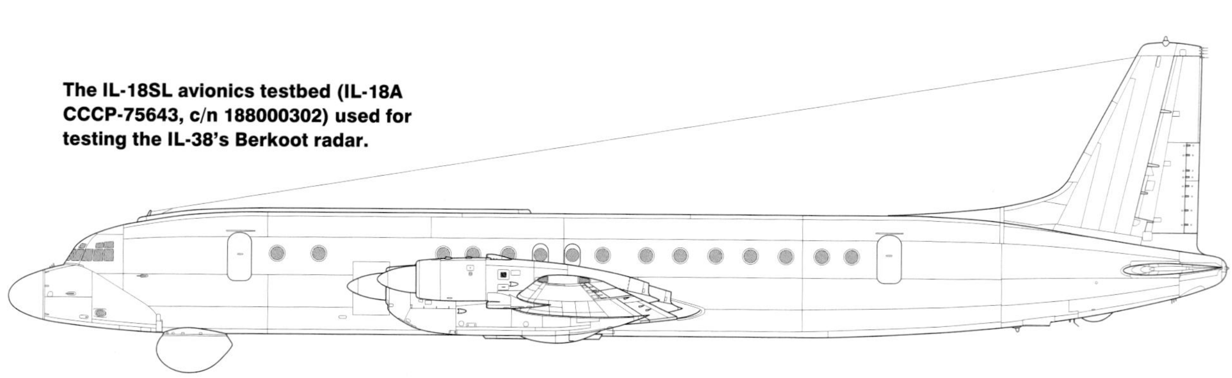

The IL-18SL avionics testbed (IL-18A CCCP-75643, c/n 188000302) used for testing the IL-38's Berkoot radar.

The IL-18 in Colour

IL-18D RA-75464 of the Rossiya State Transport Company takes off at Moscow-Vnukovo. Unlike the airline's other two communications relay-configured IL-18s, this one wears the stylish livery introduced in 1995. Yuriy Kirsanov

P-835, a very late-production IL-18D (c/n 188011205) belonging to Chosonminhang (Civil Aviation Administration of Korea). Yefim Gordon archive

Seen in front of the old wooden hangar at Chkalovskaya AB on 15th August 1999 during the Aviation Day celebrations, Russian Air Force/223rd Flight Unit IL-18D RA-75478 (c/n 189011302) is a staff transport with an IL-22 style TA-6A APU (note the exhaust under the port stabiliser). The aircraft has been jacked up for landing gear operation checks. Mikhail Yevdokimov

Unlike the other IL-24N, the second example (CCCP-75466) was stripped of its mission equipment, including the famous pod, even before it was transferred to the Russian register. It is seen here at the GosNII GA apron, still in full Aeroflot Polar colours. Yefim Gordon archive

IL-18B CCCP-75658 (c/n 188000602) was preserved in a children's playground in Zaporozhye, the Ukraine. This early aircraft stayed in service long enough to receive the 1973-standard livery. via Yefim Gordon

The IL-20M ELINT aircraft is instantly recognisable by the ventral SLAR pod and lateral fairings. Note the mudguard on the nosewheels to avoid damaging the pod and the accordingly bulged nose gear doors. Yefim Gordon

Top and above: **Two views of IL-20M '90 Red' (c/n 173011501) at Shaikovka AB. This example is unusual in having a Perfectly Green colour scheme; most *Coot-As* were painted grey.** Yefim Gordon

This head-on view emphasises the elliptical section of the IL-20M's SLAR pod. Yefim Gordon

Some IL-20Ms, including '90 Red' (c/n 173011501), displayed the construction number on the tail, IL-18 style. Yefim Gordon

An IL-20M parked on the taxiway of a Russian Naval Air Arm base. Note the red metal covers over the lateral antennas of the Romb-4 SIGINT system; these carry the aircraft's tactical code which is not applied to the airframe itself! A support has been placed under the tail – just in case. Sergey Skrynnikov

This well-weathered IL-20M (c/n 173011502) illustrates the finish usually worn by *Coot-As*. The tactical code '20 Red' is just discernible aft of the wing; this aircraft was on temporary deployment to Sperenberg AB, East Germany, in the 1980s and early 1990s. Yefim Gordon archive

Above and left: **RA-75923, the only civil-registered IL-20M, pictured at the open doors day at Chkalovskaya AB on 14th August 1999. Note the absence of the rear dorsal aerial and the rear ventral blister radomes. An IL-22M-11 *Coot-B* ABCP with the same registration was also present on the same day!** Yefim Gordon

Top: **A very weathered CCCP-75481 (c/n 173011503), the second production IL-20RT, at Ostrov AB. By 1996 the aircraft had been stripped of mission equipment, serving with the Russian Navy's Combat & Conversion Training Centre as an IL-38 crew trainer and camera platform.** Yefim Gordon

Above and right: **Two more views of CCCP-75481 at Ostrov AB in the summer of 1996, showing to advantage the dorsal 'hump' and rear thimble radome.** Yefim Gordon

Right: **Front view of the IL-20RT.** Yefim Gordon

Top: **Weathered and tattered IL-20RT CCCP-75481 may appear, it was fully operational in 1996. Here an APA-5DM power cart based on the ubiquitous Ural-4320 6x6 army lorry is about to start up the APU prior to another training flight. Note the bicycles stacked near the main landing gear; they save the airmen a lot of legwork on the vast territory of Ostrov AB.** Yefim Gordon

Centre and above: **CCCP-75903 No 1 represents the first production version of the IL-22 airborne command post (aka IL-18D-36 'Bizon'). Based at Kubinka AB, this particular *Coot-B* retained the CCCP- prefix and Soviet flag until 1998.** Yefim Gordon

Right: **The tail of CCCP-75903 No 1 at Kubinka AB. Note the twin APU intake doors above and below the cheatline, the large APU exhaust just visible under the port stabiliser and the c/n (0393610235) on the tail.** Yefim Gordon

Top: **IL-22M-11 RA-75913 (c/n 0394011098) at Chkalovskaya AB on 15th August 1999. This base boasts the largest concentration of *Coot-Bs* anywhere.** Mikhail Yevdokimov

Above left: **Snowbound and wrapped in tarpaulins, an 8th ADON IL-22M-11 (CCCP-75927, c/n 2964017558) sits amidst a tangle of barbed wire fences in early 1996. This was probably the last *Coot-B* to be manufactured as such.** Yefim Gordon

Above right: **For various reasons the IL-22s are not especially active in recent years – although in Soviet times you could see them flying around Chkalovskaya AB almost daily. On this stored example the propellers have been carefully wrapped up in canvas covers to protect them from the elements.** Yefim Gordon

Lower three pictures: **The need for IL-22 ABCPs proved so high that, when no more new *Coot-Bs* were forthcoming, several IL-18Ds were requisitioned from Aeroflot for conversion. IL-22M-11 RA-75917 is one of them, having previously flown as IL-18D CCCP-75451 No 2 (c/n 187010105).** Mikhail Yevdokimov

RA-75923 No 1 (ex-CCCP-75923), another IL-22M-11 converted from an IL-18D. This aircraft started life as CCCP-75463 and retained its original c/n (187010305) after conversion. Mikhail Yevdokimov

This view shows the lack of cabin windows ahead of the starboard wing, a characteristic feature of the IL-22M-11. Mikhail Yevdokimov

Do not believe the 'IL-18D' nose titles – YL-LAO seen in the static park at the MAKS-99 airshow in the stylish livery of the Latvian charter airline Concors is a demilitarised IL-22M-11 (ex-Belorussian Air Force CCCP-75916, c/n 2964017102), as revealed by the characteristic 'cigar' atop the fin and the portside APU exhaust. Note the sticker of the Sky High Pilots' Club near the forward entry door. Yefim Gordon

Top and above: **'20 Red', one of several *Mays* operated by the 240th GvOSAP at the Russian Naval Aviation Combat & Conversion Training Centre, Ostrov AB.** Yefim Gordon

Above and below: **Landing gear tucking away, an IL-38 leaves a heavy smoke trail as it takes off on a training mission.** Yefim Gordon

Above and left: **IL-38 '06 Red' at Ostrov AB. Note that the paintwork displays varying degrees of weathering on different parts of the airframe.** Yefim Gordon

'22 Red' (c/n 08...011006), a 240th GvOSAP IL-38 which took part in the RIAT'96 show at RAF Fairford. Note the tail artwork and the bright blue propellers, an unusual shade. The sticker on the nose is a memento from RIAT'96. Yefim Gordon

Part of the hardstand at Ostrov, showing two of the Russian Navy Combat & Conversion Training Centre's IL-38s. The glossy finish on the 'airshow' example makes a striking contrast with the other aircraft. Yefim Gordon

A Pacific Fleet IL-38 taxies out for take-off. Yefim Gordon

A North Fleet IL-38 on a patrol mission over the Arctic Ocean. via Yefim Gordon

This page:

The *May* operates in close co-operation with 'friendly' ships, as seen here. via Yefim Gordon

240th GvOSAP IL-38 '20 Red' rests between missions, with the obligatory fire extinguisher parked in front. Here the spinners have red tips. Yefim Gordon

Opposite page:

The paintwork on this IL-38 is almost obliterated by the elements (or deliberately removed prior to an overhaul), revealing interesting details such as the skin reinforcement panel in the propellers' plane of rotation, traces of antenna fairings and wiring conduits associated with the Vishnya communications intelligence (COMINT) system and the square dielectric panel of unknown purpose. via Yefim Gordon

The prototype of the IL-38N (alias IL-38SD) upgraded version with the Morskoy Zmey (Sea Dragon) search & targeting suite at Pushkin during trials. The 'suitcase' on top of the forward fuselage houses a signals intelligence system antenna array. Yefim Gordon archive

Algerian Government IL-18V 'Salon' 7T-VRA seen in the late 1960s. Sergey and Dmitriy Komissarov

EC-BS
7T-VRA
REPUBLIQUE ALGERIENNE

IL-18D LZ-BEA operated in Balkan Bulgarian Airlines' old colours as an IL-18Gr freighter. Sergey and Dmitriy Komissarov collection

An Air Cess IL-18V (c/n 184006903) enjoying life on the Equatorial Guinean register as 3C-KKJ. The airline's aircraft have been known to operate from other 'ports of registry'. Peter Davison

Another Air Cess IL-18V, 3C-KKR, in a slightly different colour scheme. This aircraft was sold to Kazakhstan in early 2003 and reregistered UN 75002. Peter Davison

The end of the line is near. IL-18V D-AOAO, still in Interflug colours, used to fly as DDR-STF before German reunification, as DM-STF No 2 before 1981 and originally as CCCP-75553. On 7th February 1991 Interflug went bankrupt and this aircraft was sold to Air Cargo (the future BerLine) – only to run full circle and become RA-75553 with Viola Avia Trans in 1995. Yefim Gordon archive

Above: **As a Russian adage goes, 'if you see "Buffalo" written on an elephant's cage, don't believe your eyes'. Anonymous-looking EL-ALW (though the green cheatline is a giveaway that the aircraft belongs to Santa Cruz Imperial) wears 'IL-18B' nose titles, but this is not true – the cabin window arrangement identifies it as an IL-18V, and so does the construction number 182004904.** Peter Davison

Below: **LOT Polish Airlines IL-18V SP-LSD in 1977-standard livery.** Yefim Gordon archive

We hope you enjoyed this book . . .

Midland Publishing titles are edited and designed by an experienced and enthusiastic team of specialists.

Further titles are in preparation but we always welcome ideas from authors or readers for books they would like to see published.

In addition, our associate, Midland Counties Publications, offers an exceptionally wide range of aviation, military, naval and transport books and videos for sale by mail-order around the world.

For a copy of the appropriate catalogue, or to order further copies of this book, and any of many other Midland Publishing titles, please write, telephone, fax or e-mail to:

Midland Counties Publications
4 Watling Drive, Hinckley,
Leics, LE10 3EY, England

Tel: (+44) 01455 254 450
Fax: (+44) 01455 233 737
E-mail: midlandbooks@compuserve.com
www.midlandcountiessuperstore.com

US distribution by Specialty Press – see page 2.

Aerofax
TUPOLEV Tu-134
The USSR's Short-Range Jet Airliner

Dmitriy Komissarov

The Tu-134 has seen passenger service for over 35 years in 42 countries. Its multifarious other activities include VIP transportation, support of Air Force, Army and Navy headquarters and research and test work. The type has helped train thousands of military pilots and navigators for the Soviet Air Forces' tactical and long-range bomber forces.

Compiled from first-hand Russian sources, the book gives a full account of the Tu-134 and the type's design, test and operational history.

Sbk, 280 x 215 mm, c176pp, c200 col, c90 b/w photos, dwgs. Spring 2004
1 85780 159 8 **c£19.99/US $34.95**

Aerofax
ILYUSHIN IL-76
Russia's Versatile Jet Freighter

Yefim Gordon and Dmitriy Komissarov

The Soviet Union's answer to the Lockheed Starlifter first flew in 1971 and has become familiar both in its intended military guise and as a commercial freighter. It has also been developed as the IL-78 for aerial refuelling, and in AEW and other versions.

There is not only a full development history and technical description, but extensive tables detailing each aircraft built, with c/n, serial and so on, and detailed notes on every operator, both civil and military, and their fleets.

Softback, 280 x 215 mm, 160 pages
c250 b/w and colour photos, drawings
1 85780 106 7 **£19.95/US $34.95**

Aerofax
TUPOLEV Tu-95/Tu-142 'BEAR'

Yefim Gordon and Vladimir Rigmant

During the 'Cold War' Tupolev's Tu-95 'Bear' strategic bomber provided an awesome spectacle. It was the mainstay of the USSR's strike force, a reliable and adaptable weapons platform. Additional roles included electronic/photographic reconnaissance and maritime patrol, AEW and command and control.

The author has had unparalleled access to the Tupolev OKB archives, taking the lid off a story previously full of speculation to produce the most comprehensive study to date.

Softback, 280 x 216 mm, 128 pages
236 b/w, 24 col photos, 12 diagrams
1 85780 046 X **£14.95/US $24.95**

Aerofax
MIKOYAN-GUREVICH MiG-17

Yefim Gordon

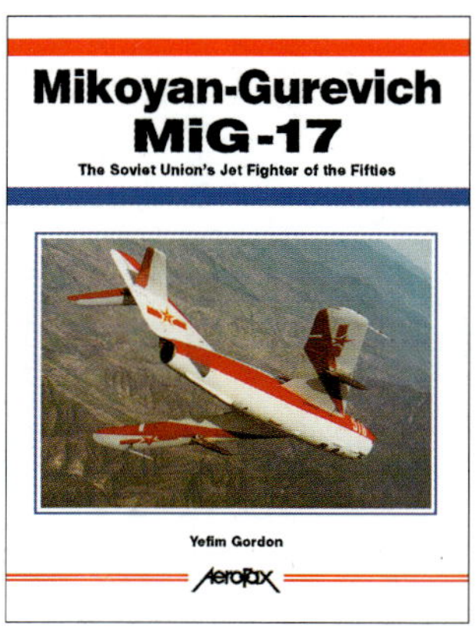

The Soviet Union produced and used around 9,000 MiG-17s. First flown in January 1950, it is an extensively upgraded MiG-15 with a redesigned 'scimitar' wing and lengthened fuselage.

The type was built under various desig-nations including the Polish Lim-5P and Lim-6bis and the Czech S-105, and served not only with the Soviet armed forces but with other Warsaw Pact nations, seeing combat in the Middle East, in North Vietnam and in Nigeria.

Softback, 280 x 215 mm, 144 pages
172 b/w, 32 colour photos, col artworks
1 85780 107 5 **£18.99/US $27.95**

Aerofax
MIKOYAN-GUREVICH MiG-19

Yefim Gordon

The MiG-19 'Farmer' represented a major technological leap for the VVS as it was one of their first fighters to accommodate AAMs, ground control intercept system and other advances. By the end of the 1950s it became their standard fighter and was integrated into the inventories of most Warsaw Pact countries and Soviet allies. It was produced in thousands in the USSR, Czechoslovakia and China (J-6, JJ-6 & Q-5); export customers included Cuba, Egypt, Albania, Vietnam and Pakistan.

Softback, 280 x 215 mm, 160 pages,
194 b/w photos, 16pp of colour, dwgs.
1 85780 149 0 **£19.99/US $32.95**

Aerofax
SAAB GRIPEN
Sweden's 21st Century Multi-role Aircraft

Gerard Keijsper

Today's Swedish Air Force has as its spearhead the Saab Gripen. A fourth-generation fighter that embraces state-of-the-art technology, the Gripen has an impressive multi-role capability, making it a more than worthy successor to the Viggen and Draken.

First flown in 1988, operational capability with the Swedish Air Force was achieved in October 1997. Production continues and sales are being made to South Africa, Hungary, Poland and the Czech Republic.

Softback, 280 x 215 mm, 144 pages
230 colour photographs, plus dwgs
1 85780 137 7 **£19.99/US $32.95**

Aerofax
LOCKHEED'S SR-71 'BLACKBIRD' FAMILY

James Goodall and Jay Miller

Though only 50 of these craft were built, everything about them was unique. The stories of the development program, the General Dynamics 'Kingfish' competition, the M-21 and D-21 effort, the F-12 saga, and the operational history of the A-12 and SR-71 under the auspices of the CIA and the USAF are all covered in detail. The high-speed, high-altitude recce overflights performed by SR-71As from bases in the US, Japan and the UK during the Cold War are also covered.

Softback, 280 x 215 mm, 128 pages,
205 b/w, 43 colour photos, plus illusts
1 85780 138 5 **£15.99/US $24.95**